Random Matrices and Non-Commutative Probability

Random Matrices and Non-Commutative Probability

Arup Bose

CRC Press is an imprint of the
Taylor & Francis Group, an **informa** business
A CHAPMAN & HALL BOOK

First edition published 2022
by CRC Press
6000 Broken Sound Parkway NW, Suite 300, Boca Raton, FL 33487-2742

and by CRC Press
2 Park Square, Milton Park, Abingdon, Oxon, OX14 4RN

CRC Press is an imprint of Taylor & Francis Group, LLC

ISBN: 9780367700812 (hbk)
ISBN: 9780367705008 (pbk)
ISBN: 9781003144496 (ebk)

DOI: 10.1201/9781003144496

Typeset in Latin Modern font
by KnowledgeWorks Global Ltd.

To my teacher GJB (Gutti Jogesh Babu)

Contents

Preface

Non-commutative probability or free probability was invented by Dan Voiculescu in the late 1980s and grew initially as a part of pure mathematics. Soon, one of its central concepts, namely free independence, attracted the attention of classical probabilists. Scientists and researchers from other areas such as probability, statistics and wireless communications have also found it to be a natural ally, especially for its connections with the behavior of large dimensional random matrices.

There are high level books connecting random matrices and non-commutative probability, for instance, *An Introduction to Random Matrices* by G. Anderson, A. Guionnet and O. Zeitouni; *Lectures on the Combinatorics of Free Probability* by A. Nica and R. Speicher; and the most recent one, *Free Probability and Random Matrices* by J.A. Mingo and R. Speicher.

Those outside pure mathematics may find it daunting or time consuming to get into this area due to the technical barriers, both in terms of the mathematics that is needed as well as the language in which it is carried out. Colleagues and students have shared their wish for an easily accessible resource.

The structure of this book has grown out of courses on random matrices and on the combinatorial approach to non-commutative probability that I taught at Master's and PhD levels. Interactions with the students through these courses was a joy. They include Riddhipratim Basu, Shirshendu Ganguly, Monika Bhattacharjee, Ayan Bhattacharya, Soumendu Sundar Mukherjee, Debapratim Banerjee, Sohom Bhattacharya, Apratim Dey, Sukrit Chakraborty, Biltu Dan and Priyanka Sen. They kept me focused with their enthusiastic participation. I also gained greatly from interactions with my post-doctoral visitor, Kartick Adhikari.

The book offers an elementary introduction to non-commutative probability, with a focus on free independence and links to random matrices. Many ideas of non-commutative probability are developed by analogy with concepts from classical probability. It is deliberately kept at an easy level, sacrificing only a little in terms of rigor, and is self-contained. Some familiarity with the basic convergence concepts in probability and a bit of mathematical maturity are helpful for the reader to make smooth progress. The highlight of the book is probably the chapter on asymptotic freeness results for random matrices.

I hope the reader will enjoy the first look at the basics and will be inspired to dig deeper into non-commutative probability and large random matrices.

The curious who would like to know how free probability is similar to and different from classical probability may also find the book attractive.

I am grateful to Włodek Bryc, Iain Johnstone and Roland Speicher for encouragement and suggestions. I benefited immensely from the insightful and encouraging comments of the referees who reviewed the original book proposal, and I have tried to implement most, if not all, of them.

Madhuchhanda Bhattacharjee volunteered to create all the simulations and plots. Arijit Chakrabarty has contributed to several early chapters, and in particular, significantly to the chapter on C^*-probability spaces. The material on the embedding-based proof for the asymptotic freeness of independent sample covariance matrices arose from work with Monika, and the material on cross-covariance matrices is based on work with Apratim and Monika. The free cumulant-based construction of free products of $*$-probability spaces presented in the final chapter was developed through discussions with Soumendu. He, along with Kartick and Monika, have helped me bring the chapter on asymptotic freeness to its present shape. Soumendu has also diligently proofread the manuscript. Monika and Priyanka provided additional checking. I am greatly indebted to all of them.

Partha Pratim Halder worked with my raw cover photo and produced a wonderful finished picture. Sashi Kumar solved all my LaTeX issues in no time. Finally, it is always a great joy to work with the Acquiring Editor John Kimmel. I understand that he is retiring soon and the entire community of authors will miss him sorely.

Arup Bose
March 16, 2021
Kolkata, India

About the Author

Arup Bose is on the faculty of the Theoretical Statistics and Mathematics Unit, Indian Statistical Institute, Kolkata, India. He has research contributions in statistics, probability, economics and econometrics. He is a Fellow of the Institute of Mathematical Statistics (USA), and of all three Indian national science academies. He is a recipient of the S.S. Bhatnagar Prize and the C.R. Rao Award and holds a J.C. Bose Fellowship. He has been on the editorial board of several journals. He has authored four books—*Patterned Random Matrices*, *Large Covariance and Autocovariance Matrices* (with Monika Bhattacharjee), *U-Statistics, M_m-Estimators and Resampling* (with Snigdhansu Chatterjee) and *Random Circulant Matrices* (with Koushik Saha).

Notation

Real and complex numbers:

$\mathbb{Z}_n$, $\mathbb{N}$: the sets $\{1, 2, \ldots, n\}$ and $\{0, 1, 2, \ldots\}$.

$\mathbb{R}$, $\mathbb{C}$: set of real and complex numbers.

$\bar{z}$, $\mathcal{I}(z)$, $\mathcal{R}(z)$: conjugate, imaginary part, and real part of z; $\iota = \sqrt{-1}$.

$\mathbb{C}^+$ and $\mathbb{C}^-$: set of complex numbers z with $\mathcal{I}(z) > 0$ and $\mathcal{I}(z) < 0$.

δ_{xy}: 1 or 0 according as $x = y$ or $x \neq y$; $x \wedge y$: minimum of x and y.

$a_n \approx b_n$, $a_n = O(b_n)$, $a_n = o(b_n)$: $a_n/b_n \to 1$; a_n/b_n is bounded; $a_n/b_n \to 0$ as $n \to \infty$.

Counts:

$\#A$ or $|A|$: number of elements in the set A.

C_n: the n-th Catalan number $(n+1)^{-1}\binom{2n}{n}$.

Partitions:

w: word or partition.

$\mathcal{C}(2k)$: set of Catalan words of length $2k$.

$\mathcal{P}_n$: set of partitions of $\{1, \ldots, n\}$; $\mathcal{P} = \cup_{n \geq 1} \mathcal{P}_n$.

$\mathcal{P}_2(2k)$: set of pair-partitions in $\mathcal{P}_{2k}$ or pair-matched circuits of length $2k$.

$NC(k)$: set of non-crossing partitions of $\{1, \ldots, k\}$.

$NC_2(2k)$: set of non-crossing pair-partitions of $\{1, \ldots, 2k\}$.

π, τ, σ: partitions or circuits, also viewed as permutations.

$a \sim b$: a and b belong to the same block of a partition.

$\pi \wedge \sigma$ in $NC(n)$, $\pi \vee \sigma$ in $NC(n)$: the largest partition smaller than both π and σ, and the smallest partition larger than both π and σ.

$|\pi|$, $K(\pi)$: number of blocks of π; Kreweras complement of π.

$\mathbf{0}_n$, $\mathbf{1}_n$: the n-block and the 1-block partitions of $\{1, \ldots, n\}$.

$[\pi,\ \sigma]$, $P^{(2)}$: set (interval) of all partitions τ, $\pi \leq \tau \leq \sigma$; set of all intervals.

$\mu[\cdot,\ \cdot]$, $\zeta[\cdot,\ \cdot]$: Möbius function and its inverse for non-crossing partitions.

γ_n: cyclic permutation $1 \to 2 \to \cdots \to n \to 1$ of $\{1, \ldots, n\}$.

$\pi\gamma_n$: composition of the two permutations π and γ.

Probability and non-commutative probability:

E, Var: expectation and variance.

δ_x: point mass at x; MP_y: Marčenko-Pastur law with parameter y.

$\boxplus$ and $\boxtimes$: free additive and multiplicative convolutions.

F_A: empirical distributions of A.

$m_k(\mu)$, $m_k(A)$: h-the moment of the probability measure μ and F_A.

$\{x_\pi\}$: multiplicative extension of $\{x_n\}$ on the non-crossing partitions.

$c_{k_1,\ldots,k_n}$: cumulants

κ_n, $\kappa_\pi(\cdot)$: free cumulant and its multiplicative extension.

$\Rightarrow$: weak convergence of measures.

Algebras and matrices:

$\mathcal{A}$, $\mathbf{1}_\mathcal{A}$: unital algebra/$*$-algebra/C^*-algebra, identity/unity of $\mathcal{A}$.

$\mathcal{M}_n(\mathbb{C})$: $*$-algebra of $n \times n$ matrices with complex entries.

Tr: trace; tr: average trace; E tr: expectation of the average trace.

$\mathtt{sp}(a)$: spectrum of the operator a, eigenvalues of the matrix a.

φ: state on a unital algebra/$*$-algebra/C^*-algebra.

W_n: Wigner matrix. A real symmetric matrix with independent entries.

E_n: Elliptic matrix $((x_{ij}))$, the pairs (x_{ij}, x_{ji}) are i.i.d.

IID_n: Matrix with i.i.d. entries.

S-matrix: Sample covariance matrix, XX^*, entries of X are independent.

C_n: Cross-covariance matrix XY^*, entries of X and Y are pair-correlated.

T_n, H_n: Toeplitz matrix, $T_n = ((x_{|i-j|}))$, Hankel matrix, $H_n = ((x_{i+j-2}))$.

SC_n, RC_n: Symmetric circulant matrix with one right and one left shift.

s, s_1, s_2: standard semi-circular variables.

e: elliptic variable.

c: (standard) circular variable; v: cross-covariance variable.

Transforms:

s_μ and s_a: Stieltjes transform of the measure μ and the variable a.

$\mathcal{R}_a$ and $\mathcal{R}_\mu$: R-transform of the variable a and the measure μ.

S_a: S-transform of the variable a.

$f^{\langle -1 \rangle}$: inverse of f under composition.

Equations are numbered within a chapter. Definitions, theorems, examples, remarks, lemmas, corollaries are numbered within a section, within a chapter.

Introduction

Non-commutative probability or free probability was invented in the late 1980s by the mathematician Dan Voiculescu, initially in an attempt to solve important questions in operator algebras. Hence it grew as a part of pure mathematics. Soon, a central idea in non-commutative probability, that of free independence, caught the attention of probabilists. Surprising and interesting connections were discovered between free independence and large dimensional random matrices. These have now led to significant applications in statistics, wireless communication and other areas.

This book attempts to provide an elementary introduction to non-commutative probability, with a focus on free independence and links to random matrices. Many ideas of non-commutative probability are developed by analogy with concepts from classical probability. Having some familiarity with the basic convergence concepts in probability, such as weak and almost sure convergence, and a bit of mathematical maturity, will be helpful for the reader. Here is a description of how the material of the book develops.

Chapter 1 (*Classical independence, moments and cumulants*). For a vector of real random variables, the logarithm of the moment generating function (assumed finite) is the cumulant generating function. Probabilistic independence of the components is equivalent to the vanishing of all *mixed cumulants* and also to any joint product moment being the product of the marginal moments.

The moments and cumulants of random variables have an interesting one-to-one combinatorial relation. We explain it for a single random variable X. Let $\mathcal{P} = \cup \mathcal{P}_n$ be the set of all partitions of $\{1, \ldots, n\}, n \geq 1$. For any sequence of numbers $\{a_n : n \geq 1\}$ and any $\pi \in \mathcal{P}$, define

$$a_\pi = a_{|V_1|} \cdots a_{|V_k|}, \quad \text{where } \pi = \{V_1, \ldots, V_k\}.$$

The family $\{a_\pi : \pi \in \mathcal{P}\}$ is the *multiplicative extension* of $\{a_n : n \geq 1\}$. Let $\{m_n\}$ and $\{c_n\}$ be the moment and cumulant sequences of X. Then

$$m_\pi = \sum_{\sigma \in \mathcal{P}_n : \sigma \leq \pi} c_\sigma \text{ for all } \pi \in \mathcal{P}_n,\ n \geq 1.$$

This is a linear relation. Its inversion involves the *Möbius function*, say μ which is defined for all pairs (σ, π), $\sigma \leq \pi$ and we obtain

$$c_\pi = \sum_{\sigma \in \mathcal{P}_n : \sigma \leq \pi} m_\sigma \mu(\sigma, \pi) \text{ for all } \pi \in \mathcal{P}_n,\ n \geq 1.$$

The general moment-cumulant formula has nice applications. If $(X_1, \ldots, X_n)$ is Gaussian with mean zero, then only second order cumulants, which are variances and covariances, survive. This gives the formula of Isserlis (1918)[59]:

$$\mathrm{E}(X_1 X_2 \cdots X_n) = \begin{cases} 0 & \text{if } n \text{ is odd,} \\ \sum \prod \mathrm{E}(X_{i_k} X_{j_k}) & \text{if } n \text{ is even} \end{cases}$$

where the sum is over the product of all pair-partitions $\{i_k, j_k\}$ of $\{1, \ldots, n\}$.

We also explore the relation between weak convergence and convergence of moments or cumulants. As a simple application, a quick proof of the central limit theorem is given. This relation becomes useful when dealing with convergence of spectral distribution of random matrices.

Chapter 2 (*Non-commutative probability*). A *non-commutative probability space* is a pair $(\mathcal{A}, \varphi)$ where $\mathcal{A}$ is a *unital algebra* and the *state* φ is a linear functional on $\mathcal{A}$ such that $\varphi(\mathbf{1}_{\mathcal{A}}) = 1$ where $\mathbf{1}_{\mathcal{A}}$ is the identity of $\mathcal{A}$. The state acts like an expectation operator. If $\mathcal{A}$ is a $*$-algebra and φ is positive ($\varphi(aa^*) \geq 0$ for all $a \in \mathcal{A}$) then $(\mathcal{A}, \varphi)$ is called a *$*$-probability space*. The set of $n \times n$ random matrices $\mathcal{M}_n(\mathbb{C})$ with complex entries and usual conjugation operation along with the state $\varphi(\cdot) = n^{-1} \mathrm{E}\,\mathrm{Trace}(\cdot)$ is a $*$-probability space.

The analog of classical random variables are the variables of the algebra. The *distribution* of variables is the collection of their moments, defined via the state. The moments of an arbitrary self-adjoint variable may not define a unique probability law. Instead of the lattice of all partitions, the sub-lattice of all *non-crossing partitions* now becomes the central object. The corresponding Möbius function and the moments are used to define *free cumulants*.

Two important probability laws are introduced. The *standard semi-circular law* is symmetric about 0 and its even moments are the Catalan numbers. Its second free cumulant is 1 and all other free cumulants are 0. Hence it is also known as the *free Gaussian* law. Recall that for the classical Poissonian law with mean 1, all its cumulants are 1. A (compactly supported) probability law μ on $\mathbb{R}$ is called *free Poisson* with parameter 1 if all its free cumulants are 1.

Chapter 3 (*Free independence*). Amongst the available notions of independence in the non-commutative set up, *free independence*, discovered by Voiculescu (1991)[100] is the most useful. In analogy with classical independence, we define free independence by the vanishing of all mixed free cumulants. In Chapter 13 we show that this definition is equivalent to the traditional moment-based definition. Free independence serves as a rich source of interesting and important non-commutative variables and probability laws. It is also a crucial tool to explore relations between distributions, variables and random matrices when their dimensions grow. Existence of freely independent variables with specified distributions is guaranteed by a construction of the *free product* of $*$-probability spaces, analogous to the construction of product probability spaces. The details of this construction is given Chapter 13.

The *free binomial* and the *families* of *semi-circular, circular* and *elliptic* variables are defined. The semi-circular family is the non-commutative analog of the multivariate Gaussian random vector and the free analog of Isserlis' formula is established.

Suppose μ_1 and μ_2 are two compactly supported probability measures. The *free additive convolution* $\mu_1 \boxplus \mu_2$ is the probability measure whose free cumulants are the sum of the free cumulants of μ_1 and μ_2.

A crucial tool in the study of the combinatorics of non-crossing partitions is the *Kreweras complementation map* $K : NC(n) \to NC(n)$ where $NC(n)$ is the set of non-crossing partitions of $\{1, \ldots, n\}$. We provide some basic properties of this map and use them to establish relations between joint moments and joint free cumulants of polynomials of free variables and also to identify the distribution of variables from given joint moments and free cumulants formulae. The chapter ends with an introduction to the *compound free Poisson* variable, an analog of the classical compound Poisson random variable.

Chapter 4 (*Convergence*). Two classical probability convergence concepts can be retained in the non-commutative set up. The first is the *weak convergence* of probability laws of self-adjoint variables whenever these probability laws exist. The second is the convergence of joint moments or equivalently of free cumulants. Let $(\mathcal{A}_n, \varphi_n)$, $n \geq 1$ be a sequence of $*$-probability spaces and let $(\mathcal{A}, \varphi)$ be another $*$-probability space. Variables $\{a_i^{(n)} : i \in I\}$ from $\mathcal{A}_n$ are said to converge jointly to $\{a_i : i \in I\}$ from $\mathcal{A}$ if,

$$\varphi_n\big(\Pi(\{a_i^{(n)}, a_i^{*(n)} : i \in I\}\big) \to \varphi\big(\Pi(\{a_i, a_i^* : i \in I\}\big) \text{ for all polynomials } \Pi.$$

This is called *algebraic convergence* and we write $\{a_i^{(n)} : i \in I\} \xrightarrow{*} \{a_i : i \in I\}$.

The two notions are related. Suppose self-adjoint variables $a^{(n)} \xrightarrow{*} a$. If the moments of $a^{(n)}$ identify unique probability laws $\{\mu_n\}$ and if the limit moments define a unique probability law μ, then μ_n converges weakly to μ.

The *free multivariate central limit theorem*, the convergence of an appropriate sequence of free binomial laws to a free Poisson law, and more generally the convergence of the free convolution of mixtures of free binomial laws to a compound free Poisson law are established. The concept of *asymptotic freeness*, heavily used in Chapters 9 and 11, is also introduced.

Chapter 5 (*Transforms*). The three most important transforms in non-commutative probability, namely the $\mathcal{R}$, S and Stieltjes transforms, are introduced for the single variable case, and their interrelations are given. The $\mathcal{R}$ transform has the additive property for free additive convolution. The S-transform enjoys the multiplicative property for free multiplicative convolutions. The Stieltjes transform has many uses, in particular in the study of weak convergence and convergence of spectral distribution of non-Hermitian matrices. The concept of *free infinite divisibility* is described very briefly and the semi-circular and free Poisson laws serve as examples.

Chapter 6 (*C^*-probability space*). [The reader may skip the details of this chapter at the first reading.] A $*$-algebra $\mathcal{A}$ is a *C^*-algebra*, if it is equipped with a norm $||\cdot||$ which satisfies

$$||ab|| \leq ||a||||b|| \text{ and } ||a^*a|| = ||a||^2 \text{ for all } a, b \in \mathcal{A}.$$

A state φ is called *tracial* if $\varphi(ab) = \varphi(ba)$ for all $a, b \in \mathcal{A}$. If $(\mathcal{A}, \varphi)$ is a $*$-probability space where $\mathcal{A}$ is a C^*-algebra and φ is tracial, then it is a *C^*-probability space*. For example, $\mathcal{M}_n(\mathbb{C})$ is a C^*-algebra with the norm

$$||A|| = \sup_{x \in \mathbb{R}^n : ||x||=1} |x' A^* A x|, \text{ for all } A \in \mathcal{M}_n(\mathbb{C}).$$

In Chapter 2, we have seen examples of self-adjoint variables in $*$-probability spaces, whose moments give rise to probability laws on the real line. The advantage of working with a C^*-probability space is that the moments of *any* self-adjoint variable from such a space identify a probability law. The price to pay is that all these probability laws have compact support.

If a_1 and a_2 are self-adjoint in a C^*-probability space, with probability laws μ_1 and μ_2, then $a_1 + a_2$ also has a probability law. This law is $\mu_1 \boxplus \mu_2$ if a_1 and a_2 are free. This provides a more rigorous foundation to free additive convolution that was introduced in Chapter 3. Free multiplicative convolution is a bit more complicated. If in addition, a_1 is non-negative (so $a^{1/2}$ is well-defined) then the probability law of the self-adjoint variable $a^{1/2} a_2 a^{1/2}$ exists and is the free multiplicative convolution of the probability laws of a_1 and a_2 if these are free.

Chapter 7 (*Random matrices*). A matrix whose elements are random variables is a *random matrix.* The famous Wishart distribution (Wishart (1928)[107]) arose as the distribution of the entries of the *sample variance-covariance matrix* when the observations are from a normal distribution. Wigner (1955)[105] introduced a matrix, now called the *Wigner matrix.* Since then, there has been a flurry of activity and gradually links have been established between random matrices and numerous areas of sciences.

Suppose A_n is an $n \times n$ random matrix with eigenvalues $\lambda_1, \ldots, \lambda_n$. Its empirical spectral measure or distribution is given by

$$\mu_n = \frac{1}{n} \sum_{i=1}^{n} \delta_{\lambda_i},$$

where δ_x is the Dirac delta measure at x. Note that it is a *random distribution function.* Its expectation is a non-random distribution function and is called the *expected ESD* (EESD). The *Limiting Spectral Distribution (LSD)* is the weak limit, if it exists, of the EESD or of the ESD *in probability* or *almost surely.* In this book, we mostly deal with the convergence of the EESD.

Suppose the entries of the matrix come from an independent identically distributed *input* sequence with mean zero and variance one, and are placed in some *pattern* in the matrix. We impose the following restriction on the pattern: *Property B: the maximum number of times any entry is repeated in a row remains uniformly bounded across all rows as $n \to \infty$.*

We provide a quick and unified treatment of patterned random matrices, as the dimension grows. We show that for **symmetric** patterned matrices with Property B, the EESD is a tight sequence and all sub-sequential limits have moments bounded by some Gaussian moments. Some specific patterned matrices which satisfy Property B are listed below.

The *Wigner matrix* W_n (with scaling) is of the form $n^{-1/2}x_{i,j}$ where $x_{i,j} = x_{j,i}$ and the entries are otherwise independent. The *elliptic matrix* is a generalization of the Wigner matrix where the symmetry condition is dropped and it is assumed that $x_{i,j}$ and $x_{j,i}$ have a common correlation across i, j. Another generalization is the *IID matrix* where all entries are assumed to the i.i.d. Suppose X and Y are two $p \times n$ random matrices. Then $S = n^{-1}XX^*$ is called a *sample covariance matrix.* The matrix $C = n^{-1}XY^*$ is called a *cross-covariance matrix.* The random symmetric *Toeplitz and Hankel matrices* are $T_n = n^{-1/2}((x_{|i-j|}))_{1\leq i,j\leq n}$ and $H_n = n^{-1/2}((x_{i+j}))_{1\leq i,j\leq n}$. The *Symmetric Circulant matrix* (with scaling) is the usual right-circular shift circulant matrix with added condition of symmetry. The *Reverse Circulant matrix* (with scaling) is also a circulant but with a left-circular shift. It is also symmetric.

Chapter 8 (*Convergence of some important matrices*). We use the main result of Chapter 7 to establish the weak convergence of the EESD for several symmetric random matrices:

(i) Wigner matrix–semi-circular law;

(ii) Symmetric Circulant–standard Gaussian law.

(iii) Reverse Circulant–symmetrized Rayleigh law.

(iv) The limit laws for Toeplitz and Hankel exist, but are not known explicitly.

(v) S-matrix. If $p \to \infty, p/n \to y \neq 0$, the LSD is the Marčenko-Pastur law. If $p \to \infty, p/n \to 0$, the LSD of $\sqrt{np^{-1}}(S - pI_p)$ is the semi-circular law.

It is briefly explained how the weak convergence of the EESD can be upgraded to almost sure weak convergence of the ESD. These matrices also converge as elements of the $*$-probability spaces $(\mathcal{M}_n(\mathbb{C}), \mathrm{E\,tr})$ The LSD of non-Hermitian matrices, specifically the IID, the elliptic and the cross-covariance matrices, are not discussed, but their algebraic convergence is proved.

Chapter 9 (*Joint convergence I: single pattern*). If a sequence of patterned random matrices converges as elements of $(\mathcal{M}_n(\mathbb{C}), \frac{1}{n}\mathrm{E\,Tr})$ then their independent copies also converge jointly. In particular, independent copies of each of the matrices listed above converge jointly. Independent copies of Wigner, circular, elliptic, cross-covariance and S-matrices are asymptotically free. Symmetric Circulants are asymptotically (classically) independent. The Reverse Circulant limits exhibit the so-called *half independence.* The Toeplitz and Hankel limits do not exhibit any known notions of independence or dependence.

Chapter 10 (*Joint convergence II: multiple patterns*). In this chapter we show joint convergence when the sequences involved are not necessarily of the same pattern. A general sufficient condition for joint algebraic convergence of independent copies of matrices with different patterns is given. This condition is then verified for the following five matrices, two patterns at a time: Wigner, Toeplitz, Hankel, Reverse Circulant and Symmetric Circulant. As a consequence, the EESD of any symmetric matrix polynomial of multiple copies of the above five matrices converges weakly. For more than two patterns at a time, the proof is notationally complex and we do not pursue this.

Chapter 11 (*Asymptotic freeness of random matrices*). This chapter focuses on presenting different asymptotic freeness results for random matrices. We have already seen a few asymptotic freeness results in Chapter 9. We give some more results here.

Independent elliptic matrices are asymptotically free and are also free with other appropriate non-random matrices. In particular, the same is true of independent IID matrices and independent Wigner matrices. This result is used, along with embedding, to give a second proof of the asymptotic freeness of independent S matrices. Independent cross-covariance matrices are also asymptotically free. Some other asymptotic freeness results from the literature are mentioned but not proved.

Chapter 12 (*Brown measure*). For any element a in a C^*-probability space, there is an associated measure, introduced by Brown (1986)[28]. It is a probability measure that serves as a generalization of the ESD of matrices, to the eigenvalue distribution of operators in a von Neumann algebra which is equipped with a trace. We provide a very brief look at this concept. For more information please refer to Mingo and Speicher (2017)[72].

For many non-Hermitian matrices, the LSD and the Brown measure are identical. The Brown measure of an element is not always easy to compute, unless it is an R-diagonal variable. In that case, its Brown measure can be expressed in terms of the S-transform of aa^*. We use this result without proof and compute the Brown measure of some variables that arose naturally in the previous chapters.

Chapter 13 (*Tying three loose ends*). We finally tie up the the three loose ends that we have left so far. First, we establish a crucial bound for the Möbius function of $NC(n)$ which has been used in several proofs in the book.

Then we give the standard definition of free independence via properties of the moments and prove that this is equivalent to the free cumulant definition that we have given in Chapter 3 and have used throughout the book.

Theorem 3.2.1 of Chapter 3 on the free product construction of $*$-probability spaces is a vital result. In particular it guarantees the existence of free self-adjoint variables with given probability laws. This was not proved earlier and now we give a step-by-step proof. The development differs from the traditional proof in that it uses free cumulants instead of moments to carry out the construction. We leave out the analogous construction for C^*-probability space and the reader can find it in Nica and Speicher (2006)[74].

1

Classical independence, moments and cumulants

Classical independence is one of the fundamental concepts in probability theory. For random variables which have finite moment generating functions, independence is equivalent to the vanishing of all mixed cumulants. In general we can switch between the moments and cumulants of random variables via the Möbius function on the set of all partitions. This linear relation between multiplicative extensions of moments and cumulants leads us to Isserlis' formula, popularly known as Wick's formula, to compute mixed moments of Gaussian random variables. The idea of the Möbius function will be crucial to us later too when we define the analog of cumulants and moments for non-commutative variables. It will lead us quickly to the concept of free independence. Later we shall also establish a "free version" of Isserlis' formula.

1.1 Classical independence

Suppose $\{X_i : 1 \leq i \leq n\}$ is a collection of real valued random variables defined on some probability space $(\Omega, \mathcal{A}, P)$.

Definition 1.1.1. (Independence) Random variables $\{X_i : 1 \leq i \leq n\}$ are said to be *independent* if for all Borel sets $B_1, \ldots, B_n$,

$$P(X_i \in B_i,\ 1 \leq i \leq n) = \prod_{i=1}^{n} P(X_i \in B_i).$$

◇

The joint characteristic function of $\{X_i : 1 \leq i \leq n\}$ is defined as

$$\phi_{X_1,\ldots,X_n}(t_1, \ldots, t_n) := \mathrm{E}\left[\exp\left\{\iota \sum_{j=1}^{n} t_j X_j\right\}\right] \text{ for all } t_1, \ldots, t_n \in \mathbb{R}.$$

A commonly used equivalent criterion for independence is

$$\phi_{X_1,\ldots,X_n}(t_1, \ldots, t_n) = \prod_{j=1}^{n} \phi_{X_j}(t_j) \text{ for all } t_1, \ldots, t_n \in \mathbb{R}.$$

DOI: 10.1201/9781003144496-1

The joint *moment generating function* (m.g.f.) is defined as

$$M_{X_1,\ldots,X_n}(t_1,\ldots,t_n) = \mathrm{E}\left[\exp\left\{\sum_{j=1}^{n} t_j X_j\right\}\right],\ t_1,\ldots,t_n \in \mathbb{R}.$$

In this book we shall always assume that the m.g.f. is finite for all $t_1,\ldots,t_n$ in a neighborhood N of the origin in $\mathbb{R}^n$. In that case M has the following power series expansion:

$$M_{X_1,\ldots,X_n}(t_1,\ldots,t_n) = \sum_{k_1,\ldots,k_n=0}^{\infty} \frac{t_1^{k_1}\cdots t_n^{k_n}}{k_1!\cdots k_n!}\,\mathrm{E}\left(X_1^{k_1}\ldots X_n^{k_n}\right),\ (t_1,\ldots,t_n)\in N.$$

The following lemma is often useful in proving independence.

Lemma 1.1.1. Let the m.g.f. $M_{X_1,\ldots,X_n}(t_1,\ldots,t_n)$ exist in a neighborhood N of the origin. Then, $\{X_i : 1 \le i \le n\}$ are independent if and only if

$$M_{X_1,\ldots,X_n}(t_1,\ldots,t_n) = \prod_{j=1}^{n} M_{X_j}(t_j) \text{ for all } (t_1,\ldots,t_n)\in N.$$

♦

The above product rule can be converted to an addition rule by taking the logarithm. For convenience, let us write $t = (t_1,\ldots,t_n)$.

Definition 1.1.2. (Cumulant generating function) If $M_{X_1,\ldots,X_n}$ is finite in an open neighborhood N of the origin, then the joint *cumulant generating function* (c.g.f.), is defined as

$$C_{X_1,\ldots,X_n}(t_1,\ldots,t_n) := \log M_{X_1,\ldots,X_n}(t_1,\ldots,t_n),\ t\in N$$

which also has a power series expansion of the form

$$C_{X_1,\ldots,X_n}(t_1,\ldots,t_n) = \sum_{k_1,\ldots,k_n=0}^{\infty} \frac{t_1^{k_1}\cdots t_n^{k_n}}{k_1!\cdots k_n!} c_{k_1,\ldots,k_n}(X_1,\ldots,X_n),\ t\in N.$$

The real numbers $c_{k_1,\ldots,k_n}(X_1,\ldots,X_n)$ are called the cumulants of $\{X_i : 1 \le i \le n\}$. If $k_j \neq 0$ for at least two indices j, then $c_{k_1,\ldots,k_n}(X_1,\ldots,X_n)$ is called a *mixed cumulant* of $\{X_i : 1 \le i \le n\}$. ◇

It is easy to convince oneself that $c_{k_1,\ldots,k_n}$ involves only those moments $\mathrm{E}(X_1^{t_1}\cdots X_n^{t_n})$ where $t_j \le k_j$ for all j.

A restatement of Lemma 1.1.1 in terms of the cumulants is the following. Its proof is easy and is left as an exercise.

Lemma 1.1.2. Suppose that $\{X_i : 1 \le i \le n\}$ have a finite m.g.f. in a neighborhood N of the origin. Then they are independent if and only if

$$C_{X_1,\ldots,X_n}(t_1,\ldots,t_n) = \sum_{j=1}^{n} C_{X_j}(t_j), \quad \text{for all } t_1,\ldots,t_n.$$

That is, they are independent if and only if all mixed cumulants are zero. ♦

Example 1.1.1. A random variable X follows the Gaussian law or the normal probability law with mean μ and variance σ^2 (we write $X \sim N(\mu,\ \sigma^2)$) if,

$$M_X(t) = \exp\left(t\mu + \frac{1}{2}t^2\sigma^2\right),\ t \in \mathbb{R}$$

or, equivalently,

$$C_X(t) \quad = \quad t\mu + \frac{1}{2}t^2\sigma^2,\ t \in \mathbb{R}.$$

In that case

$$c_1(X) = \mu, \quad c_2(X) = \sigma^2 \quad \text{and} \quad c_n(X) = 0,\ n \geq 3. \tag{1.1}$$

Incidentally, there exists no other non-trivial probability law which has only finitely many non-vanishing cumulants. This follows from a theorem of Marcinkiewicz (1938)[70]. See Lukacs (1970, page 213)[66] for a proof of this difficult result. ▲

Example 1.1.2. The pair (X, Y) has a bivariate Gaussian law if and only if

$$c_{k_1,k_2}(X_1, X_2) = 0 \quad \text{whenever} \quad k_1 + k_2 \geq 3\,.$$

A similar property also holds for Gaussian vectors of any dimension. Moreover, using an appropriate linearity property of the cumulants, it is easy to see that any linear transformation of a Gaussian vector is again Gaussian. The proof of this is left as an exercise. ▲

Example 1.1.3. If X follows the Poisson law with mean λ (we write $X \sim Poi(\lambda)$), then, for all $t \in \mathbb{R}$,

$$M_X(t) \quad = \quad \exp\left(\lambda(e^t - 1)\right),\ \text{and}$$

$$\begin{aligned} C_X(t) \quad &= \quad \lambda(e^t - 1) \\ &= \quad \lambda\sum_{j=1}^{\infty}\frac{t^j}{j!}. \end{aligned}$$

Hence

$$c_j(X) = \lambda \ \text{ for all } j \geq 1. \tag{1.2}$$

Conversely, a random variable that satisfies (1.2) is distributed as $Poi(\lambda)$. ▲

Example 1.1.4. By using the cumulants in Examples 1.1.1 and 1.1.3 above, it can be easily shown that sums of independent Gaussian random variables are again Gaussian and likewise for Poisson. ▲

1.2 CLT via cumulants

The following lemma will be useful to us. We denote weak convergence of probability measures by $\Rightarrow$.

Lemma 1.2.1. (Weak convergence via moments and cumulants) (a) Suppose that $\{Y_n : n \in \mathbb{N}\}$ is a sequence of random variables with distribution functions $\{G_n\}$ such that for all $k \in \mathbb{N}$,

$$\lim_{n\to\infty} \mathrm{E}(Y_n^k) = m_k \text{ (finite)}.$$

Further, suppose that there is a unique distribution function G whose k-th moment is m_k for every k. Then, $G_n \Rightarrow G$.

(b) Suppose that $\{Y_n : n \in \mathbb{N}\}$ is a sequence of random variables with distribution functions $\{G_n\}$ such that for all $k \in \mathbb{N}$,

$$\lim_{n\to\infty} c_k(Y_n) = c_k \text{ (finite)}.$$

Further suppose that there is a unique distribution function G whose k-th cumulant is c_k for all $k \geq 1$. Then, $G_n \Rightarrow G$. ♦

Proof. (a) Since all moments converge, all powers of $\{Y_n\}$ are uniformly integrable. Further, $\{G_n\}$ is a tight sequence. Consider any sub-sequence of $\{G_n\}$. By tightness, there is a further sub-sequence which converges weakly to, say, G. Then, by uniform integrability, all moments of G exist and

$$\int x^k dG(x) = m_k \text{ for all } k.$$

Now by our assumption, $\{m_k\}$ determines G uniquely. That is, the limit does not depend on the chosen sub-sequence. Hence the original sequence $\{G_n\}$ converges weakly to G. This completes the proof of (a). Proof of Part (b) can be done easily by using Lemma 1.3.1 and Part (a). This is left as an exercise. ■

Remark 1.2.1. (Moments and uniqueness of measures) Suppose μ is a probability law with the moment sequence $\{m_k\}$. There are many sufficient conditions known for μ to be the unique probability law with the moment sequence $\{m_k\}$. Here are three of them in increasing order of generality. While (a) is easy to prove using the Stone-Weierstrass theorem, for proofs of (b) and (c), see Bose (2018)[20] and Bai and Silverstein (2010)[5].

(a) *Compact support.* If μ has compact support, then it is unique.

(b) *Riesz's condition.* (Riesz (1923)[80]) If $\liminf \frac{1}{k} m_{2k}^{\frac{1}{2k}} < \infty$, then μ is unique.

(c) *Carleman's condition.* (Carleman (1926)[32]) If $\sum_{i=1}^{\infty} m_{2k}^{-\frac{1}{2k}} = \infty$, then μ is unique. ●

Let us show how the classical central limit theorem (CLT) can be proved using cumulants or moments by an application of Lemma 1.2.1.

Theorem 1.2.2. (CLT via cumulants) (a) Suppose that $\{Y_n : n \in \mathbb{N}\}$ is a sequence of i.i.d. bounded random variables with mean zero and variance one. Then,

$$n^{-1/2} \sum_{i=1}^{n} Y_i \Rightarrow Y \,,$$

where $Y \sim N(0, 1)$.
(b) The above remains true if the boundedness condition is dropped. ◆

Proof. First note that the Gaussian probability law is uniquely determined by its moments. This can be concluded by verifying that the Gaussian moments satisfy Riesz's condition.
(a) Define

$$Z_n := n^{-1/2} \sum_{i=1}^{n} Y_i,\ n \geq 1\,.$$

Since the variables are bounded, all moments are finite. Hence it suffices to show that, for all $k \in \mathbb{N}$,

$$\lim_{n\to\infty} c_k(Z_n) = \begin{cases} 1 & k = 2\,, \\ 0 & k \neq 2\,. \end{cases} \tag{1.3}$$

We shall use the following two facts. Their proofs are left as an exercise.

Fact 1. For any random variable X, and any real constants a and b,

$$c_j(aX + b) = a^j c_j(X) \ \text{ for all } \ j \geq 2.$$

Fact 2. For independent random variables X and Y,

$$c_j(X + Y) = c_j(X) + c_j(Y) \ \text{ for all } \ j \geq 1.$$

Using the above facts and the fact that Y_i are identically distributed, it follows that, for any $k \in \mathbb{N}$,

$$\begin{aligned} c_k(Z_n) &= nc_k(n^{-1/2}Y_1) \\ &= n^{1-k/2}c_k(Y_1) \to 0 \ \text{ if } \ k \geq 3. \end{aligned}$$

On the other hand, it is easy to check that

$$\mathrm{E}(Y_1) = c_1(Y_1)\,, \tag{1.4}$$

$$\mathrm{E}(Y_1^2) = c_2(Y_1) + (c_1(Y_1))^2\,. \tag{1.5}$$

Therefore, $c_2(Y_1) = 1$ which implies that $c_2(Z_n) = 1$ for all n. Finally,

$c_1(Z_n) = 0$ because $c_1(Y_1) = 0$. This establishes (1.3) and the proof in this case is complete by invoking Lemma 1.2.1.

(b) We can first truncate the random variables at a fixed level B and invoke Part (a) to show that its standarized sum obeys the CLT. Then we can show that the variance of the difference is upper bounded (across n) by a quantity which depends on B and which tends to 0 as $B \to \infty$. Then the proof is completed by using a limit argument. The details are left as an exercise. ■

1.3 Cumulants to moments

The formula for deriving moments from cumulants is relatively easier than going in the reverse direction. So we do this first. We focus on the case when we have a single random variable. Suppose X is a random variable with m.g.f. $M_X(\cdot)$ and c.g.f. $C_X(\cdot)$ that are finite in a neighborhood N of the origin. Let us begin by understanding the relation between the first few moments and cumulants. Let

$$m_n = \mathrm{E}(X^n),\ n \in \mathbb{N}.$$

Therefore,

$$\begin{aligned} M_X(t) &= 1 + \sum_{n=1}^{\infty} \frac{m_n}{n!} t^n,\ t \in N, \\ C_X(t) &= \log\big(1 + \sum_{n=1}^{\infty} \frac{m_n}{n!} t^n\big),\ t \in N. \end{aligned}$$

The relation between m_1, m_2, $c_1(X)$ and $c_2(X)$ is as in (1.4) and (1.5). A similar calculation will show that

$$m_3 = c_3(X) + 3c_1(X)c_2(X) + (c_1(X))^3. \tag{1.6}$$

The set $\{1,2\}$ has two partitions, namely $\{\{1\},\{2\}\}$ and $\{\{1,2\}\}$. Likewise, the set $\{1,2,3\}$ has five partitions, $\{\{1\},\{2\},\{3\}\}$, $\{\{1,2\},\{3\}\}$, $\{\{1,3\},\{2\}\}$, $\{\{1\},\{2,3\}\}$ and $\{\{1,2,3\}\}$. Each of these five partitions contributes a term, in a *multiplicative* way, in the formula (1.6).

This happens to be a general feature that captures the dependence between moments and cumulants. We now elaborate on this.

Define

$$\begin{aligned} \mathcal{P}_n &:= \text{Set of all partitions of } \{1,\ldots,n\}, \\ \mathcal{P} &:= \cup_{n=1}^{\infty} \mathcal{P}_n. \end{aligned}$$

Any partition π can be written as $\pi = \{V_1, \ldots, V_k\}$ where $V_1, \ldots, V_k$ are the *blocks* of π, and $|\pi|$ will denote the number of blocks of π. The number of elements in a block V of π will be denoted by $\#V$ or $|V|$. For example, the partition $\pi = \{\{1,2\},\{3\}\}$ of $\{1,2,3\}$ has two blocks, $V_1 = \{1,2\}$ and $V_2 = \{3\}$ Then $|\pi| = 2$, $|V_1| = 2$ and $\#V_1 = 1$. A partition π is called a *pair-partition* if $|V_i| = 2$ for all i. For any $\pi, \sigma \in \mathcal{P}_n$, we say that $\pi \leq \sigma$ if every block of π is contained in some block of σ. It is easy to check that this defines a partial order and is called the *reverse refinement partial order.* The partitions $\{\{1\},\ldots,\{n\}\}$ and $\{1,\ldots,n\}$ shall be denoted by $\mathbf{0}_n$ and $\mathbf{1}_n$ respectively. Clearly, they are the smallest and the largest elements of $\mathcal{P}_n$.

Before we express the moments in terms of the cumulants and the partitions, we need the idea of a *multiplicative extension* of a sequence.

Definition 1.3.1. (Multiplicative extension) Suppose that $\{a_n : n \geq 1\}$ is any sequence of complex numbers. For any $\pi \in \mathcal{P}$, define

$$a_\pi := a_{|V_1|} \cdots a_{|V_k|}\,, \quad \text{whenever} \quad \pi = \{V_1, \ldots, V_k\}.$$

Then $\{a_\pi : \pi \in \mathcal{P}\}$ is the *multiplicative extension* of $\{a_n : n \geq 1\}$. ◇

Now we can state our formula which gives a *linear relation* between $\{m_\pi\}$ and $\{c_\pi\}$.

Lemma 1.3.1. Suppose that X is a random variable with moment and cumulant sequence $\{m_n\}$ and $\{c_n\}$ respectively. Then,

$$m_\pi = \sum_{\sigma \in \mathcal{P}_n:\ \sigma \leq \pi} c_\sigma \text{ for all } \pi \in \mathcal{P}_n,\ n \geq 1\,, \tag{1.7}$$

where $\{c_\sigma :\ \sigma \in \mathcal{P}\}$ is the multiplicative extension of $\{c_n\}$. In particular,

$$m_n = \sum_{\sigma \in \mathcal{P}_n} c_\sigma,\ n \geq 1\,. \tag{1.8}$$

♦

Proof. Let M and C denote the m.g.f. and the c.g.f. of X respectively, defined in a neighborhood N of the origin. Note that

$$M(t) = \exp\left(C(t)\right),\ t \in N\,.$$

Hence, the first derivative of M is given by

$$M^{(1)}(t) = C^{(1)}(t) \exp\left(C(t)\right),\ t \in N\,.$$

Differentiating once again, we get

$$\begin{aligned} M^{(2)}(t) &= C^{(2)}(t) \exp\left(C(t)\right) + \left(C^{(1)}(t)\right)^2 \exp\left(C(t)\right),\ t \in N \\ &= \exp\left(C(t)\right) \left[C^{(2)}(t) + \left(C^{(1)}(t)\right)^2\right]. \end{aligned}$$

Proceeding similarly, it can be proved by induction that for all $n \geq 1$,

$$M^{(n)}(t) = \exp(C(t)) \sum_{\pi=\{V_1,\ldots,V_k\}\in\mathcal{P}_n} C^{(|V_1|)}(t)\cdots C^{(|V_k|)}(t),\ t \in N\,.$$

Using the above equation with $t = 0$, (1.8) follows. The proof of (1.7) is left as an exercise. ■

Example 1.3.1. Suppose that $X \sim N(0,1)$. Then, by using its m.g.f., it is easy to check that

$$c_2(X) = 1, \text{ and } c_j(X) = 0 \text{ for all } j \neq 2\,.$$

Hence

$$c_\sigma = \begin{cases} 1 & \text{if } \sigma \text{ is a pair-partition,} \\ 0 & \text{otherwise.} \end{cases}$$

Therefore

$$\begin{aligned} m_n &= \text{the number of pair-partitions in } \mathcal{P}_n \\ &= \begin{cases} \frac{(2k)!}{2^k k!} & \text{if } n = 2k \text{ is even,} \\ 0 & \text{if } n \text{ is odd.} \end{cases} \end{aligned} \tag{1.9}$$

▲

Example 1.3.2. Suppose that $X \sim Poi(\lambda)$. Then, we have already seen that

$$c_j = \lambda,\ j = 1, 2, \ldots.$$

Hence,

$$c_\sigma = \lambda^{|\sigma|} \text{ for any partition } \sigma.$$

As a consequence,

$$m_n = \sum_{j=1}^{n} \lambda^j S(n,j)\,,$$

where

$$S(n,j) := \#\{\pi \in \mathcal{P}_n : |\pi| = j\}.$$

Here and elsewhere, for any set A, $\#A$ denotes the number of elements of A. In particular, if $\lambda = 1$, then

$$m_n = \#\mathcal{P}_n.$$

The numbers $S(n,j)$ are the *Stirling numbers of the second kind* and $\sum_{j=1}^{n} S(n,j)$ are the *Bell numbers.* Stanley (2011)[96] and Stanley (2001)[95] are definitive sources for number sequences arising in combinatorics. ▲

1.4 Moments to cumulants, the Möbius function

Since it is obvious that the moments and cumulants are in one to-one correspondence, the linear relation (1.7) is invertible. This inverse relation may be expressed via the Möbius function. We need some preliminaries before we can define this function.

Let P be any finite partially ordered set (POSET) with the partial order $\leq$. For any two elements π and τ, consider the sets

$$V_1 := \{\sigma : \pi \geq \sigma \text{ and } \tau \geq \sigma\},\ V_2 := \{\sigma : \pi \leq \sigma \text{ and } \tau \leq \sigma\}.$$

If there is a unique largest element of V_1 then it will be denoted by $\pi \wedge \tau$. Similarly, if there is a unique smallest element of V_2 then it will be denoted by $\pi \vee \tau$.

Definition 1.4.1. (Lattice) The POSET $(P,\ \leq)$ is called a *lattice* if both $\pi \wedge \tau$ and $\pi \vee \tau$ exist for all π and τ in P. ◇

For example, it is evident that $\mathcal{P}_n$ is a lattice for every n.

For any two elements $\pi \leq \sigma$, $(\pi,\ \sigma)$ denotes the (ordered) pair π, σ. The *interval* $[\pi,\ \sigma]$ is defined in the natural way:

$$[\pi,\ \sigma] := \{\tau : \pi \leq \tau \leq \sigma\}.$$

Let the set of all intervals of P be denoted by

$$P^{(2)} := \{[\pi,\ \sigma] : \pi, \sigma \in P, \pi \leq \sigma\}. \tag{1.10}$$

For any two complex valued functions $F, G : P^{(2)} \to \mathbb{C}$, their *convolution* $F * G : P^{(2)} \to \mathbb{C}$ is defined by:

$$F * G\ [\pi,\ \sigma] := \sum_{\substack{\rho \in P \\ \pi \leq \rho \leq \sigma}} F[\pi,\ \rho]G[\rho,\ \sigma]. \tag{1.11}$$

It is easy to see that $*$ is not commutative but is associative.
F is said to be *invertible* if there exists a (unique) G (called the inverse of F), for which

$$F * G\ [\pi,\ \sigma] = G * F\ [\pi,\ \sigma] = \mathbb{I}(\pi = \sigma),\ \forall \pi \leq \sigma \in P \tag{1.12}$$

where $\mathbb{I}$ is the indicator function.

We need the following lemma on the existence of the inverse.

Lemma 1.4.1. Suppose $(P,\ \leq)$ is a finite lattice. Then the function F on $P^{(2)}$ is invertible if and only if $F[\pi,\ \pi] \neq 0$ for every $\pi \in P$. ♦

Proof. We will use the following crucial observation to prove the lemma. Suppose $n = \#P$. Then P can be enumerated as $E = \{\pi_1, \pi_2, \ldots, \pi_n\}$ where, for any $i < j$, either π_i, π_j are not ordered or $\pi_i \le \pi_j$. We leave the proof of this fact as an exercise for the reader. There may be more than one such enumeration but we will choose and fix one such and use this particular enumeration of P for the rest of the proof.

Now, with the function F, associate the $n \times n$ matrix $F_E = ((F_{ij}))$ where

$$F_{ij} = \begin{cases} F[\pi_i,\ \pi_j] & \text{if} \quad i \le j \text{ and } \pi_i \le \pi_j, \\ \\ 0 & \text{otherwise.} \end{cases} \tag{1.13}$$

Then F_E is an upper triangular matrix and its diagonal elements are $\{F[\pi_i, \pi_i]\}$.

Now suppose $F[\pi,\ \pi] \neq 0$ for all π. Then all the diagonal elements of F_E are non-zero and hence F_E is invertible and its inverse is also upper triangular. Call the inverse G_E. Now construct a function G on $P^{(2)}$ using the matrix G_E by reversing the process described above. It is then easy to see that the matrix relation $F_E G_E = G_E F_E = I$ is equivalent to saying that the function G is the inverse of the function F.

The converse is easy. If F has an inverse G, then by definition of inverses,

$$F * G\ [\pi,\ \pi] = F[\pi,\ \pi]G[\pi,\ \pi] = \mathbb{I}(\pi = \pi) = 1, \text{ for all } \pi \in P.$$

Hence $F[\pi,\ \pi] \neq 0$ for all π. ■

Definition 1.4.2. (Zeta and Möbius functions) Suppose $(P,\ \le)$ is a finite lattice. The *Zeta function* ξ of P is defined by

$$\xi[\pi,\ \sigma] := 1 \text{ for all } [\pi,\ \sigma] \in P^{(2)}. \tag{1.14}$$

The *Möbius function* μ of P is the inverse of ξ. ◇

Note that the Möbius function exists by Lemma 1.4.1. Moreover, it is unique (does not depend on the enumeration that we used in the proof of Lemma 1.4.1). As a consequence,

$$\mu * \xi\ [\pi,\ \sigma] = \xi * \mu\ [\pi,\ \sigma] = I(\pi = \sigma) \text{ for all } \pi \le \sigma \in P. \tag{1.15}$$

Example 1.4.1. (From moments to cumulants) Recall $\mathcal{P}_n$, the set of all partitions of the set $\{1, 2, \ldots, n\}$ with the reverse refinement partial order. Clearly $(\mathcal{P}_n,\ \le)$ is a lattice. As a consequence, it is equipped with its own Möbius function. The moment-cumulant relation can be written in terms of the Möbius function as follows. We have already seen the linear relation

$$m_\pi = \sum_{\sigma \in \mathcal{P}_n:\ \sigma \le \pi} c_\sigma \quad \text{for all} \quad \pi \in \mathcal{P}_n. \tag{1.16}$$

This relation is a convolution relation: for each fixed n, identify $m_\pi, \pi \in \mathcal{P}_n$

with the function M on $\mathcal{P}_n^{(2)}$ which is given by $M[\mathbf{0}_n, \pi] = m_\pi$ and zero otherwise. We can do likewise for c_π. Then the above relation can be written as

$$c * \zeta = m.$$

Recall that $\mathcal{P}_n$ is a lattice. Let μ_n be the Möbius function on $\mathcal{P}_n^{(2)}$. Convolving both sides of the above equation with μ_n on the right, we get

$$c_\pi = \sum_{\sigma \leq \pi} m_\sigma \mu_n[\sigma, \pi], \ \pi \in \mathcal{P}_n. \tag{1.17}$$

This, when specialized to $\pi = \mathbf{1}_n$, gives c_n in terms of m_k, $k \leq n$, and the Möbius function μ_n. Incidentally, μ_n is defined on $\mathcal{P}_n^{(2)}$ for each fixed n. We can define μ on $\cup_n \mathcal{P}_n^{(2)}$ in the obvious way. ▲

1.5 Classical Isserlis' formula

Now suppose $X_1, \ldots, X_n$ are any real-valued random variables. Note that now the moments and the cumulants can be mixed and we need multi-indexes to write these down, keeping track of which random variable corresponds to which index. We may then define the multiplicative extension of the moment and cumulant sequences as before and relation (1.18) given below holds.

More generally, Lemma 1.3.1 can be easily extended to many random variables. The reader may formulate and prove this extension as an exercise. We shall encounter a much more involved situation that deals with non-commutative variables in the next chapter. There we establish a similar formula. This is the reason we omit the details in the present commutative case.

$$\mathrm{E}(X_1 X_2 \cdots X_n) = \sum_{\sigma \in \mathcal{P}_n} c_\sigma, \tag{1.18}$$

where c_σ are defined in a multiplicative way, using the cumulants of the random variables $X_1, \ldots, X_n$, taking care to preserve the appropriate indices. Moreover, if we use the Möbius function on $\mathcal{P}_n^{(2)}$, we can write any joint cumulant in terms of the moments. Note that if we take $X_i \equiv X$ for all i then we get back (1.8).

This result, when specialized to Gaussian random variables, yields an important formula, which was discovered by Isserlis (1918)[59], to compute moments of products of any collection of Gaussian random variables in terms of their covariances. Suppose that $X_1, \ldots, X_n$ are jointly Gaussian with mean zero. Observe that the joint m.g.f. of a Gaussian vector is of the form $\exp\{Q(\cdot)\}$ where Q is a quadratic function. Hence the c.g.f. is a quadratic function. Thus all cumulants (mixed or pure) of order larger than two are zero. Moreover, since the means are zero, all the first order cumulants are also zero. As a consequence, specializing relation (1.18) to the present situation,

$$\mathrm{E}(X_1X_2\cdots X_n) = 0 \quad \text{(if } n \text{ is odd)},$$

and

$$\mathrm{E}(X_1X_2\cdots X_n) = \sum\prod \mathrm{E}(X_{i_k}X_{j_k}) \quad \text{(if } n \text{ is even)},$$

where the sum is over the product of all *pair-partitions* $\{\{i_k, j_k\}\}$ of $\{1,\ldots,n\}$. The above formula is commonly referred to as *Wick's formula* after the work of Wick (1950)[104] in particle physics.

1.6 Exercises

1. Prove Lemma 1.1.1.
2. Show the following:

   (a) For any random variable X, and any real constants a and b,

   $$c_j(aX+b) = a^j c_j(X) \quad \text{for all} \quad j \geq 2.$$

   (b) For independent random variables X and Y,

   $$c_j(X+Y) = c_j(X) + c_j(Y) \quad \text{for all} \quad j \geq 1.$$

3. Identify the cumulants of a binomial random variable.
4. Find the cumulants of the exponential random variable.
5. Extend Example 1.1.2 to dimensions greater than 2.
6. Suppose μ is a probability measure on $\mathbb{R}$. Show that μ is compactly supported if and only if there exists a finite constant C such that the moments of μ satisfy $|m_k(\mu)| \leq C^k$ for all non-negative integers k. Further, the support of μ is contained in the interval $[-C,\ C]$.
7. Show that the moments of the Gaussian probability law satisfy Riesz's condition.
8. Show that Riesz's condition implies Carleman's condition.
9. Suppose for some constant C, the cumulant sequence $\{c_k\}$ satisfies $|c_k| \leq C^k$ for all $k \in \mathbb{N}$. Show that the moments also satisfy this bound (with a possibly different C).
10. Suppose X is a random variable with all moments finite. Show that all its odd cumulants are 0 if and only if all its odd moments are 0.

11. Show that if X is a random variable then its n-th cumulant is finite if and only if its n-th moment is finite. Moreover in this case, the moment-cumulant relation holds for all moments and cumulants of order n or less.
12. Show that any mixed cumulant $c_{k_1,\ldots,k_n}(X_1,\ldots,X_n)$ involves only those moments $\mathrm{E}(X_1^{t_1}\cdots X_n^{t_n})$ where $t_j \leq k_j$ for all j.
13. Show by using cumulants that the sum of independent Gaussian random variables is again Gaussian.
14. Show by using cumulants that any linear combination of jointly Gaussian random variables is again Gaussian.
15. Suppose X_n is a random variable which obeys the binomial law with parameters n and p_n such that $np_n \to \lambda$ as $n \to \infty$. By using cumulants show that the law of X_n converges to the Poisson law with parameter λ.
16. Prove Lemma 1.2.1 (b).
17. In Lemma 1.3.1, establish (1.7) using (1.8).
18. The d_2 metric is defined on the space of all probability laws with finite second moment as:

$$d_2(F,\ G) := \inf_{X\sim F, Y\sim G} [\mathrm{E}(X-Y)^2]^{1/2}.$$

It is known that $d_2(F_n, F) \to 0$ if and only if F_n converges to F weakly and the second moment of F_n converges to the second moment of F. Use d_2 to complete the proof of Theorem 1.2.2 (b).

19. Using Lemma 1.2.1 (b) and the above metric, establish the following CLT for U-statistics. Suppose h is a real-valued function from $\mathbb{R}^m \to \mathbb{R}$ which is symmetric in its arguments. Suppose $\{X_i\}$ is a sequence of i.i.d. random variables. Suppose that $\mathrm{E}[h(X_1,\ldots,X_m)] = 0$ and $\mathrm{Var}[h(X_1,\ldots,X_m)] < \infty$. Define

$$U_n := \frac{1}{\binom{n}{m}} \sum_{1\leq i_1<i_2<\cdots<i_m\leq n} h(X_{i_1},\ldots,X_{i_m}).$$

Show that as $n \to \infty$, $n^{1/2}U_n$ converges to the normal distribution with mean 0 and variance

$$\sigma_1^2 := m^2\,\mathrm{Var}[\mathrm{E}(h(X_1,\ldots,X_m)|X_1)].$$

20. Show that the number of pair-partitions of $\{1,\ldots 2n\}$ equals $\frac{(2n)!}{2^n n!}$.
21. State and prove an extension of Lemma 1.3.1 to several random variables so that a mixed moment of any order is expressed as a sum of the product of suitable cumulants of equal or lower order. Use this relation and the Möbius function on $\mathcal{P}_n$ to express any cumulant in terms of the moments of equal and lower order.

22. A complex random variable $Z = X + \iota Y$ is called *circularly symmetric central complex Gaussian*, if X and Y are independent real Gaussian with zero means. Show that Isserlis's formula remains true for complex Gaussian random variables.

23. Prove that the enumeration claimed in the beginning of the proof of Lemma 1.4.1 can always be achieved. Show by example that this enumeration need not be unique. Prove that nevertheless, the Möbius function is independent of the specific choice of the enumeration.

24. Suppose P is a finite lattice with the Möbius function μ. Show that for any π and σ such that $\pi \leq \sigma$,

$$\sum_{\pi \leq \tau \leq \sigma} \mu[\pi, \tau] = \sum_{\pi \leq \tau \leq \sigma} \mu[\tau, \sigma] = \begin{cases} 1 & \text{if } \pi = \sigma, \\ 0 & \text{if } \pi < \sigma. \end{cases} \tag{1.19}$$

25. Find the Möbius function for $\mathcal{P}_3$, the set of all partitions of $\{1, 2, 3\}$.

26. Suppose P_1 and P_2 are finite lattices and μ_1 and μ_2 are their associated Möbius functions. Suppose $f : P_1 \to P_2$ is a lattice isomorphism. [Unfamiliar reader, the definition of lattice isomorphism is available in Definition 13.1.1]. Show that

$$\mu_2[f(\pi), f(\sigma)] = \mu_1[\pi, \sigma], \quad \text{for all } \pi, \sigma \in P_1,\ \pi \leq \sigma.$$

27. The Möbius function was used to move between moments and free cumulants in the two formulae (1.16) and (1.17). Show that the following general version can be proved by the same arguments. Let P be a finite lattice and let μ be the Möbius function on $P^{(2)}$. Suppose $f, g : P \to \mathbb{C}$ are two functions. Then the following two relations are equivalent.

$$f(\pi) = \sum_{\sigma \in P, \sigma \leq \pi} g(\sigma) \tag{1.20}$$

$$g(\pi) = \sum_{\sigma \in P, \sigma \leq \pi} f(\sigma)\mu[\sigma, \pi]. \tag{1.21}$$

2

Non-commutative probability

A non-commutative probability space is an algebra equipped with a unital linear functional, called a state. The analogs of classical random variables are the non-commutative variables/elements of the algebra. We have to do away with the idea of a joint probability law, it being replaced by the linear functional, the analog of the expectation operator. The moments are defined via this linear functional. The distribution of variables is defined as the collection of their moments. Instead of the lattice of all partitions, the sub-lattice of all *non-crossing partitions* now becomes the central object. The corresponding Möbius function and the moments are used to define *free cumulants*, yielding a one to-one correspondence.

Two important probability laws, namely the *free Gaussian* and the *free Poisson* are also introduced in this chapter.

There are many books that cover more advanced topics which may be of interest to the reader. Speicher (1997)[92] and Nica and Speicher (2006)[74] are excellent references for a combinatorial approach to non-commutative probability or free probability. Much of the material covered in Chapters 2–6 are adapted from the latter. Mingo and Speicher (2017)[72] and Anderson, Guionnet and Zeitouni (2010)[1] are advanced monographs on free probability and random matrices. Other books on free probability include Voiculescu, Dykema and Nica (1992)[102] and the highly technical Hiai and Petz (2000)[58].

2.1 Non-crossing partition

Let S be a finite totally ordered set. For any partition $\pi = \{V_1, V_2, \ldots, V_r\}$ of S, we call $V_1, V_2, \ldots, V_r$ the *blocks* of π. Given two elements $p, q \in S$, we write $p \sim_\pi q$ if p and q belong to the same block of π.

Definition 2.1.1. (Non-crossing partition) A partition π of S is called *crossing* if there exists $p_1 < q_1 < p_2 < q_2$ in S such that $p_1 \sim_\pi p_2$ and $q_1 \sim_\pi q_2$ but $\{p_1, p_2, q_1, q_2\}$ are not in the same block. If π is not crossing, then it is called *non-crossing*. ◇

We shall be mostly interested in the case where $S = \mathbb{Z}_n = \{1, 2, \ldots, n\}$ with the natural ordering. The set of all non-crossing partitions of $\mathbb{Z}_n$ will be

DOI: 10.1201/9781003144496-2

denoted by $NC(n)$. Clearly it inherits the reverse refinement partial order of $\mathcal{P}_n$ described in Chapter 1.

Example 2.1.1. The partitions $\pi_1 = \{\{1,4,7\},\{2,3\}\},\{5,6\}\}$ and $\pi_2 = \{\{1,3,5\},\{2,6\},\{4\},\{7\}\}$ are respectively non-crossing and crossing partitions of $\mathbb{Z}_7$. ▲

Consider $\mathbb{Z}_k = \{1,2,\ldots,k\}$. Arrange these integers on a circle sequentially. Consider any partition of this set and draw edges between all points of each partition block. Then the partition is *non-crossing* if and only if edges of different blocks do not cross. A pictorial illustration of the non-crossing and crossing partitions π_1 and π_2 of Example 2.1.1 is given in Figure 2.1.

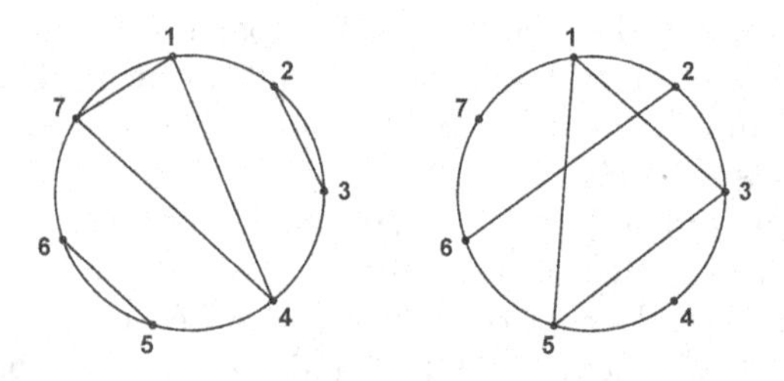

FIGURE 2.1
Non-crossing and crossing partitions π_1 and π_2 of Example 2.1.1.

Lemma 2.1.1. The partially ordered set $NC(n)$ is a lattice. ♦

Proof. First take any two elements $\pi = \{D_1,\ldots,D_t\}$ and $\sigma = \{W_1,\ldots,W_s\}$ in $NC(n)$.

Consider the partition $\tau = \{D_i \bigcap W_j, 1 \le i \le t, 1 \le j \le s\}$. It is easy to check that τ is a non-crossing partition. Clearly, it is the largest partition which is smaller than both π and σ. This shows that $\tau = \pi \wedge \sigma$ is in $NC(n)$.

Now consider π, σ in $NC(n)$ and consider the set

$$U := \{\tau : \tau \ge \pi, \tau \ge \sigma\}.$$

This set is non-empty since $\mathbf{1}_n$ is in U. List the elements of U as $U = \{\sigma_1, \sigma_2, \ldots, \sigma_k\}$. Then let

$$\eta := \sigma_1 \wedge \sigma_2 \wedge \cdots \wedge \sigma_k.$$

By using the previous step inductively, it follows that $\eta \in NC(n)$. Clearly η is the smallest element of U and hence $\eta = \pi \vee \sigma$. This completes the proof. ■

2.2 Free cumulants

Since $NC(n)$ is a finite lattice for every $n \geq 1$, there is a (unique) Möbius function (see Definition 1.4.2). We shall use a common notation μ for these functions. We must emphasize that this Möbius function is completely different from the one connected to $\mathcal{P}_n$, $n \geq 1$.

Using this function, we are now ready to define free cumulants of a random variable or a probability law whose moments are all finite. While this definition may not have much intuitive appeal right away, as we proceed, we shall realize how and why this is a very meaningful concept.

Definition 2.2.1. (Free cumulant) Suppose X is a random variable with moments $\{m_n\}$. Then for every n, the n-th *free cumulant* of X or of its probability law is defined by

$$\kappa_n = \sum_{\pi \in NC(n)} m_\pi \mu[\pi, \mathbf{1}_n], \; n \geq 1 \tag{2.1}$$

where μ is the Möbius function of $NC(n)$, $n \geq 1$. ◇

Since the Möbius function of $NC(n)$ is entirely different from the Möbius function of $\mathcal{P}_n$, it follows that the set of free cumulants is *not* the same as the set of cumulants.

Let us see some quick consequences of Definition 2.2.1. As in Chapter 1, we can obtain the *multiplicative extension* of the sequence $\{\kappa_n\}$: for any $\pi = \{V_1, V_2, \ldots, V_k\} \in NC(n)$, $n \geq 1$, define

$$\kappa_\pi = \prod_{i=1}^{k} \kappa_{|V_i|}.$$

Then using (2.1), we can say that the following convolution equation holds:

$$k = m * \mu,$$

in the sense that, for any $\pi \in NC(n)$, $n \geq 1$,

$$\kappa_\pi = \sum_{\sigma \in NC(n):\, \sigma \leq \pi} m_\sigma \mu[\sigma, \pi]. \tag{2.2}$$

Hence, by the inversion Lemma 1.4.1, we can say

$$m_\pi = \sum_{\sigma \in NC(n):\, \sigma \leq \pi} \kappa_\sigma.$$

Example 2.2.1. In particular, the n-th order moment is given by

$$m_n = m_{1_n} = \sum_{\pi \in NC(n)} \kappa_\pi.$$

Recall the similar relation given in (1.8) of Chapter 1:

$$m_n = m_{1_n} = \sum_{\pi \in \mathcal{P}_n} c_\pi.$$

Thus the moments can be expressed in terms of either the cumulants or the free cumulants. If we use $n = 1$ in these relations, we get

$$m_1 = \kappa_1 = c_1.$$

For $n = 2$, $NC(2)$ equals $\mathcal{P}_2$ and has two elements:

$$\pi_1 = \{\{1\}, \{2\}\} \text{ and } \pi_2 = \{1, 2\}.$$

Thus we have

$$m_2 = \kappa_2 + \kappa_1^2 = c_2 + c_1^2.$$

As a consequence

$$\kappa_2 = m_2 - m_1^2 = c_2.$$

So, the first two cumulants and free cumulants coincide and are the mean and the variance respectively. Higher order free cumulants are increasingly harder to express in terms of moments. Detailed information about the Möbius function μ of the lattice of non-crossing partitions is needed for that. ▲

As in the classical case, given a sequence of numbers, it may require significant effort to decide if there is a unique probability law whose free cumulant sequence is the given sequence. Moreover, as we shall see later, moments in the non-commutative setup need not be attached to a probability law.

2.3 Free Gaussian or semi-circular law

Recall that a random variable X is standard Gaussian if and only if its second cumulant is 1 and all other cumulants are 0. We wish to define the *free Gaussian* law by demanding the same for free cumulants.

Before we can show that such a probability law exists, we need a result on counts of non-crossing partitions. The numbers

$$C_n := \frac{1}{n+1}\binom{2n}{n}, \quad n \geq 1,$$

TABLE 2.1
The first few Catalan numbers

n	C_n
1	1
2	2
3	5
4	14
5	42

are known as *Catalan numbers.* These numbers arise as counts of many useful objects in mathematics and other sciences. Table 2.1 gives the values of C_n for a few values of n. In particular, we shall make heavy use of their relation to counts of non-crossing partitions. Let

$$NC_2(2n) := \{\pi : \pi \in NC(2n) \text{ and it is a pair-partition}\}. \tag{2.3}$$

The next lemma shows that $NC(k)$ and $NC_2(2k)$ are in bijection and their cardinality is C_k. *Later in Lemma 8.2.2, we provide a refined form of this bijection.*

Lemma 2.3.1. For every integer $k \geq 1$,

$$\#NC(k) = \#NC_2(2k) = C_k. \tag{2.4}$$

So the number of non-crossing partitions of $\mathbb{Z}_k$ and the number of non-crossing pair-partitions of $\mathbb{Z}_{2k}$ are the same and equal the k-th Catalan number. ♦

Proof. We first sketch a proof of the second equality. For any non-crossing pair-partition, for each partition block, mark its first and second occurrences by $+1$ and -1 respectively.

For example, the non-crossing partitions $\{\{1,6\},\{2,5\},\{3,4\}\}$ and $\{\{1,2\},\{3,4\}\}$ are represented respectively by $(1,1,1,-1,-1,-1)$ and $(1,-1,1,-1)$.

In general, consider the last $+1$. It must be followed by a -1. Pair these two. Then move to the previous $+1$ and pair it with the next -1 available. Continue the process until all $+1$ and -1 are exhausted.

It is not hard to check that this provides a bijection between the non-crossing pair-partitions of $\mathbb{Z}_{2k}$ and sequences $\{u_l\}_{1\leq l\leq 2k}$ which satisfy: each $u_l = \pm 1$, $S_l = \sum_{j=1}^{l} u_j \geq 0$ for all $l \geq 1$, and $S_{2k} = 0$. By the *reflection principle* (recall the well-known ballot problem), the total number of such paths is easily seen to be $\frac{(2k)!}{(k+1)!k!} = C_k$. We omit the details. This proves the second equality.

Now we prove that $\#NC(k) = C_k$. Let $D_n = \#NC(n)$ for $n \geq 1$ and $D_0 = 1$. We must prove that $D_n = C_n$ for all $n \geq 1$. Consider a partition $\pi \in NC(n)$. Suppose that j is the last element in $\mathbb{Z}_n$ such that j and 1 are in the same block Note that any element in the set $\mathbb{Z}_j$ cannot be in the same partition block with any element in the set $\{j+1, \ldots, n\}$. Thus the blocks of π consist of a non-crossing partition of $\mathbb{Z}_j$ and of $\{j+1, \ldots, n\}$. By varying j, we get all the non-crossing partitions of $\mathbb{Z}_n$. Hence we get

$$D_n = D_0 D_{n-1} + D_1 D_{n-2} + \cdots + D_{n-1} D_0, \quad n \geq 2. \tag{2.5}$$

It is not too hard now to show that the unique solution to this equation is $D_n = C_n$. The details are left as an exercise. This proves the lemma completely. ■

Definition 2.3.1. (Semi-circular law) The *standard semi-circular law* is defined by the probability density

$$f(x) := \begin{cases} \frac{1}{2\pi}\sqrt{4-x^2} & \text{if } \; -2 < x < 2, \\ 0 & \text{otherwise.} \end{cases} \tag{2.6}$$

◇

Its moment sequence is given by

$$m_h = \begin{cases} C_n & \text{if } h = 2n, \\ 0 & \text{if } h \text{ is odd.} \end{cases} \tag{2.7}$$

Figure 2.2 illustrates the density (2.6). Even though the graph is not a semi-circle, it has become the norm to call it the semi-circular law. If a random variable X follows the semi-circylar law, then, for any $\sigma > 0$, the law of the random variable σX is also called the semi-circular law, with variance σ^2. Usually "semi-circular" will refer to the standard semi-circular.

The connection between the semi-circular law and the non-crossing partitions goes deeper. In Chapters 8 and 9 we shall see how the semi-circular law arises naturally in the study of convergence of Wigner random matrices.

To verify the moment formula (2.7), first note that since the law is symmetric about 0, all odd moments are 0. The even moment of order $2k$ equals

$$\begin{aligned} \frac{1}{2\pi}\int_{-2}^{2} x^{2k}\sqrt{4-x^2}dx &= \frac{1}{\pi}\int_{0}^{2} x^{2k}\sqrt{4-x^2}dx \\ &= \frac{2^{2k+1}}{\pi}\int_0^1 y^{k-1/2}(1-y)^{1/2}dy \quad (\text{setting } x = 2\sqrt{y}) \\ &= \frac{2^{2k+1}}{\pi}\frac{\Gamma(k+1/2)\Gamma(3/2)}{\Gamma(k+2)} \quad (\text{beta integral}) \\ &= \frac{1}{k+1}\binom{2k}{k} = C_k. \end{aligned}$$

Definition 2.3.2. (Free Gaussian) A random variable X or its probability law μ on $\mathbb{R}$ is said to be (standard) *free Gaussian* if and only if its second free cumulant is 1 and all other free cumulants are 0. ◇

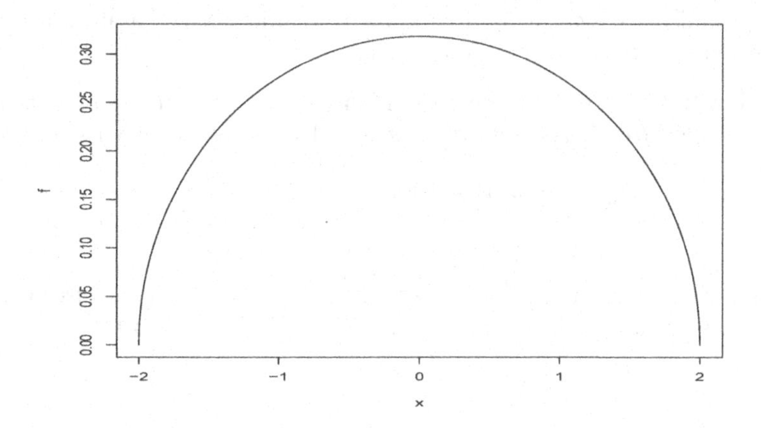

FIGURE 2.2
Density of the standard semi-circular law.

The above definition makes sense only if we are able to show that there indeed is a unique probability measure with these free cumulants. Define $\{m_n\}$ by the relation

$$m_n := \sum_{\pi \in NC(n)} \kappa_\pi.$$

Since $\kappa_2 = 1$, and all other free cumulants are 0, the only non-crossing partitions that can contribute to the right side of the above equation are those for which all blocks are of size two. Hence, if n is odd, $m_n = 0$. For the even cases, by Lemma 2.3.1,

$$m_{2n} = \#NC_2(2n) = C_n.$$

Hence $\{m_n\}$ are the moments of the standard semi-circular law. Since the latter is the unique law with these moments, we can conclude that the (standard) free Gaussian law is nothing but the (standard) semi-circular law.

If the random variable X has the standard free Gaussian law, then for any $\sigma > 0$, σX is also called a free Gaussian random variable, with variance σ^2.

2.4 Free Poisson law

Recall that X is a Poisson random variable with parameter λ if and only if all its cumulants are λ. We can now define a *free Poisson* random variable by demanding the same condition for the free cumulants of X.

Here we define the free Poisson law only with parameter 1. The general definition and discussion of the free Poisson with arbitrary parameter $\lambda > 0$ are postponed until we acquire more concepts.

Definition 2.4.1. (Free Poisson) A random variable X or its probability law μ on $\mathbb{R}$ is called *free Poisson* with parameter 1 if its free cumulants satisfy

$$\kappa_n(\mu) = 1 \text{ for all } \ n \geq 1.$$

◇

To show that such a probability law exists, let Y be a standard free Gaussian random variable. Then the moments of $X = Y^2$ are given by

$$\begin{aligned} m_n(X) &= \mathrm{E}(Y^{2n}) \\ &= \frac{1}{n+1}\binom{2n}{n} \\ &= \#NC(n) = \sum_{\pi \in NC(n)} 1. \end{aligned} \tag{2.8}$$

From (2.8) and the moment-free cumulant relation (1.16), we see that $\kappa_n(X) = 1$ for all n. So, X has the free Poisson law with parameter 1. Since this law has support $[0,\ 4]$, it is the unique law with these moments or free cumulants.

It is interesting to observe that while, in the classical case, the Gaussian and the Poisson laws are of different nature—absolutely continuous and discrete respectively, the standard free Gaussian Y and the free Poisson X with parameter 1 are both random variables with absolutely continuous probability laws with compact supports, and are related as $Y^2 = X$ in law.

2.5 Non-commutative and $*$-probability spaces

We now define the basic ingredients of a non-commutative setup. The basic structure on which our variables will live shall be either an algebra or a $*$-algebra.

Definition 2.5.1. (Non-commutative algebra) A collection $\mathcal{A}$ is called a (complex) *algebra* if it is a vector space (with addition $+$) over the complex numbers endowed with a multiplication with respect to which, for all $x, y, z \in \mathcal{A}$ and $\alpha \in \mathbb{C}$,

(i) $x(yz) = (xy)z$,

(ii) $(x + y)z = xz + yz$,

(iii) $x(y + z) = xy + xz$, and

(iv) $\alpha(xy) = (\alpha x)y = x(\alpha y)$.

This algebra is called *unital* if there exists an identity/unit element with respect to multiplication. We use $1_{\mathcal{A}}$ to denote this identity element. ◇

Definition 2.5.2. (Non-commutative probability space) Let $\mathcal{A}$ over $\mathbb{C}$ be a unital algebra with identity/unity $1_{\mathcal{A}}$. Let $\varphi : \mathcal{A} \to \mathbb{C}$ be a linear functional which is unital (that is $\varphi(\mathbf{1}_{\mathcal{A}}) = 1$.) Then $(\mathcal{A}, \varphi)$ is called a *non-commutative probability space* (NCP) and φ is called a *state.* ◇

Note that there are no probability laws involved. However, it is important as well as helpful to observe that φ is really the analog of the (classical) expectation operator. We shall refer to members of $\mathcal{A}$ as *elements* or *variables* as opposed to random variables in classical probability.

A state φ is said to be *tracial* if

$$\varphi(ab) = \varphi(ba), \quad \text{for all} \quad a, b \in \mathcal{A}. \tag{2.9}$$

Example 2.5.1. Suppose μ is a probability law on $\mathbb{R}$ with finite moments $\{m_k\}$. Then we can construct a unital algebra $\mathcal{A}$ generated by an indeterminate x and define φ on this algebra by declaring $\varphi(x^k) = m_k$. Then $(\mathcal{A}, \varphi)$ is an NCP and φ is tracial. ▲

We shall need a broader framework and this is given below.

Definition 2.5.3. ($*$-algebra) An algebra $\mathcal{A}$ is called a *$*$-algebra* if there exists a mapping $x \to x^*$ from $\mathcal{A} \to \mathcal{A}$ such that, for all $x, y \in \mathcal{A}$ and $\alpha \in \mathbb{C}$,

(i) $(x + y)^* = x^* + y^*$,

(ii) $(\alpha x)^* = \bar{\alpha} x^*$,

(iii) $(xy)^* = y^* x^*$, and

(iv) $(x^*)^* = x$. ◇

A variable a of a $*$-algebra $\mathcal{A}$ is called *self-adjoint* if $a = a^*$. Any variable a can always be written as $a = x + \iota y$ where x and y are self-adjoint—simply define

$$x := \frac{a + a^*}{2}, \quad \text{and} \quad y := \frac{\iota(a^* - a)}{2}.$$

A unital linear functional φ on a $*$-algebra is said to be *positive* if

$$\varphi(a^* a) \geq 0 \quad \text{for all} \quad a \in \mathcal{A}.$$

Definition 2.5.4. ($*$-probability space) Let $\mathcal{A}$ be a unital $*$-algebra with a state φ that is positive. Then $(\mathcal{A}, \varphi)$ is called a *$*$-probability space.* ◇

Note that a $*$-probability space is an NCP if we simply ignore the existence of the $*$-operation (and hence the positivity of the state).

We may often refer to both types of spaces as NCP unless we specifically need to make a distinction.

Example 2.5.2. Suppose $(\Omega, \mathcal{F}, P)$ is a classical probability space. Let $\mathcal{A} = L^\infty(\Omega, P)$ = set of all measurable and essentially bounded (bounded up to a set of measure zero) complex-valued functions. Define $\varphi(\cdot)$ by

$$\varphi(a) := \int_\Omega a(\omega) dP(\omega), \quad a \in \mathcal{A}.$$

Then clearly $(\mathcal{A}, \varphi)$ is a $*$-probability space. Actually, in this case, $\mathcal{A}$ is commutative, and, for any $a \in \mathcal{A}$, a^* is its complex conjugate.

The algebra $\mathcal{B} = L^{\infty,-}(\Omega, P)$ of all complex valued random variables with all moments finite is also a $*$-probability space with the same state. This state is tracial and positive. ▲

Example 2.5.3. Consider a classical standard normal random variable X on some probability space. Consider the algebra $\mathcal{A}$ of all polynomials $p(X)$ in X with complex coefficients and define the state φ as

$$\begin{aligned} \varphi(p(X)) &:= \mathrm{E}(p(X)) \\ &= \int p(x) d\mu(x), \end{aligned}$$

where μ denotes the standard Gaussian measure. It is easy to check that $(\mathcal{A}, \varphi)$ is a (commutative) $*$-probability space. ▲

Example 2.5.4. For any positive integer d, let

$$\mathcal{M}_d(\mathbb{C}) = \text{ set of all } \; d \times d \; \text{ matrices with complex entries}$$

with the multiplication and addition operation defined in the usual way. The $*$-operation is the same as taking conjugate transpose. Let $\mathrm{tr} : \mathcal{M}_d(\mathbb{C}) \to \mathbb{C}$ be the normalized trace defined by,

$$\mathrm{tr}(a) := \frac{1}{d}\mathrm{Trace}(a) = \frac{1}{d}\sum_{i=1}^{d} \alpha_{ii} \quad \text{for all} \quad a = ((\alpha_{ij}))_{i,j=1}^{d} \in \mathcal{M}_d(\mathbb{C}).$$

Then $(\mathcal{M}_d(\mathbb{C}), \mathrm{tr})$ is a (genuinely non-commutative) $*$-probability space. Moreover, the state tr is tracial and positive.

If, in the above matrices we allow entries to be (random) variables from one of the algebras $\mathcal{A}$ or $\mathcal{B}$ described in Example 2.5.2, then $(\mathcal{M}_d(\mathbb{C}), \mathrm{E}\,\mathrm{tr})$ also forms a $*$-probability space. Here

$$\mathrm{E}\,\mathrm{tr}(a) = \frac{1}{d}\sum_{i=1}^{d} \mathrm{E}(\alpha_{ii}) \quad \text{for all} \quad a = ((\alpha_{ij}))_{i,j=1}^{d} \in \mathcal{M}_d(\mathbb{C}).$$

This state is also tracial and positive. We will be encountering these $*$-probability spaces quite frequently in this book. ▲

2.6 Moments and probability laws of variables

We now proceed to define moments of non-commutative variables.

Moments of a single variable. Let $(\mathcal{A}, \varphi)$ be a $*$-probability space. First consider a single variable $a \in \mathcal{A}$. The numbers $\{\varphi(a^{\epsilon_1} a^{\epsilon_2} \cdots a^{\epsilon_n}), n \geq 1\}$, where $\epsilon_i \in \{1, *\}$ for all $1 \leq i \leq n$, are the $*-$*moments* of a. Since we are in the non-commutative setup, these moments need not define any probability law.

If $(\mathcal{A}, \varphi)$ is only an NCP, and not necessarily a $*$-probability space, then the numbers $\{\varphi(a^k)\}_{k=1}^{\infty}$ are called the *moments of* a. Again, this sequence need not define any probability law.

Example 2.6.1. A variable s on a $*$-probability space $(\mathcal{A}, \varphi)$ is said to be *semi-circular* if it is self-adjoint and

$$\varphi(s^h) = \begin{cases} C_n & \text{if } h = 2n, \\ 0 & \text{if } h \text{ is odd.} \end{cases} \tag{2.10}$$

Note that these moments define the free Gaussian probability law.

The *existence* of such a variable can be demonstrated in an obvious way. Consider the NCP defined in Example 2.5.1 with μ as the semi-circular law. Recall that $\mathcal{A}$ was generated by a single indeterminate x. We can convert $\mathcal{A}$ into a $*$-algebra by declaring $(\alpha x)^* = \bar{\alpha} x$ for all $\alpha \in \mathbb{C}$ and then extending this operation to all of $\mathcal{A}$ in a natural way. This converts $(\mathcal{A}, \varphi)$ into a $*$-probability space. By construction x is a self-adjoint semi-circular variable. ▲

Definition 2.6.1. (Probability law of a variable) Let $(\mathcal{A}, \varphi)$ be a $*$-probability space. Suppose $a \in \mathcal{A}$ is a *normal element*, that is, $aa^* = a^*a$. Suppose there is a *unique* probability law μ_a on $\mathbb{C}$ such that

$$\varphi(a^m (a^*)^n) = \int_{\mathbb{C}} z^m \bar{z}^n d\mu_a(z) \quad \text{for all} \quad m, n \in \mathbb{N}. \tag{2.11}$$

Then μ_a is called the *probability law of* a. ◇

Note that if μ_a satisfies (2.11) and is compactly supported, then it is indeed unique. If, in addition, a is self-adjoint, then μ_a is supported on (a compact subset of) $\mathbb{R}$. To see this, observe that, using the above equation and $a = a^*$,

$$\begin{aligned} \int_{\mathbb{C}} |z - \bar{z}|^2 d\mu(z) &= \int_{\mathbb{C}} (z - \bar{z})(\bar{z} - z) d\mu_a(z) \\ &= 2\varphi(aa^*) - \varphi(a^2) - \varphi((a^*)^2) \\ &= 0. \end{aligned}$$

Hence the support of μ_a is contained in $\{z : z = \bar{z}\} \subset \mathbb{R}$.

Also note that, if a is self-adjoint, and φ is positive on the sub-algebra generated by a (that is, $\varphi(xx^*) \geq 0$ for all x in this sub-algebra), then $\{m_k := \varphi(a^k)\}$ is indeed a classical moment sequence. To see this, by Herglotz's theorem (or Bochner's theorem, see Loéve (1963)[65]), it is enough to show that the sequence is positive semi-definite. We proceed as follows: fix any $k \geq 1$ and complex numbers $\{c_1, \ldots, c_k\}$. Now

$$\begin{aligned}
\sum_{i,j=1}^{k} c_i \bar{c}_j m_{i+j} &= \varphi\Big(\sum_{i,j=1}^{k} c_i \bar{c}_j a^{i+j}\Big) \\
&= \varphi\Big(\sum_{i=1}^{k} c_i a^i \sum_{j=1}^{k} \bar{c}_j a^j\Big) \\
&= \varphi\Big(\Big(\sum_{i=1}^{k} c_i a^i\Big)\Big(\sum_{j=1}^{k} c_j a^j\Big)^*\Big) \quad \text{(since } a \text{ is self-adjoint)} \\
&\geq 0 \quad \text{(by positivity).}
\end{aligned}$$

Thus $\{m_k\}$ is a moment sequence. If it identifies a *unique* probability law, then this law is μ_a. In Remark 1.2.1, we have given sufficient criteria for this uniqueness.

Joint moments of several variables. The above definition of moments and distribution for a single variable extends easily to more than one variable. Consider any collection of variables $\{a_i : i \in I\}$ from $\mathcal{A}$. Then $\{\varphi(\Pi(a_i, a_i^* : i \in I)) : \Pi$ is a finite degree monomial$\}$ are called their *joint $*$-moments.* For simplicity, we shall often write moments instead of $*$−moments.

Multiplicative extension. So far we have mostly worked with a single random variable. We shall now start working with several variables together. Since the variables are non-commutative, we shall always carefully keep track of the order in which they appear in our formulae and expressions.

Let $(\mathcal{A}, \varphi)$ be an NCP (or a $*$-probability space). Define a sequence of *multi-linear functionals* (that is, linear in each co-ordinate) $(\varphi_n)_{n \in \mathbb{N}}$ on $\mathcal{A}^n$ via

$$\varphi_n(a_1, a_2, \ldots, a_n) := \varphi(a_1 a_2 \cdots a_n). \tag{2.12}$$

We extend $\{\varphi_n, n \geq 1\}$ to $\{\varphi_\pi, \pi \in NC(n), n \geq 1\}$ *multiplicatively* in a recursive way by the following formula. If $\pi = \{V_1, V_2, \ldots, V_r\} \in NC(n)$, then

$$\varphi_\pi[a_1, a_2, \ldots, a_n] := \varphi(V_1)[a_1, a_2, \ldots, a_n] \cdots \varphi(V_r)[a_1, a_2, \ldots, a_n], \tag{2.13}$$

where

$$\varphi(V)[a_1, a_2, \ldots, a_n] := \varphi_s(a_{i_1}, a_{i_2}, \ldots, a_{i_s}) = \varphi(a_{i_1} a_{i_2} \cdots a_{i_s})$$

for $V = \{i_1, i_2, \ldots, i_s\}$ with $i_1 < i_2 < \cdots < i_s$. Observe how the order of the variables has been preserved and the different uses of the two types of braces ( ) and [ ] in (2.12) and (2.13). In particular,

$$\varphi_{1_n}[a_1, a_2, \ldots, a_n] = \varphi_n(a_1, a_2, \ldots, a_n) = \varphi(a_1 a_2 \cdots a_n). \tag{2.14}$$

Definition 2.6.2. (Joint, mixed and marginal free cumulants) The *joint free cumulant* of order n is defined on $\mathcal{A}^n$ by

$$\kappa_n(a_1, a_2, \ldots, a_n) = \sum_{\sigma \in NC(n)} \varphi_\sigma[a_1, a_2, ..., a_n]\mu[\sigma, \mathbf{1}_n], \ (a_1, \ldots a_n) \in \mathcal{A}^n, \tag{2.15}$$

where μ is the Möbius function of $NC(n)$. The quantity $\kappa_n(a_1, a_2, \ldots, a_n)$ is called a *mixed free cumulant* if for some $i \neq j$, $a_i \neq a_j$, or $a_i \neq a_j^*$. For any $\epsilon_i \in \{1, *\}, 1 \leq i \leq n$, $\kappa_n(a^{\epsilon_1}, a^{\epsilon_2}, \ldots, a^{\epsilon_n})$ is called a *marginal free cumulant* of order n of $\{a, a^*\}$. If a is self-adjoint, then

$$\kappa_n(a) := \kappa_n(a, a, \ldots, a)$$

is called the n-th free cumulant of a. ◇

Incidentally, the earliest combinatorial description of the relations between moments and free cumulants was given in Speicher (1994)[91], which was later fully developed in Krawcyk and Speicher (2000)[62].

Recall that $\{\varphi_n\}$ are multi-linear functionals. Since the free cumulants $\{\kappa_n\}$ in (2.15) are linear combinations of $\{\varphi_\pi\}$, they are also multi-linear. In particular, for any variables $\{a_i, b_i\}$ and constants $\{c_i\}$,

$$\begin{aligned}
\kappa_n(a_1 + b_1, \ldots, a_n + b_n) &= \kappa_n(a_1, \ldots, a_n) + \kappa_n(a_1, b_2, a_3, \ldots, a_n) \\
&\quad + \cdots + \kappa_n(b_1, \ldots, b_n), \quad \text{and} \\
\kappa_n(c_1 a, c_2 a, \ldots, c_n a) &= c_1 c_2 \cdots c_n \kappa_n(a, a, \ldots, a) \\
&= c_1 c_2 \cdots c_n \kappa_n(a).
\end{aligned}$$

Just as in (2.13), the numbers $\{\kappa_n(a_1, a_2, \ldots, a_n) : n \geq 1\}$ have a multiplicative extension $\{\kappa_\pi : \pi \in NC(n), n \geq 1\}$. Moreover, by using Lemma 1.4.1, we can say that, for any $\pi \in NC(n), \ n \geq 1$,

$$\kappa_\pi[a_1, a_2, ..., a_n] := \sum_{\sigma \in NC(n): \ \sigma \leq \pi} \varphi_\sigma[a_1, a_2, \ldots, a_n]\mu[\sigma, \pi], \tag{2.16}$$

where μ is the Möbius function of $NC(n)$. Note that

$$\begin{aligned}
\kappa_{1_n}[a_1, a_2, \ldots, a_n] &= \kappa_n(a_1, a_2, \ldots, a_n), \quad \text{and} \\
\kappa_2(a_1, a_2) &= \varphi(a_1 a_2) - \varphi(a_1)\varphi(a_2).
\end{aligned}$$

The number $\kappa_2(a_1, a_2)$ is called the *covariance* between a_1 and a_2. Note that, in general,

$$\kappa_2(a_1, a_2) \neq \kappa_2(a_2, a_1).$$

Equality holds for all a_1 and a_2 if and only if φ is tracial.
Using the Möbius function of $NC(n)$, $n \geq 1$, it is easily seen that

$$\varphi(a_1 a_2 \cdots a_n) = \sum_{\sigma \in NC(n):\ \sigma \leq 1_n} \kappa_\sigma[a_1, a_2, \ldots, a_n]. \tag{2.17}$$

Moreover, for any $\pi \in NC(n)$,

$$\varphi_\pi[a_1, a_2, \ldots, a_n] = \sum_{\sigma \in NC(n):\ \sigma \leq \pi} \kappa_\sigma[a_1, a_2, \ldots, a_n]. \tag{2.18}$$

In particular, (2.15)–(2.18) establish a one-to-one correspondence between free cumulants and moments. These relations will be frequently used and we shall refer to them collectively also as the *moment-free cumulant relation.*

Example 2.6.2. (a) Since φ is a state, clearly all moments of the variable $1_{\mathcal{A}}$ equal 1. By using the moment-free cumulant relations, it is easy to see that

$$\kappa_1(1_{\mathcal{A}}) = 1, \quad \text{and} \quad \kappa_n(1_{\mathcal{A}}) = 0 \text{ for all } \ n \geq 2.$$

(b) Since inserting the identity into any monomial does not change its moments, it is easy to see, by using the moment-free cumulant relations, that all mixed free cumulants of $1_{\mathcal{A}}$ and $a_1, \ldots, a_n$, where $1_{\mathcal{A}}$ appears at least once, are zero for any variables $a_1, \ldots, a_n$. ▲

Example 2.6.3. (Free Poisson variable) Consider an algebra $\mathcal{A}$ generated by a single indeterminate a and define φ on $\mathcal{A}$ by

$$\varphi(a^n) := \sum_{\pi \in NC(n)} \lambda^{|\pi|},$$

where $\lambda > 0$ is a constant. Note that using (2.17), the free cumulants of a are given by

$$\kappa_n(a) = \lambda \ \text{ for all } \ n.$$

We say that a is a free Poisson variable with parameter λ. In Section 2.4 we have already seen that this sequence of moments defines a unique probability law when $\lambda = 1$. *We shall show later through Theorem 4.3.1 and Example 6.5.2, that this continues to be true for any* $\lambda > 0$. ▲

Example 2.6.4. For any constants $c_1, \ldots c_n$, all mixed free cumulants of the variables $\{c_i 1_{\mathcal{A}}\}$ vanish. This is easy to verify by first computing their moments and then using the moment-free cumulant relation. ▲

Example 2.6.5. For any constants $c_1, \ldots c_n$, the mixed free cumulants involving the variables $\{c_i 1_{\mathcal{A}}\}$ and any other set of variables $\{a_j\}$ also vanish. This is easy to verify by first identifying how their mixed moments look and then using the moment-free cumulant relation. ▲

2.7 Exercises

1. Show that the Catalan numbers C_n satisfy $C_n \leq 4^n$ for all $n \geq 1$.
2. Show that $C_n \approx \dfrac{4^n}{n^{3/2}\sqrt{\pi}}$ as $n \to \infty$.
3. If the random variable X is distributed as the standard semi-circular law then show that $\mathrm{E}(2+X)^n = C_{n+1}$.
4. Suppose π is a partition of $\{1, \ldots, n\}$. Show that $\pi \in NC(n)$ if and only if (a) and (b) hold:

   (a) there is a block V of π which consists solely of neighboring integers, say $\{i, i+1, \ldots i+p\}$ for some $p \geq 0$,

   (b) if we remove V, then the remaining blocks of π form a non-crossing partition of the remaining integers. Now there will a block whose elements, when ordered, will not have elements in between from any of the remaining blocks. Removing this block and repeating this scheme of removal eventually leave us with the empty partition.
5. Show that if $\pi \in NC(n)$ for some $n \geq 2$ and if all blocks of π have at least two elements, then there has to be at least one block, say V, with at least two consecutive integers.
6. Show that the unique solution to the recursive equation

   $$D_{n+1} = D_0 D_n + D_1 D_{n-1} + \cdots + D_n D_0, \quad n \geq 1$$

   where $D_0 = D_1 = 1$ and D_i are non-negative integers is $D_n = C_n$. Hint: You may like to use generating functions.
7. (The ballot problem). Consider the usual random walk. Show that the number of paths from $(0,0)$ to $(0,2n)$ such that they all lie on or above the x-axis is the same as C_n.
8. For every $k \geq 1$, construct a direct bijection between $NC(k)$ and $NC_2(2k)$.
9. Show that the decomposition $a = x + \iota y$ where x and y are self-adjoint is unique.
10. If φ is a positive linear functional on a $*$-algebra $\mathcal{A}$, then show that for every $a, b \in \mathcal{A}$,

    $$|\varphi(b^* a)|^2 \leq \varphi(a^* a)\varphi(b^* b).$$

    Hint: Adapt the proof of Cauchy-Schwartz inequality for real numbers.

11. Check the details of the claim that $(\mathcal{A}, \varphi)$ is a $*$-probability space in Example 2.6.1.

12. Show that tr and E tr are both tracial.

13. Verify that φ_n defined in (2.12) is indeed a multi-linear functional on $\mathcal{A}^n$.

14. Show that κ_n defined in (2.15) is indeed a multi-linear functional on $\mathcal{A}^n$.

15. Verify (2.16).

16. Suppose a is a variable in some NCP. Show that all odd moments of a are 0 if and only if all its odd free cumulants are 0.

17. Verify that

$$NC(3) = \{\mathbf{0}_3, \pi_1, \pi_2, \pi_3, \mathbf{1}_3\}$$

where π_j denotes the two-block partition one of whose blocks is the singleton $\{j\}$. Using (1.19), show that the Möbius function of $NC(3)$ is given by

$$\begin{aligned} \mu[\mathbf{0}_3, \pi_j] &= \mu[\pi_j, \mathbf{1}_3] = -1 \\ \mu[\mathbf{0}_3, \mathbf{1}_3] &= 2. \end{aligned}$$

18. Show that (2.2) holds.

19. Suppose $(\mathcal{A}, \varphi)$ is a $*$-probability space. Suppose $u \in \mathcal{A}$ is such that $uu^* = u^*u = 1_{\mathcal{A}}$ and

$$\varphi(u^n) = \varphi(u^{*n}) = 0 \ \text{ for all } \ n \in \mathbb{N}.$$

Then show that the moments of $u + u^*$ are given by

$$\varphi((u+u^*)^k) = \begin{cases} \binom{2j}{j} & \text{if } k = 2j \text{ for some } j, \\ 0 & \text{otherwise.} \end{cases} \tag{2.19}$$

These moments determine the *arc-sine* probability law μ whose density is given by

$$f_\mu(x) = \begin{cases} \frac{1}{\pi\sqrt{4-t^2}} & \text{if } |x| \leq 2, \\ 0 & \text{otherwise.} \end{cases} \tag{2.20}$$

20. Suppose U_1 and U_2 are i.i.d. random variables distributed uniformly on the interval $[0,\ 1]$. Show that the probability law of the random variable $W := 2\sqrt{U_1}\cos(2\pi U_2)$ is the standard semi-circular law.

3

Free independence

There are several notions of independence in the non-commutative setup. The most useful amongst these is *free independence* which was discovered by Voiculescu (1991)[100]. Traditionally free independence is defined through certain requirements on the moments. However, by analogy with classical independence, we define free independence by the vanishing of all mixed free cumulants. In Chapter 13 we shall establish the equivalence of the two definitions.

Existence of freely independent variables with specified distributions is guaranteed by a construction of *free product* of $*$-probability spaces, analogous to the construction of product probability spaces.

The *free binomial* and the *families* of *semi-circular, circular* and *elliptic* variables are defined. A semi-circular family is the non-commutative analog of a multivariate Gaussian random variable. The free analog of Isserlis' formula is established.

Given two compactly supported probability measures μ_1 and μ_2, their *free additive convolution*, $\mu_1 \boxplus \mu_2$, is a probability measure whose free cumulants are the sums of the free cumulants of μ_1 and μ_2. That this yields a valid probability measure is shown in Chapter 5.

The *Kreweras complementation map* is used to establish relations between joint moments and joint free cumulants of polynomials of free variables and also to identify the distribution of variables from given joint moments and free cumulants. The chapter ends with an introduction to compound free Poisson distribution.

3.1 Free independence

For classical independence, constants are independent of each other and of any other variables. We also know that (measurable) functions of different sets of independent random variables are again independent.

For free independence, while the first condition is true, the second will not make sense since there is no measurability concept. However, all variables must reside on the same NCP. We now give the formal definition of free independence.

DOI: 10.1201/9781003144496-3

Definition 3.1.1. (Free independence) Suppose $(\mathcal{A}, \varphi)$ is a $*$-probability space. Then the $*$-sub-algebras $(\mathcal{A}_i)_{i \in I}$ of $\mathcal{A}$ are said to be $*$-freely independent (or simply $*$-free) if, for all $n \geq 2$, and all $a_1, a_2, \ldots, a_n$ from $(\mathcal{A}_i)_{i \in I}$, $\kappa_n(a_1, a_2, \ldots, a_n) = 0$ whenever at least two of the a_i are from different $\mathcal{A}_i$. In particular, any collection of variables is said to be $*$-free if the sub-algebras generated by these variables are $*$-free. ◇

This concept remains valid if the underlying NCP is not necessarily a $*$-probability space, in which case, it is a convention to call the sub-algebras freely independent or just free. We shall use the terms free or freely independent to denote both free independence and $*$-free independence, it being clear from the context which one we are specifically referring to.

We emphasize that freeness is tied to the state φ. Algebras or variables which are free with respect to a state φ need not remain free with respect to a different state.

Nevertheless, the calculations in Example 2.6.2 imply that "constants are free of everything," irrespective of what the state is. So, at least for starters, we know that free variables exist and the concept is not vacuous.

For any collection of free variables, since their mixed free cumulants vanish, any mixed moment can be expressed in terms of the marginal free cumulants, which, in turn, can be expressed in terms of the corresponding marginal moments. However, it is not at all clear if this final expression can be written down easily. Let us see some examples first.

Example 3.1.1. If a and b are free in $(\mathcal{A}, \varphi)$, then for all non-negative integers n and m,

$$\varphi(a^n b^m) = \varphi(a^n)\varphi(b^m) = \varphi(b^m a^n).$$

In particular, $\varphi(ab) = \varphi(ba)$. This can be proved by using the moment-free cumulant relation and the freeness of a and b as follows:

$$\begin{aligned}
\varphi(a^n b^m) &= \sum_{\pi \in NC(m+n)} \kappa_\pi[a, \ldots, a, b, \ldots, b] \text{ (moment-free cumulant relation)} \\
&= \sum_{\pi_1 \in NC(n),\ \pi_2 \in NC(m)} \kappa_{\pi_1}[a, a, \ldots, a]\kappa_{\pi_2}[b, b, \ldots, b] \quad \text{(freeness)} \\
&= \sum_{\pi_1 \in NC(n)} \kappa_{\pi_1}[a, a, \ldots, a] \sum_{\pi_2 \in NC(m)} \kappa_{\pi_2}[b, b, \ldots, b] \\
&= \varphi(a^n)\varphi(b^m) \\
&= \varphi(b^m)\varphi(a^n) \\
&= \varphi(b^m a^n), \quad \text{by the same arguments as presented above.}
\end{aligned}$$

▲

When a and b are free, even though any of their mixed moments can be expressed in terms of their marginal moments, which marginal moments are involved depends on the specific mixed moment we are dealing with.

Example 3.1.2. Suppose a and b are free in $(\mathcal{A}, \varphi)$. Then,

$$\begin{aligned}
\varphi(aba) &= \kappa_1(a)\kappa_1(a)\kappa_1(b) + \kappa_2(a)\kappa_1(b), \quad \text{other terms being } 0 \\
&= \kappa_1(b)[\kappa_2(a) + \kappa_1^2(a)] \\
&= \varphi(b)\varphi(a^2) \\
&= \varphi(ba^2) \quad \text{(by Example 3.1.1).}
\end{aligned}$$

Similarly, it can be checked that

$$\begin{aligned}
\varphi(abab) &= \varphi^2(a)\varphi^2(b), \quad \text{and} \\
\varphi(abba) &= \varphi(a^2)\varphi(b^2) \\
&= \varphi(a^2b^2) \\
&= \varphi(b^2a^2) = \varphi(bbaa).
\end{aligned}$$

Note that if a and b were classically independent, then the above two mixed moments would be identical. Thus, it is clear that free independence and classical independence are very different. ▲

The following extremely useful result is an immediate consequence of the linearity of free cumulants and freeness. Its proof is left as an exercise.

Theorem 3.1.1. (Free cumulant of a free sum) If a and b are freely independent variables, then, for all $n \geq 1$,

$$\kappa_n(a + b, \ldots, a + b) = \kappa_n(a, \ldots, a) + \kappa_n(b, \ldots, b)\,.$$

◆

Example 3.1.3. Suppose $(\mathcal{A}, \varphi)$ is an NCP on which we have *two* free standard semi-circular variables s_1 and s_2. Note that the existence of two such variables is *not* guaranteed by the developments made so far. We shall address this issue in Theorem 3.2.1 below. *Later in Chapter 8 we shall see how free semi-circular variables arise naturally in the context of large dimensional symmetric random matrices.*

Consider the variable

$$s = \frac{1}{\sqrt{2}}(s_1 + s_2).$$

We claim that s is a standard semi-circular variable. Clearly s is self-adjoint. Now, by freeness and linearity of free cumulants,

$$\begin{aligned}
\kappa_1(s) &= 0, \\
\kappa_2(s, s) &= \kappa_2\Big(\frac{s_1}{\sqrt{2}}\Big) + \kappa_2\Big(\frac{s_2}{\sqrt{2}}\Big) \\
&= \frac{1}{2} + \frac{1}{2} = 1, \text{and} \\
\kappa_n(s) &= 0 \text{ for } n \geq 3 \text{ since the same is true for } s_1 \text{ and } s_2.
\end{aligned}$$

This is commonly referred to by the phrase "free additive convolution of semi-circulars is again a semi-circular." The reader is invited to formulate and prove a more general claim that involves more than two free semi-circular variables which may not all be standardized to have variance 1. ▲

Example 3.1.4. Suppose a and b are two free Poisson variables with parameters λ_1 and λ_2 respectively. Again, existence of such variables is guaranteed by Theorem 3.2.1 stated below. Then $a+b$ is free Poisson with parameter $\lambda_1+\lambda_2$. To verify this calculate the free cumulants using linearity and freeness. ▲

3.2 Free product of $*$-probability spaces

How do we guarantee that there are enough non-trivial collections of free variables? Recall that, in the classical case, the product probability space construction guarantees the existence of classically independent variables with given marginals.

There is a similar construction of free products of non-commutative and $*$-probability spaces. In Theorem 3.2.1, we state the result for $*$-probability spaces. *We postpone its proof to Section 13.3 of Chapter 13.*

We mention the following facts without proof, for the understanding of the theorem.

Construction of free product. Given algebras (respectively $*$-algebras) $\{\mathcal{A}_i,\ i \in I\}$ where I is some index set, we can construct an algebra (respectively a $*$-algebra) $\mathcal{A}$, called the *free product* of $\{\mathcal{A}_i\}$, as follows:

(a) Identify all the identitites/unities $\{\mathbf{1}_{\mathcal{A}_i}\}$ of $\{\mathcal{A}_i\}$ and declare it to be the identity/unity $\mathbf{1}_{\mathcal{A}}$ of $\mathcal{A}$.

(b) Form a family of unital homomorphisms $\{V_i : \mathcal{A}_i \to \mathcal{A},\ i \in I\}$, such that the following universal property holds: whenever $\mathcal{B}$ is a unital algebra and $\{T_i : \mathcal{A}_i \to \mathcal{B},\ i \in I\}$ is a family of unital homomorphisms, there exists a unique unital homomorphism $T : \mathcal{A} \to \mathcal{B}$ such that $TV_i = T_i$ for all $i \in I$.

Each $V_i : \mathcal{A}_i \to \mathcal{A}$ is one-to-one, and hence we may view *each* $\mathcal{A}_i$ *as a sub-algebra of* $\mathcal{A}$ for every $i \in I$. We can demonstrate this with the following diagram:

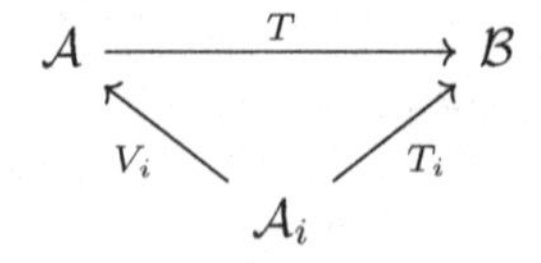

We now state the theorem. A version of the result is also true if we have only NCP without a $*$-operation.

Theorem 3.2.1. (Free and $*$-free product) Let $(\mathcal{A}_i, \varphi^{(i)})_{i\in I}$ be a family of $*$-probability spaces. Then there exists a $*$-probability space $(\mathcal{A}, \varphi)$, called the $*$-*free* product of $(\mathcal{A}_i, \varphi^{(i)})_{i\in I}$, such that there is a copy of $\mathcal{A}_i$ in $\mathcal{A}$, $i \in I$ which are freely independent in $(\mathcal{A}, \varphi)$ and φ restricted to $\mathcal{A}_i$ equals $\varphi^{(i)}$ for all i. ◆

This theorem equips us with rich collections of variables and distributions. In particular, given variables a_i on $(\mathcal{A}_i, \varphi_i)$, $i \geq 1$, there is an NCP $(\mathcal{A}, \varphi)$ with variables $\{b_i\}$ such that $\{b_i\}$ are free, and the distribution of a_i is the same as that of b_i for all i. In particular, $\varphi_i(a_i^k) = \varphi(b_i^k)$ for all i and k.

3.3 Free binomial

Example 3.3.1. A self-adjoint variable x in a $*$-probability space $(\mathcal{A}, \varphi)$ is said to be a Bernoulli variable with parameter p if its moments are given by

$$\varphi(x^n) = p \quad \text{for all} \quad n \geq 1.$$

Note that the sequence $\{\varphi(x^n)\}$ defines a unique probability law μ_x on $\mathbb{R}$, namely the classical Bernoulli law, with these as the moments.

Now suppose $a_1, \ldots, a_n$ are freely independent Bernoulli variables each with parameter p on some $*$-probability space $(\mathcal{A}, \varphi)$. Then $a = a_1 + \cdots + a_n$ is said to be a *free binomial* variable.

The moments $\{\varphi(a^n)\}$ *determine a unique probability law* on $\mathbb{R}$, which is naturally called the *free additive convolution* of Bernoulli laws. This probability law is called the *free binomial law* with parameters n and p. We shall investigate this law further in Example 5.3.2. ▲

3.4 Semi-circular family

Recall the multivariate normal vector. We now wish to define its analog in the non-commutative case. First an example.

Example 3.4.1. Suppose s_1 and s_2 are standard free semi-circular variables on some NCP. Let $-1 \leq \rho \leq 1$ be a real number. Define $x := \rho s_1 + \sqrt{1-\rho^2} s_2$ and consider the distribution of $\{x, s_1\}$. Clearly, all pure and mixed free cumulants of x and s_1 of order greater than two and of order one vanish. The free cumulants of order two are given by

$$\kappa_2(x) = 1, \ \kappa_2(x, s_1) = \kappa_2(s_1, x) = \rho.$$

In particular, x is a semi-circular variable.

Now suppose s is a *vector* whose n components are free semi-circular variables in some NCP. Suppose $\Sigma = ((\sigma_{ij}))$ is an $n \times n$ positive semi-definite matrix. Let A be the $n \times n$ matrix with real entries such that $AA^\top = \Sigma$, where $A^\top$ denotes the transpose of A.

Define the n vector x by $x = As$. Then each component x_i is a semi-circular variable with variance σ_{ii}. All mixed free cumulants of order greater than two vanish for $\{x_i\}$. The mixed free cumulants of order two are given by $\kappa_2(x_i, x_j) = \sigma_{ij}$. ▲

Definition 3.4.1. (Semi-circular family) Self-adjoint variables $\{s_i, 1 \leq i \leq n\}$ in an NCP $(\mathcal{A}, \varphi)$ are said to form a *semi-circular family* if every s_i is a semi-circular variable (perhaps with different variances) and all mixed free cumulants of order greater than two vanish. Moreover, the matrix $((\kappa_2(s_i, s_j)))$ is positive semi-definite. Such a family is also called a *free Gaussian family.* The matrix $((\kappa_2(s_i, s_j)))$ is called the *variance-covariance matrix* or the *dispersion matrix* of $\{s_i\}$. ◇

Note that we have not required the dispersion matrix to be symmetric.

Example 3.4.2. Suppose $\{s_1, s_2\}$ is a semi-circular family. Define

$$a_1 = s_1 + s_2,\ a_2 = s_1 - s_2.$$

Then $\{a_1, a_2\}$ is a semi-circular family. If, further, $\kappa_2(s_1, s_1) = \kappa_2(s_2, s_2)$ and $\kappa_2(s_1, s_2) = \kappa_2(s_2, s_1)$, then a_1 and a_2 are free.

More generally, suppose $\{s_i\}$ is a semi-circular family with dispersion matrix Σ. Let s be the column vector of the $\{s_i\}$. Then for any linear transformation A, the components of As form a semi-circular family. If, further, $A\Sigma A^\top = I$ (the identity matrix), then the components of As are free. ▲

3.5 Free Isserlis' formula

Recall Isserlis' formula for classical Gaussian random variables from Section 1.5. It is natural to expect a similar formula for semi-circular families.

(Free) Isserlis' formula. Suppose $\{s_i, 1 \leq i \leq k\}$ is a semi-circular family. Then

$$\varphi(s_1 s_2 \cdots s_k) = \sum_{\pi \in NC_2(k)} \prod_{\{r,s\} \in \pi, r<s} k_2(s_r, s_s) = \sum_{\pi \in NC_2(k)} \prod_{\{r,s\} \in \pi, r<s} \varphi(s_r s_s)$$

where $\{r, s\}$ denotes a typical block of π.

To prove this, first invoke the moment-free cumulant relation, by which the left side is the sum over all possible free cumulants. But then all free cumulants of order one and of order greater than two vanish. Hence free cumulants only of order two survive. This justifies the first equality. The second equality then follows since $\varphi(s_i) = 0$ for all i.

A consequence of this formula is that if k is odd, then the above moment is zero. If all s_i are identical, say s, then we recover the moments of s from the above formula.

3.6 Circular and elliptic variables

In classical probability, consider a standard complex Gaussian variable $Z = X + \iota Y$. Recall that this means X and Y are independent and identically distributed Gaussian random variables with mean zero and variance $1/2$ each. Let $Z^* = X - \iota Y$ be the complex conjugate of Z. It is then easy to verify that $\text{Cov}(Z, Z^*) = \text{E}(ZZ^*) = 1$ and all other mixed cumulants of $\{Z, Z^*\}$ are zero. This motivates the following analog in the non-commutative case.

Definition 3.6.1. (Circular variable) A variable c in some $*$-probability space is said to be a *circular variable* if, for some $\alpha > 0$, $\kappa_2(c, c^*) = \kappa_2(c^*, c) = \alpha$ and all other free cumulants of $\{c, c^*\}$ are 0. If $\alpha = 1$, then c is said to be a standard circular variable. ◇

Note that, if c is a circular variable, then c^* is also a circular variable.

Example 3.6.1. Suppose s_1 and s_2 are free semi-circular variables. Define

$$c := \frac{s_1 + \iota s_2}{\sqrt{2}}.$$

Then it is easy to check that c is a standard circular variable. ▲

Example 3.6.2. Suppose c is a standard circular variable. Define

$$s_1 := \frac{c + c^*}{\sqrt{2}} \quad \text{and} \quad s_2 := \frac{c - c^*}{\sqrt{2}\iota}.$$

Then s_1 and s_2 are self-adjoint, free, and each has a semi-circular distribution. Moreover, $c = \frac{s_1 + \iota s_2}{\sqrt{2}}$.

Because of the above and Example 3.6.1, we can *define* a circular element c by the relation $c = \frac{s_1 + \iota s_2}{\sqrt{2}}$, where s_1 and s_2 are standard free semi-circular variables. ▲

An extension of the above ideas leads to the following definition.

Definition 3.6.2. (Elliptic variable) Suppose $(\mathcal{A}, \varphi)$ is a $*$-probability space. An element $e \in \mathcal{A}$ is said to be an *elliptic* variable with parameter ρ, $0 \leq \rho \leq 1$

if the free cumulants of e of order greater than two are zero and the second order free cumulants of e are given by

$$\begin{aligned}\kappa_2(e,e) &= \kappa_2(e^*,e^*) = \rho,\\ \kappa_2(e,e^*) &= \kappa_2(e^*,e) = 1.\end{aligned}$$

◇

Note that $\rho = 1$ and $\rho = 0$ yield respectively the semi-circular variable and the circular variable.

As in the case for circular variables, an elliptic variable can be *defined* as

$$e = \sqrt{\frac{1+\rho}{2}}s_1 + i\sqrt{\frac{1-\rho}{2}}s_2, \tag{3.1}$$

where s_1 and s_2 are free standard semi-circular variables. This representation allows us to derive a formula for the moments of e. Define

$$\delta_{xy} := \begin{cases} 1 & \text{if } x = y,\\ 0 & \text{if } x \neq y. \end{cases} \tag{3.2}$$

Lemma 3.6.1. (Moments of an elliptic variable) An element e in a $*$-probability space $(\mathcal{A}, \varphi)$ is an elliptic variable with parameter ρ if and only if for $\epsilon_1, \ldots, \epsilon_p \in \{1, *\}$,

$$\varphi(e^{\epsilon_1}e^{\epsilon_2}\cdots e^{\epsilon_p}) = \begin{cases} \sum_{\pi \in NC_2(2k)} \rho^{T(\pi)} & \text{if } p = 2k,\\ 0 & \text{if } p = 2k+1, \end{cases} \tag{3.3}$$

where

$$T(\pi) = \#\{\{r,s\} \text{ is a block of } \pi \ : \ \delta_{\epsilon_r\epsilon_s} = 1\}. \tag{3.4}$$

♦

Proof. Suppose e is elliptic with parameter ρ. Then using the moment-free cumulant relation,

$$\varphi(e^{\epsilon_1}e^{\epsilon_2}\cdots e^{\epsilon_p}) = \sum_{\pi \in NC(p)} \prod_{V \in \pi} \kappa_{|V|}[e^{\epsilon_1}, \ldots, e^{\epsilon_p}].$$

Now, using the representation (3.1), only pair-partitions will contribute since other free cumulants are zero. Therefore, if p is odd, then the right side of the last equation is zero as no pair-partition is possible. If $p = 2k$, we get

$$\begin{aligned}\varphi(e^{\epsilon_1}e^{\epsilon_2}\cdots e^{\epsilon_{2k}}) &= \sum_{\pi \in NC_2(2k)} \prod_{\{r,s\}\in\pi, r<s} \kappa_2(e^{\epsilon_r}, e^{\epsilon_s})\\ &= \sum_{\pi \in NC_2(2k)} \rho^{T(\pi)}.\end{aligned}$$

Hence the result. ■

Note that in Lemma 3.6.1, if $\rho = 1$, then we get back the moments of the standard semi-circular variable. If $\rho = 0$, then (3.3) yields the moments of a circular variable.

Lemma 3.6.2. (Moments of a circular variable) Suppose c is a standard circular variable in a $*$-probability space $(\mathcal{A}, \varphi)$. Then

$$\varphi(c^{\epsilon_1} c^{\epsilon_2} \cdots c^{\epsilon_p}) = \begin{cases} \sum_{\pi \in NC_2(2k)} \prod_{\{r,s\} \in \pi, r<s} (1 - \delta_{\epsilon_r, \epsilon_s}) & \text{if } p = 2k, \\ 0 \text{ if } p = 2k+1. \end{cases} \tag{3.5}$$

♦

In Chapter 8, we shall see how circular and elliptic variables arise naturally in the context of high dimensional random matrices.

3.7 Free additive convolution

Suppose X and Y are two classically independent random variables with probability laws μ_1 and μ_2. Then the law of $X + Y$ is the additive convolution of μ_1 and μ_2.

We can consider an analogous notion for variables which are free. This can be done conveniently if we set up the background machinery of C^*-algebras. However, that needs a bit of advanced mathematics which we shall postpone to Chapter 6. For now, we shall be less ambitious and consider the familiar $*$-algebra set up.

Let $(\mathcal{A}, \varphi)$ be a $*$-probability space. Suppose we have two self-adjoint free variables a and b in this NCP. Consider a third variable $a+b$ and observe that it is also self-adjoint.

As noted after Definition 2.6.1, $\{\varphi(a^k)\}$, $\{\varphi(b^k)\}$ and $\{\varphi((a+b)^k)\}$ are moment sequences. Suppose they define unique probability laws μ_a, μ_b and μ_{a+b} respectively. Then μ_{a+b} is called the *free additive convolution* of μ_a and μ_b. We write this measure as $\mu_a \boxplus \mu_b$, and it is the same as $\mu_b \boxplus \mu_a$. The n-fold free additive convolution of μ with itself will be denoted by $\mu^{\boxplus n}$.

We shall see later through the approach via C^*-algebra that the free additive convolution $\mu_1 \boxplus \mu_2$ of *any* two compactly supported probability measures μ_1 and μ_2 is always well-defined. Moreover, it is not too hard to show that $\mu_1 \boxplus \mu_2$ is also compactly supported. The notion of free additive convolution is easily extended to any finite number of compactly supported probability measures.

Example 3.7.1. (a) If μ_1 and μ_2 are the semi-circular laws with variances σ_1^2 and σ_2^2 respectively, then $\mu_1 \boxplus \mu_2$ is the semi-circular law with variance $\sigma_1^2 + \sigma_2^2$.

(b) If μ_1 and μ_2 are the free binomial laws with parameters $(n,\ p)$ and $(m,\ p)$ respectively, then $\mu_1 \boxplus \mu_2$ is the free binomial law with parameters $(n+m, p)$.

(c) If μ_1 and μ_2 are the free Poisson laws with parameters λ_1 and λ_2 respectively, then $\mu_1 \boxplus \mu_2$ is the free Poisson law with parameter $\lambda_1 + \lambda_2$. ▲

Free additive convolution leads to a natural notion of *free infinite divisibility*. From Example 3.7.1, it should be clear to the reader that the semi-circular law and the free Poisson law are both free infinitely divisible. We shall discuss free infinite divisibility briefly in Chapter 5.

3.8 Kreweras complement

We know that mixed moments of free variables can be calculated in terms of their marginal moments. However, it is not easy to write a general formula for this. The *Kreweras complementation map* $K : NC(n) \to NC(n)$ is an extremely useful tool in this regard. It is defined as follows.

Consider additional elements $\bar{1}, \bar{2}, \ldots, \bar{n}$ with the ordering $\bar{1} < \bar{2} < \cdots < \bar{n}$. Then $NC(\bar{1}, \bar{2}, \ldots, \bar{n}) = NC(\bar{n})$ denotes their non-crossing partitions and is also identified with $NC(n)$ in an obvious way.

Now *interlace* the numbers $\bar{1}, \bar{2}, \ldots, \bar{n}$ and $1, 2, \ldots, n$ with the ordering:

$$1 < \bar{1} < 2 < \bar{2} < \cdots < n < \bar{n}.$$

The set of non-crossing partitions of the above set is identified with $NC(2n)$ in the most natural way.

Definition 3.8.1. (Kreweras complement) For any $\pi \in NC(n)$, its Kreweras complement, $K(\pi) \in NC(\bar{n}) \cong NC(n)$, is defined to be the *biggest* element among those $\sigma \in NC(\bar{n})$ which have the property that

$$\pi \cup \sigma \in NC(1, \bar{1}, 2, \bar{2}, \ldots, n, \bar{n}) \cong NC(2n).$$

◇

Example 3.8.1. Consider $\pi_1 = \{\{1\}, \{2\}, \{3\}, \{4\}\}$ and $\pi_2 = \{\{1, 2\}, \{3, 4\}\}$. Then $K(\pi_1) = \{1, 2, 3, 4\}$ and $K(\pi_2) = \{\{1\}, \{2, 4\}, \{3\}\}$. ▲

Before we present some simple but useful properties of the Kreweras complement, we need the idea that any partition π of $\mathbb{Z}_n$ may be considered as a *permutation* of $\mathbb{Z}_n$ as follows: order the elements within each block of π in an increasing way. Then within each block consider the *cyclic permutation* which shifts every element to the next right element of the block, and the last element of the block is shifted to the first element of the block. This gives the permutation on $\mathbb{Z}_n$ whose cycles are the permutations within each block of π.

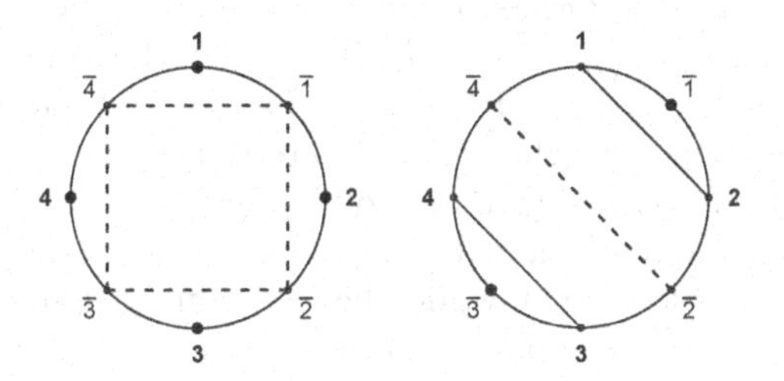

FIGURE 3.1
$K(\pi_1)$ and $K(\pi_2)$ for Example 3.8.1.

Example 3.8.2. If $\pi_1 = \{\{1\}, \{2\}, \ldots, \{n\}\}$, then since each block is a singleton, we get the *identity permutation.* If $\pi_2 = \{\{1, 2, \ldots n\}\}$, then the corresponding permutation is the *cyclic permutation* $1 \to 2 \to \cdots \to n \to 1$. Suppose $\pi_3 = \{\{1, 2\}, \{3, 4\}\}$. Then the corresponding permutation of $1, 2, 3, 4$ is $2, 1, 4, 3$ with two cycles $1 \to 2 \to 1$ and $3 \to 4 \to 3$. ▲

The cyclic permutation $1 \to 2 \to \cdots \to n \to 1$ shall be denoted by γ_n, or simply by γ, if the value of n is clear from the context. It is clear that γ_n^n is the identity permutation. The composition $\pi_1\pi_2$ of two permutations is defined in the natural way—first apply π_2 and then apply π_1. The composition $\pi\gamma$ shall be especially important to us. Recall that for any partition π, $|\pi|$ denotes the number of blocks of π.

Lemma 3.8.1. (Properties of the Kreweras complement)

(a) $K(\mathbf{0}_n) = \mathbf{1}_n$ and $K(\mathbf{1}_n) = \mathbf{0}_n$.

(b) If $\pi \leq \sigma$, then $K(\pi) \geq K(\sigma)$.

(c) For any $\pi \in NC(n)$, we have $|\pi| + |K(\pi)| = n + 1$.

(d) If $\sigma \leq K(\pi)$ in $NC(n)$, then $\pi \cup \sigma \in NC(2n)$.

(e) For any $\pi \in NC(n)$, $\gamma_n K^2(\pi) = \pi$. Thus $K^2(\pi)$ is a cyclic permutation of π and hence has the same number of blocks as π.

(f) The composition map K^{2n} is the identity map, and hence K is a bijection on $NC(n)$.

(g) If $\pi \in NC_2(2k)$, then $K(\pi) = \pi\gamma_{2k}$, and hence $|\pi\gamma_{2k}| = k + 1$.

(h) If $\pi \in \mathcal{P}_2(2k)$, then $|\pi\gamma_{2k}| \leq k + 1$, and equality holds if and only if $\pi \in NC_2(2k)$. ♦

Proof. (a) This is trivial.

(b) Suppose $\{V_1, \ldots, V_k\}$ are the blocks of π, written in increasing order of their first elements.

First assume that σ is formed from π by combining two blocks, say $V_i = \{i_1 < \cdots < i_r\}$ and $V_j = \{j_1 < \cdots < j_s\}$ of π. Then the elements of $\{\bar{1}, \ldots, \bar{n}\}$ which are to the left of i_1 (left of 1 is defined as $\bar{n}$), and those that are to the right of $\bar{j_s}$ (element to the right of $\bar{n}$ is $\bar{1}$) are treated in the same way while building $K(\pi)$ and $K(\sigma)$ and give rise to identical blocks.

On the other hand, as V_i and V_j are combined in σ, the elements i_r and j_1 fall in the same block of σ while they remain in different blocks of π. Therefore, in $K(\pi)$, $\overline{j_1 - 1}$ and $\overline{j_s}$ are in the same block. But in $K(\sigma)$, these certainly belong to different blocks. This proves that $K(\pi) \geq K(\sigma)$ and *$K(\sigma)$ has exactly one more block than $K(\pi)$*.

Now the general case follows from the above observation since we can combine two blocks of π repeatedly to obtain σ from π in finitely many steps.

(c) This is trivially true when $\pi = \mathbf{0}_n$ since, in that case, $K(\pi) = \mathbf{1}_n$. Now the rest of the proof can be completed by using the italicized observation in the proof of Part (b). For instance, consider any partition π such that $|\pi| = n-1$. Then π is obtained by combining two blocks of $\mathbf{0}_n$. Hence $|K(\pi)| = 2$. The reader can now complete the argument for any partition π of any size.

(d) Note that $\pi \cup \sigma$ is a partition of $\{1, \bar{1}, \ldots, n, \bar{n}\}$. We need to argue that it is non-crossing. Since $\sigma \leq K(\pi)$, it splits the blocks of $K(\pi)$ in a non-crossing manner. On the other hand, $\pi \cup K(\pi) \in NC(2n)$. Clearly, splitting the blocks of $K(\pi)$ cannot destroy the non-crossing nature. This establishes (d).

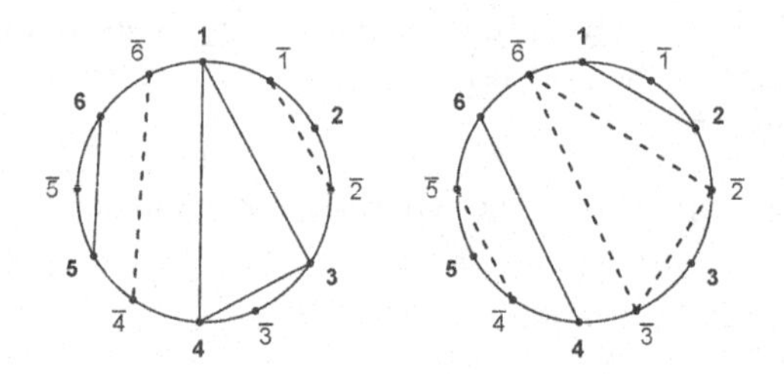

FIGURE 3.2
$K(\pi)$ and $K(K(\pi))$ for $\pi = \{\{1,3,4\},\{2\},\{5,6\}\}$.

(e) First, we give an illustrative example. Consider the partition $\pi = \{\{1,3,4\},\{2\},\{5,6\}\} \in NC(6)$. Then

$$K(\pi) = \{\{\bar{1},\bar{2}\},\{\bar{3}\},\{\bar{4},\bar{6}\},\{\bar{5}\}\} \cong \{\{1,2\},\{3\},\{4,6\},\{5\}\}.$$

Then

$$K(K(\pi)) = \{\{\bar{1}\},\{\bar{2},\bar{3},\bar{6}\},\{\bar{4},\bar{5}\}\} \cong \{\{1\},\{2,3,6\},\{4,5\}\}$$

and applying γ on $K(K(\pi))$, we clearly get back π. See Figure 3.2.

For the general case, consider the ordered numbers

$$\mathbb{I}_{2n} := \{1, \bar{1}, \ldots n, \bar{n}\},$$

marked on a circle sequentially and with the obvious ordering $1 < \bar{1} < 2 < \cdots < n$. It is most convenient to keep Figure 3.2 in mind while we provide the proof.

Recall that any partition is non-crossing if and only if edges from different blocks do not cross each other. Let π be a non-crossing partition of $\mathbb{Z}_n$. Its Kreweras complement is first formed as a partition, say $\bar{K}(\pi)$ of $\{\bar{1}, \ldots, \bar{n}\}$.

We now find the Kreweras complement of $\bar{K}(\pi)$. Consider the cyclic permutation γ_{2n} of $\mathbb{I}_{2n}$, which yields $\{\bar{1}, 2, \bar{2}, \ldots, \bar{n}, 1\}$, with the obvious new ordering. We can think of this as an interlacing of the two sets $\{\bar{1}, \bar{2}, \ldots, \bar{n}\}$ $\{2, 3, \ldots, n, 1\}$ and $\bar{K}(\pi)$ is a non-crossing partition of the "first elements of the pairs" $\{(\bar{1}, 2), (\bar{2}, 2), \ldots, (\bar{n}, 1)\}$. Note that the non-crossing nature of any partition is not destroyed by this cyclic movement. Now note that the Kreweras complement $K(\bar{K}(\pi))$ of $\bar{K}(\pi)$ is a non-crossing partition of $\{2, 3, \ldots, n, 1\}$ and is exactly the same as π. But $K(\bar{K}(\pi)) = \gamma_n(K(K(\pi)))$. This proves (e).

(f) Using Part (e), $K^{2n}(\pi) = \gamma_n^{-n}(\pi) = \pi$ as γ_n^{-n} is the identity permutation. This shows that K is a bijection.

(g) The relation can be verified directly when $k = 2$.

(1) For $\pi = \{\{1, 2\}, \{3, 4\}\}$, we have $\pi\gamma_4 = K(\pi) = \{\{1\}, \{2, 4\}, \{3\}\}$.
(2) For $\pi = \{\{1, 4\}, \{2, 3\}\}$, we have $\pi\gamma_4 = K(\pi) = \{\{1, 3\}, \{2\}, \{\{4\}\}$.

We can prove the relation for all k by induction. Suppose it is true for all $\pi \in NC_2(2i)$, $i \leq k - 1$.

Take $\pi \in NC_2(2k)$. Now there is a pair $\{i, i + 1\} \in \pi$. Clearly, $\{i\}$ is a block of $K(\pi)$. At the same time, $\pi\gamma_{2k}(i) = i$. Now remove the cycle $\{i, i+1\}$. Then we are left with $\pi_1 \in NC(2(k - 1))$ and all the earlier cycles are intact. Now by induction the Kreweras complement $K(\pi_1)$ is $\pi_1\gamma_{2k}$. Clearly,

$$K(\pi) = K(\pi_1) \cup \{i\} = \pi\gamma_{2k}.$$

(h) From Part (g), we know that if $\pi \in NC_2(2j)$, then $|\pi\gamma_{2j}| = j + 1$. So we need to show the following:

$$\pi \in \mathcal{P}_2(2j) \cap (NC_2(2j))^c \Rightarrow |\pi\gamma_{2j}| \leq j \text{ for all } j \geq 2. \tag{3.6}$$

We now proceed inductively. First, let $k = 2$. Then the only crossing pair-partition is $\pi = \{\{1, 3\}, \{2, 4\}\}$. It can be checked that $\pi\gamma_4$ is the cycle $1 \to 4 \to 3 \to 2 \to 1$, and hence $|\pi\gamma_4| = 1$, and so (3.6) holds for $j = 2$.

Suppose (3.6) holds for all $j \leq k$. Let $\pi \in \mathcal{P}_2(2(k + 1))$ which is crossing. We must show that $|\pi\gamma_{2(k+1)}| \leq k + 1$.

First suppose there is a block $\{i, i + 1\}$ of π. Then $\pi\gamma_{2(k+1)}(i) = i$, and hence $\{i\}$ is a block of $\pi\gamma$. Now drop $\{i, i + 1\}$ from $\{1, \ldots, 2(k + 1)\}$, and relabel the remaining $2k$ numbers in the obvious natural order to get the set $\{1, 2, \ldots 2k\}$. Note that the permutations π (after dropping $\{i, i + 1\}$) and

$\pi\gamma_{2(k+1)}$ are mapped in a natural way to the permutation π_0 and $\pi_0\gamma_{2k}$, say, of $\{1,\ldots,2k\}$ in a 1-1 way. However, π_0 is a crossing pair-partition, and hence by induction hypothesis, $|\pi_0\gamma_{2k}| \leq k$. As a consequence

$$|\pi\gamma_{2(k+1)}| = |\pi_0\gamma_{2k}| + 1 \leq k + 1.$$

If π has no block of the form $\{i, i+1\}$, then $\pi\gamma_{2(k+1)}$ cannot have any singleton block. Hence $|\pi\gamma_{2(k+1)}| \leq k+1$, and the proof is complete. ■

3.9 Moments of free variables

Since mixed free cumulants of free variables vanish, in principle, every joint moment of free variables can be calculated from their marginal moments. We now derive moment formulae for mixed moments of free variables and a criterion for freeness.

Lemma 3.9.1. Let $(\mathcal{A}, \varphi)$ be an NCP.

(a) Suppose $\{a_i, 1 \leq i \leq n\}$ and $\{b_i, 1 \leq i \leq n\}$ are free. Then

$$\varphi(a_1b_1a_2b_2\cdots a_nb_n) = \sum_{\pi\in NC(n)} \kappa_\pi[a_1,\ldots,a_n]\, \varphi_{K(\pi)}[b_1,\ldots,b_n]. \tag{3.7}$$

(b) Two sub-algebras $\mathcal{A}_i$, $i = 1, 2$, of $\mathcal{A}$ are free if and only if, for all $n \geq 1$, and all choices of $\{a_1,\ldots,a_n\} \in \mathcal{A}_1$ and $\{b_1,\ldots,b_n\} \in \mathcal{A}_2$, (3.7) holds. ♦

Proof. (a) We identify the indices of the a and b variables with elements from $\{1, \bar{1}, 2, \bar{2}, \ldots, n, \bar{n}\}$ in the obvious natural way. Then using the moment-free cumulant relation, we can write

$$\varphi(a_1b_1a_2b_2\cdots a_nb_n) = \sum_{\sigma\in NC(2n)} \kappa_\sigma[a_1, b_{\bar{1}}, \ldots, a_n, b_{\bar{n}}].$$

Now note that any mixed free cumulant that involves at least one a and at least one b variable is zero. Hence there is a "separation of variables" and the above expression can be written as

$$\sum_{\pi\cup\tilde{\pi}=\sigma\in NC(2n)} \kappa_\pi[a_1,\ldots,a_n]\, \kappa_{\tilde{\pi}}[b_{\bar{1}},\ldots,b_{\bar{n}}], \tag{3.8}$$

where π and $\tilde{\pi}$ are non-crossing partitions of $\mathbb{Z}_n$ and $\{\bar{1},\ldots,\bar{n}\}$ respectively.

If we fix a π, then the largest possible $\tilde{\pi}$ such that $\pi\cup\tilde{\pi} \in NC(n)$ is nothing but the Kreweras complement of π. Moreover, by Lemma 3.8.1 (d), any $\tilde{\pi} \leq K(\pi)$ satisfies $\pi\cup\tilde{\pi} \in NC(2n)$. Hence (3.8) equals

$$\sum_{\pi\in NC(n)} \kappa_\pi[a_1,\ldots,a_n] \sum_{\tilde{\pi}\leq K(\pi)} \kappa_{\tilde{\pi}}[b_{\bar{1}},\ldots,b_{\bar{n}}]$$

and then another application of the moment-free cumulant relation on the second sum finishes the proof.

(b) If $\mathcal{A}_1$ and $\mathcal{A}_2$ are free, then (3.7) follows from Part (a). The proof of the converse is easy once we use the free product construction. We give the details for completeness. Suppose that equation (3.7) holds as stated. Construct the free product, say $(\mathcal{B}, \varphi^{(1)})$ of $(\mathcal{A}_1, \varphi)$ and $(\mathcal{A}_2, \varphi)$. Let the free cumulants on this space be identified by a superscript (1). Note that, by construction, $\varphi^{(1)}$ and $\kappa^{(1)}$ agree with φ and κ on the subspaces $\mathcal{A}_1$ and $\mathcal{A}_2$. Then since $\mathcal{A}_1$ and $\mathcal{A}_2$ are free *in this NCP*, by Part (a),

$$\begin{aligned}\varphi^{(1)}(a_1b_1a_2b_2\cdots a_nb_n) &= \sum_{\pi\in NC(n)} \kappa^{(1)}_\pi[a_1,a_2,\ldots,a_n]\ \varphi^{(1)}_{K(\pi)}[b_1,b_2,\ldots,b_n]\\ &= \sum_{\pi\in NC(n)} \kappa_\pi[a_1,a_2,\ldots,a_n]\ \varphi_{K(\pi)}[b_1,b_2,\ldots,b_n]\\ &= \varphi(a_1b_1a_2b_2\cdots a_nb_n)\ \ (\text{by } (3.7)).\end{aligned}$$

Thus $\varphi^{(1)}$ agrees with φ on the *sub-algebra of $\mathcal{A}$ generated by the two algebras $\mathcal{A}_1$ and $\mathcal{A}_2$*. But these two sub-algebras are free under the state $\varphi^{(1)}$ and hence under the state φ. ■

Example 3.9.1. If a and b are free, then (3.7) implies that

$$\begin{aligned}\varphi((ab)^n) &= \sum_{\pi\in NC(n)} \kappa_\pi[a,a,\ldots,a]\ \varphi_{K(\pi)}[b,b,\ldots,b]\\ &= \sum_{\pi\in NC(n)} \kappa_\pi[b,b,\ldots,b]\ \varphi_{K(\pi)}[a,a,\ldots,a]\\ &= \varphi((ba)^n).\end{aligned}$$

▲

Theorem 3.9.2. If a and b are free in some NCP $(\mathcal{A}, \varphi)$, then φ is tracial on the sub-algebra $\mathcal{A}_{a,b}$ generated by a and b. ◆

Proof. Incidentally, we have already seen a restricted version of this traciality in Example 3.9.1. Since φ is linear, and monomials in a and b generate $\mathcal{A}_{a,b}$, it is enough to show that for any choice of $a_i = a^{n_i}$, $b_i = b^{m_i}$, where $\{n_i, m_i\}$ are arbitrary integers,

$$\varphi(a_1b_1a_2b_2\cdots a_nb_n) = \varphi(b_1a_2b_2\cdots a_nb_na_1). \tag{3.9}$$

Note that the $\{a_i\}$ commute and so do the $\{b_i\}$. This commutativity can be used to show that (using the moment-free cumulant relation)

$$\kappa_\pi[a_1,a_2,\ldots,a_n] = \kappa_\pi[a_2,\ldots,a_n,a_1]$$

and

$$\varphi_{K(\pi)}[b_1,b_2,\ldots,b_n] = \varphi_{K(\pi)}[b_2,\ldots,b_n,b_1].$$

Now (3.9) follows from the above equalities upon invoking Lemma 3.9.1 (a). ■

3.10 Compound free Poisson

Recall that, in classical probability, a random variable X is said to have a *compound Poisson law* if its c.g.f. and m.g.f. are of the form

$$C_X(t) = \lambda \,\mathrm{E}\left[\exp(tY) - 1\right] \text{ and } M_X(t) = \exp\left[\lambda \,\mathrm{E}\left[\exp(tY) - 1\right]\right],$$

where Y is a random variable with distribution function F with a finite m.g.f. in a neighborhood of 0. We say that X follows the *compound Poisson law* with jump distribution F and rate λ. It is clear that if $P[Y = 1] = 1$ then X is a classical Poisson random variable with mean λ.

The cumulants of X are given by

$$c_n(X) = \lambda \,\mathrm{E}(Y^n), \quad n \geq 1. \tag{3.10}$$

This motivates our definition of a *compound free Poisson variable.* Recall the notation $m_n(\mu)$ for the n-th moment of a probability measure μ.

Definition 3.10.1. (Compound free Poisson variable) A self-adjoint variable a in some NCP is said to be *compound free Poisson* with rate λ and *jump distribution* ν if its free cumulants are given by

$$\kappa_n(a) = \lambda m_n(\nu), \; \forall n \geq 1.$$

◇

In Theorem 4.3.2 (b) we shall see how the moments of a compound free Poisson variable define a probability law and how this law arises as the limit law of free additive convolutions. In Remark 8.2.1 (c), we shall see that this law also arises as the limit spectral distribution of large dimensional sample covariance matrices.

Example 3.10.1. Suppose, in some $*$-probability space $(\mathcal{A}, \varphi)$, s is a semi-circular variable and is free of b which is self-adjoint. We assume that the moments of b yield a probability law μ_b with compact support. Observe that since b and s are self-adjoint, so is sbs. Let us identify its distribution.

First note that, since b and s are free, by Theorem 3.9.2, φ is tracial for the purposes of computing the moments of $\{s, b\}$. Hence $\varphi((sbs)^n) = \varphi(s^2bs^2 \cdots s^2b)$ which is in a form where Lemma 3.9.1 (a) is applicable. From Section 2.4, s^2 is free Poisson and all its free cumulants are 1. Hence

$$\begin{aligned}
\varphi((sbs)^n) &= \varphi(s^2bs^2 \cdots s^2b) \\
&= \sum_{\pi \in NC(n)} \kappa_\pi[s^2, \ldots, s^2]\varphi_{K(\pi)}[b, b, \ldots, b] \text{ (by (3.7)}, \\
&= \sum_{\pi \in NC(n)} \varphi_{K(\pi)}[b, b, \ldots, b] \\
&= \sum_{\pi \in NC(n)} \varphi_\pi[b, b, \ldots, b] \text{ as the map } \pi \to K(\pi) \text{ is one-one onto.}
\end{aligned}$$

Now, if we invoke the moment-free cumulant relation, the above equality implies that

$$\kappa_n(sbs) = \varphi(b^n) = m_n(\mu_b) \text{ for all } n.$$

That is, sbs is a compound free Poisson variable with jump distribution μ_b and rate 1.

In particular, suppose b is a Bernoulli variable with $\varphi(b^n) = p$, for all $n \geq 1$. Then all free cumulants of sbs are p. Hence sbs is a free Poisson variable with mean (rate) p. ▲

3.11 Exercises

1. Suppose $\pi \in NC(k)$. Show that $K(\pi) = \pi^{-1}\gamma_k$ where, γ_k is the cyclic permutation $1 \to 2 \to \cdots \to k \to 1$.
2. Suppose a is a variable with moments

$$\varphi(a^n) = \begin{cases} 1 & \text{if } n \text{ is even,} \\ 0 & \text{if } n \text{ is odd.} \end{cases} \tag{3.11}$$

   Show that the free cumulants of a are given by

$$\kappa_n(a) = \begin{cases} (-1)^{k-1}C_{k-1} & \text{if } n = 2k, \\ 0 & \text{otherwise,} \end{cases} \tag{3.12}$$

   where $\{C_k\}$ are the Catalan numbers.
3. Show that constants $\{c_i \mathbf{1}_{\mathcal{A}}\}$ are always free.
4. Show that constants are also free of any other variable.
5. When is a variable a free of itself in an algebra? When is a variable a $*$- free of itself in a $*$-algebra?
6. Prove Theorem 3.1.1.
7. Suppose c is a standard circular variable on some $*$-probability space $(\mathcal{A}, \varphi)$.

   (a) Show that the moments of cc^* are given by

$$\varphi((cc^*)^k) = \varphi(cc^* \cdots cc^*) = C_k$$

   where C_k is the k-th Catalan number.

   (b) Find $\kappa_n(c^{\epsilon_1}, \ldots, c^{\epsilon_n})$ for all $n \geq 1$ and $\epsilon_i \in \{1, *\}$, $1 \leq i \leq n$.
8. Suppose e is an elliptic variable with parameter $\rho \neq 0$. Show that the moments of ee^* do not depend on ρ.

9. Suppose $\{a_i, 1 \leq i \leq n\}$ are free standard semi-circular variables. Let

$$\bar{a}_n = \frac{a_1 + \ldots + a_n}{n} \text{ and } s_j = \sum_{i=1}^{j} (a_i - \bar{a}_n)^2.$$

Show that for every $n \geq 2$, $\bar{a}_n$ and $\{s_2, \ldots, s_n\}$ are free. [For a solution see Bose, Dey and Ejsmont (2018)[24]].

10. Suppose $\{a_i, 1 \leq i \leq n\}$ are free variables in some $*$-probability space $(\mathcal{A}, \varphi)$. Show that φ restricted to the $*$-algebra generated by $\{a_i, 1 \leq i \leq n\}$ is tracial.

11. Suppose a and b are free. Show that

$$\kappa_n(ab) = \sum_{\pi \in NC(n)} \kappa_\pi(a) \kappa_{K(\pi)}(b) \text{ for all } n \geq 1.$$

12. Suppose $(\mathcal{A}, \varphi)$ is a $*$-probability space. Define $\mathcal{A}_2$ as the set of all 2×2 matrices whose elements belong to $\mathcal{A}$. Show that with the natural operations, this is also a $*$-algebra. Define φ_2 on $\mathcal{A}_2$ as $\varphi_2(A) = \varphi(\text{tr}(A))$.

   (a) Show that $(\mathcal{A}_2, \varphi_2)$ is $*$-probability space.

   (b) Consider $A = ((s_{i,j})) \in \mathcal{A}_2$ where $s_{1,2} = s_{2,1}$ and $s_{i,j}, i \leq j$ are free standard semi-circular variables in $\mathcal{A}$. Find the distribution of A.

13. Suppose μ_1 and μ_2 are compactly supported probability measures on $\mathbb{R}$. Show that $\mu_1 \boxplus \mu_2$ is compactly supported.

14. Show that (3.10) indeed gives the classical cumulants of a compound Poisson random variable.

4

Convergence

Classical probability theory is rich with various types of convergence concepts. Many of these concepts cannot be paralleled in the non-commutative setup. However there are two concepts that can be retained. The notion of limits via convergence of joint moments is called algebraic convergence. At the same time, there is the notion of weak convergence of probability laws of self-adjoint variables, whenever such probability laws exist. These two notions are related, and, for self-adjoint variables, moment convergence with an added restriction implies weak convergence.

We shall use these ideas to establish the free independence analog of the classical multivariate central limit theorem, convergence of an appropriate sequence of free binomials laws to the free Poisson law, and, more generally, convergence of the free convolution of mixtures of free binomial laws to compound free Poisson laws. We shall also introduce the concept of asymptotic freeness which will be useful when we move to the study of large dimensional random matrices.

4.1 Algebraic convergence

We begin by noting that if $(\mathcal{A}, \varphi)$ is a $*$-probability space and $\mathcal{B}$ is a unital $*$-sub-algebra of $\mathcal{A}$ then $(\mathcal{B}, \varphi)$ is also a $*$-probability space. Suppose $\{a_i : i \in I\}$ are elements of a $*$-probability space $(\mathcal{A}, \varphi)$. Then the $*$-sub-algebra $\text{Span}\{a_i, a_i^* : i \in I\}$ of $\mathcal{A}$ generated by $\{a_i, a_i^* : i \in I\}$ is defined as

$$\text{Span}\{a_i, a_i^* : i \in I\} = \{\Pi(a_i, a_i^* : i \in I) : \ \Pi \text{ is a polynomial}\}. \tag{4.1}$$

We can define a non-starred version of Span in the obvious way.

Definition 4.1.1. (Convergence of variables and of $*$-probability spaces) Let $(\mathcal{A}_n, \varphi_n)$, $n \geq 1$ be a sequence of $*$-probability spaces and let $(\mathcal{A}, \varphi)$ be another $*$-probability space.

(a) (Marginal convergence) We say that $a^{(n)} \in \mathcal{A}_n$ *converges in $*$-distribution* to $a \in \mathcal{A}$ if

$$\lim \varphi_n(\Pi(a^{(n)}, a^{(n)*})) = \varphi(\Pi(a_i, a_i^*)) \ \text{ for all polynomials } \ \Pi. \tag{4.2}$$

DOI: 10.1201/9781003144496-4

We denote this convergence by $a^{(n)} \xrightarrow{*} a$. If the above convergence holds without the $*$-variables, we say that $a^{(n)}$ converges to a in distribution and write $a^{(n)} \to a$. If the context is clear, we write "convergence in distribution" and $a^{(n)} \to a$ to mean either of the above.

(b) (Joint convergence) Suppose I is an index set. The elements $\{a_i^{(n)} : i \in I\}$ from $\mathcal{A}_n$ are said to *converge (jointly)* to $\{a_i : i \in I\}$ from $\mathcal{A}$ if,

$$\Pi(\{a_i^{(n)}, a_i^{*(n)} : i \in I\}) \xrightarrow{*} \Pi(\{a_i, a_i^* : i \in I\}) \text{ for all } \Pi. \tag{4.3}$$

We write $\{a_i^{(n)} : i \in I\} \xrightarrow{*} \{a_i : i \in I\}$. We also write $\text{Span}\{a_i^{(n)}, a_i^{*(n)} : i \in I\} \to \text{Span}\{a_i, a_i^* : i \in I\}$. Again, we may consider only the non-starred version of the above.

(c) If the limit variables $\{a_i, i \in I\}$ are free or $*$-free respectively, then we say that $\{a_i^{(n)}, i \in I\}$ are free or $*$-free in the limit, or are *asymptotically free or $*$-free* respectively. ◇

In Chapter 9 we shall see examples of high dimensional random matrices which are asymptotically free or $*$-free.

Note that the limit φ automatically inherits some properties of $\{\varphi_n\}$. For instance if $\{\varphi_n\}$ are tracial or positive on $\text{Span}\{a_i^{(n)}, a_i^{*(n)} : i \in I\}$, then so is φ on $\text{Span}\{a_i, a_i^* : i \in I\}$. Moreover, joint convergence implies marginal convergence of the variables as well as the convergence of each polynomial constructed from the variables in the sequence.

Remark 4.1.1. Suppose for every $n \geq 1$, $\mathcal{A}_n = \text{Span}\{a_i^{(n)}, a_i^{*(n)} : i \in I\}$ is a unital $*$-sub-algebra of the NCP $(\mathcal{A}_n, \varphi_n)$ and, for all polynomials Π, $\lim \varphi_n(\Pi(\{a_i^{(n)}, a_i^{*(n)}) : i \in I\})$ exists. Then we can *construct* a unital polynomial $*$-algebra $\mathcal{A}$ of indeterminates $\{a_i, a_i^*\}$. We can define φ on $\mathcal{A}$ by

$$\varphi\big(\Pi(\{a_i, a_i^* : i \in I\})\big) = \lim \varphi_n\big(\Pi(\{a_i^{(n)}, a_i^{*(n)} : i \in I\})\big).$$

Then, clearly, $(\mathcal{A}, \varphi)$ is a $*$-probability space, and $(\mathcal{A}_n, \varphi_n) \xrightarrow{*} (\mathcal{A}, \varphi)$. Moreover, if $\{\varphi_n\}$ are tracial or positive, then so is φ. ●

Example 4.1.1. Let $\mathcal{A}_n$ be the $*$-algebra of all $n \times n$ diagonal matrices with real entries and the positive state $\varphi_n = \frac{1}{n}\text{Trace}$. Consider the sequence of $*$-probability spaces $(\mathcal{A}_n, \varphi_n)$. Suppose $D_n \in \mathcal{A}_n$ is a diagonal matrix with real elements $d_{i,n}$, $1 \leq i \leq n$ on the diagonal. Suppose further that

$$m_k = \lim \frac{1}{n} \sum_{i=1}^{n} d_{i,n}^k \text{ (assumed finite for all } k \in \mathbb{N}). \tag{4.4}$$

We may then define an algebra $\mathcal{A} = \text{Span}\{a\}$, where a is some indeterminate and define φ on $\mathcal{A}$ by $\varphi(a^k) = m_k$ for all $k \geq 1$. Then $(\text{Span}\{D_n\}, \varphi_n)$ converges to $(\text{Span}\{a\}, \varphi)$. ▲

In the above example, since D_n is a diagonal matrix, the numbers $d_{i,n}$ are also its eigenvalues. Even when the matrices are not necessarily diagonal, the distribution of eigenvalues is connected to the notion of algebraic convergence defined above. To see this, we first need a definition.

Definition 4.1.2. (Empirical spectral distribution) Suppose A_n is any $n \times n$ matrix with real or complex entries. Then the *empirical spectral distribution* (ESD) μ_{A_n} of A_n is the probability law which puts mass $1/n$ at each of the eigenvalues. $\diamond$

Note that the eigenvalues may be complex, and hence the ESD is a probability law on the complex plane in general.

Trace-moment formula. Suppose A_n is a real symmetric matrix with eigenvalues $\lambda_{i,n}$, $1 \leq i \leq n$. Since all eigenvalues are real, μ_{A_n} is now a probability law on $\mathbb{R}$. Let $m_k(\mu_{A_n})$ be its k-th moment. It is easy to see that

$$m_k(\mu_{A_n}) = \frac{1}{n}\sum_{i=1}^{n} \lambda_{i,n}^k = \frac{1}{n}\text{Trace}(A_n^k).$$

This is known as the *trace-moment formula.*

Example 4.1.2. Consider the algebra $\mathcal{A}_n$ of all $n \times n$ real symmetric matrices with the state $\varphi_n = \frac{1}{n}\text{Trace}$. Then a sequence of matrices $A_n \in \mathcal{A}_n$, $n \geq 1$ converges in distribution if and only if

$$\varphi_n(A_n^k) = \frac{1}{n}\text{Trace}(A_n^k) \text{ converges for all } k \in \mathbb{N}.$$

On the other hand, as we have seen above,

$$m_k(\mu_{A_n}) = \frac{1}{n}\text{Trace}(A_n^k) = \varphi_n(A_n^k).$$

If $\{A_n\}$ converges in distribution with respect to the states $\{\varphi_n\}$, then this is the same as the convergence of $\{m_k(\mu_{A_n})\}$ for every k to, say, m_k. If, further, $\{m_k\}$ determines a unique probability law, say, μ, then μ_{A_n} converges weakly to μ by Lemma 1.2.1 of Chapter 1. This line of reasoning fails if some of the eigenvalues of A_n are complex-valued. ▲

Example 4.1.3. Let $(\mathcal{A}_n, \varphi_n)$ be the $*$-probability space where $\mathcal{A}_n$ is the algebra of all $n \times n$ diagonal matrices with real random variables on the diagonal, all of whose moments are finite, and $\varphi_n = \frac{1}{n}\,\text{E}\,\text{Trace}$.

Suppose $\{X_i, i \geq 1\}$ are i.i.d. real-valued random variables with a common probablity law μ. Define the $n \times n$ diagonal random matrix D_n whose i-th diagonal entry is X_i. Let $m_k = \text{E}(X_1^k)$. Similar to Example 4.1.1, D_n converges with respect to the state φ_n to a variable a whose moments are $\varphi(a^k) = m_k$ for all k.

In this case, the ESD is the *random probability measure* μ_n which puts mass $1/n$ at each $X_i, 1 \leq i \leq n$. We know from the Glivenko-Cantelli theorem (see the classic book by Gnedenko (1997)[46]) that μ_n converges weakly to μ *almost surely* (irrespective of whether the moments of μ are finite or not). This result can also be linked to algebraic convergence as follows when all moments are finite.

Define another state $\tilde{\varphi}_n$ on $\mathcal{A}_n$ as $\frac{1}{n}$Trace. Note that φ_n and $\tilde{\varphi}_n$ are related by $\varphi_n = \mathrm{E}\,\tilde{\varphi}_n$. By the trace-moment formula and the strong law of large numbers,

$$m_k(\mu_n) = \tilde{\varphi}_n(D_n^k) = \frac{1}{n}\sum_{i=1}^{n} X_i^k \to m_k \text{ almost surely for all } k.$$

That is, D_n converges with respect to the state $\tilde{\varphi}_n$, *almost surely.* Now suppose that $\{m_k\}$ determines the probability law μ. Then again by Lemma 1.2.1, we can conclude that μ_n converges weakly to μ *almost surely.* Note that this argument required the finiteness of all moments. ▲

Example 4.1.4. Consider the $n \times n$ real symmetric matrix $T_{1,n}$ whose upper and lower first diagonals have entries equal to 1 and the rest of the entries are 0. Let $\varphi_n = \frac{1}{n}$Trace. It is easy to see that

$$\varphi_n(T_{1,n}) = 0, \text{ and } \varphi_n(T_{1,n}^2) = \frac{2(n-1)}{n} \to 2 \text{ as } n \to \infty.$$

It is not hard to show that $\varphi_n(T_{1,n}^k)$ converges for every non-negative integer k. Hence $T_{1,n}$ converges in distribution with respect to the state $\varphi_n = \frac{1}{n}$Trace. Moreover, using the trace-moment formula, it can be shown that the ESD of $T_{1,n}$ converges weakly to a compactly supported probability law. ▲

Example 4.1.5. Suppose $\{x_i,\ i = 0, 1, \ldots k\}$ is a finite sequence of real numbers. Let $T_{k,n}$ be the $n \times n$ *Toeplitz matrix* whose (i,j)-th element equals $x_{|i-j|}$ if $|i-j| \leq k$, and 0 otherwise. Then it can be shown that $\{T_{k,n}\}$ converges in distribution with respect to the states $\{\frac{1}{n}\text{Trace}\}$ as $n \to \infty$. The ESD of $T_{k,n}$ also converges weakly to a compactly supported probability law. For more information on general Toeplitz matrices, see the classic book of Grenander and Szegő (1984)[51] and, also, Gray (2006)[48]. ▲

Recall from (2.15) and (2.17), the relation between moments and free cumulants. Using this relation the following lemma can be proved easily. The proof is left as an exercise.

Lemma 4.1.1. A sequence of variables $a^{(n)} \in \mathcal{A}_n$ converges in $*$-distribution to $a \in \mathcal{A}$ if and only if all its $*$-free cumulants converge to those of a. Similar statements are true for (a) convergence in distribution, (b) joint convergence and (c) convergence of spans. ♦

Weak and algebraic convergence. Suppose for every $n \geq 1$, a_n is a *self-adjoint* variable from an NCP $(\mathcal{A}_n, \varphi_n)$. If $\varphi_n(a_n^k)$ converges for every k, say to m_k, then, according to our definition, a_n converges in distribution.

On the other hand, suppose, for each n, there exists (not necessarily unique) probability measures μ_n such that

$$\int_{\mathbb{R}} x^k d\mu_n(x) = \varphi_n(a_n^k) \text{ for all integers } k \geq 1.$$

Then it follows that $\{m_k\}$ is a moment sequence. Suppose this sequence identifies a unique probability distribution, say μ, such that

$$m_k = \int_{\mathbb{R}} x^k d\mu(x) \text{ for all integers } k \geq 1.$$

Then, by Lemma 1.2.1 (a), μ_n converges to μ weakly. Algebraic convergence and weak convergence are thus related.

4.2 Free central limit theorem

We are now ready to state and prove the basic one-dimensional free central limit theorem. We need the following notation. If μ is a probability law, then for any constant c, $\mu^{(c)}$ denotes the dilated probability law, defined by

$$\mu^{(c)}(A) = \mu(cA).$$

Recall the free additive convolution $\boxplus$, defined in Section 3.7.

Theorem 4.2.1. Suppose $\{a_i\}$ are self-adjoint free variables in some $*$-probability space $(\mathcal{A}, \varphi)$ with identical moments. Further, $\varphi(a_i) = 0$, $\varphi(a_i^2) = 1$ for all i.

(a) Then $n^{-1/2}(a_1 + \cdots + a_n) \to s$ where s is a standard semi-circular variable.

(b) Suppose further that the moments of a_i determine a probability law, say μ. Then $\left(\mu^{(n^{1/2})}\right)^{\boxplus n}$ converges weakly to the semi-circular law. ◆

Proof. (a) We have the following relations:

$$\begin{aligned}
\kappa_1\left(\frac{1}{\sqrt{n}}(a_1 + \cdots + a_n)\right) &= n^{-1/2}(\kappa_1(a_1) + \cdots + \kappa_1(a_n)) \text{ (linearity)} \\
&= 0. \\
\kappa_2\left(\frac{1}{\sqrt{n}}\sum_{i=1}^n a_i, \frac{1}{\sqrt{n}}\sum_{i=1}^n a_i\right) &= \frac{1}{n}\sum_{i=1}^n \kappa_2(a_i) \text{ (mixed free cumulants are 0)} \\
&= 1.
\end{aligned}$$

$$\begin{aligned}
\kappa_j\left(\frac{1}{\sqrt{n}}\sum_{i=1}^n a_i\right) &= n^{-j/2}\sum_{i=1}^n \kappa_j(a_i) \text{ (freeness)}, \\
&= n^{-j/2+1}\kappa_j(a_1) \qquad (4.5) \\
&\to 0 \text{ if } j \geq 3.
\end{aligned}$$

Thus all free cumulants converge to the free cumulants of the semi-circular law. By the moment-free cumulant relation, the same happens for moments. Finally, since $n^{-1/2}(a_1 + \cdots + a_n)$ is self-adjoint for every n, we can take the limit s to be also self-adjoint. This completes the proof of (a).

(b) From Part (a), all moments of $\left(\mu^{(n^{1/2})}\right)^{\boxplus n}$ converge to the moments of the semi-circular law. These moments uniquely determine the semi-circular law and hence the result follows once we invoke Lemma 1.2.1 (a). ■

Example 4.2.1. (a) If $\{a_i\}$ are free standard semi-circular variables, then we have seen in Chapter 3 that the distribution of $n^{-1/2}(a_1 + \cdots + a_n)$ is again semi-circular for every n.

(b) Suppose $\{a_i\}$ are Bernoulli with parameter p on some $*$-probability space $(\mathcal{A}, \varphi)$ and are free. Recall that then

$$\varphi(a_i^k) = p \quad \text{for all} \quad k.$$

Let $b_i = \frac{a_i - p\mathbf{1}_{\mathcal{A}}}{\sqrt{p(1-p)}}$. Then $\{b_i\}$ are free, identically distributed and $\varphi(b_i) = 0$ and $\varphi(b_i^2) = 1$. Let $x_n = a_1 + a_2 + \cdots + a_n$. Theorem 4.2.1 implies that

$$\frac{x_n - np\mathbf{1}_{\mathcal{A}}}{\sqrt{np(1-p)}} = n^{-1/2}(b_1 + \cdots + b_n)$$

converges to the standard semi-circular variable. That is, the standardized free binomial variable converges in distribution to the semi-circular variable. The probability law of this variable converges to the semi-circular law. ▲

We now state and give an outline of the proof of the multivariate version of the free CLT.

Theorem 4.2.2. Suppose for all $j \geq 1$, $\{a_{i,j}, 1 \leq i \leq k\}$ are self-adjoint variables on $(\mathcal{A}, \varphi)$, which are identically distributed and free across $j \geq 1$. Further suppose that, for all i, $\varphi(a_{i,1}) = 0$ and $\varphi(a_{i,1}^2) = \sigma_i^2$. Then

$$\{n^{-1/2}(a_{i,1} + \cdots + a_{i,n}), 1 \leq i \leq k\} \to \{s_i, 1 \leq i \leq k\}$$

which is a semi-circular family (in some $*$-probability space $(\mathcal{A}, \varphi_0)$) with $\varphi_0(s_i) = 0$, $\varphi_0(s_i^2) = \sigma_j^2$ and $\varphi_0(s_i s_j) = \varphi(a_{i,1}a_{j,1})$ for all i, j. ◆

Proof. It is enough to show the convergence of the free cumulants. We extend the proof for the single variable case. Define

$$b_{i,n} := n^{-1/2}(a_{i,1} + \cdots + a_{i,n}), \quad 1 \leq i \leq k.$$

Note that $\kappa_1(b_{i,n}) = 0$ for all i, n. Now let us verify the limit of the second order free cumulants. Fix i, j. Indeed, for every fixed n,

$$\kappa_2(b_{i,n}, b_{j,n}) = n^{-1} \sum_{k=1}^{n} k_2(a_{i,n}, a_{j,n}) = k_2(a_{i,1}, a_{j,1}).$$

If we imitate the arguments of the proof of Theorem 4.2.1, it is clear that

$$\kappa_j(b_{i_1,n},\ldots,b_{i_j,n}) \to 0 \text{ for all } j \geq 3, \text{ and } 1 \leq i_1,\ldots i_j \leq k,$$

once we use the freeness and the "identically distributed" conditions. ■

4.3 Free Poisson convergence

Suppose, for every fixed n, $X_{i,n}, 1 \leq i \leq n$, are classically i.i.d. Bernoulli random variables with probability of success p_n where $np_n \to \lambda < \infty$ as $n \to \infty$. Then we know that the probability law of $\sum_{i=1}^n X_{i,n}$ converges to the Poisson law with mean λ. We now prove a free analog of this result.

Theorem 4.3.1. Suppose, for every n, $a_{i,n}, 1 \leq i \leq n$, are free and each is a Bernoulli variable with parameter p_n in some $*$-probability space $(\mathcal{A}_n, \varphi_n)$. Suppose that $np_n \to \lambda < \infty$ as $n \to \infty$. Then the following hold:

(a) The free binomial variable $a_{1,n} + \cdots + a_{n,n}$ converges to a free Poisson variable with mean λ;

(b) The probability law of $a_{1,n}+\cdots+a_{n,n}$ converges weakly to the free Poisson law with parameter λ. ◆

Proof. (a) Write $b_n = a_{1,n} + \cdots + a_{n,n}$. It is enough to prove that all free cumulants of b_n converge to λ. Since, for every n, the mixed free cumulants of $a_{i,n}$ are all zero, and they are identically distributed, we have $k_j(b_n) = nk_j(a_{1,n})$. This immediately yields that $\kappa_1(b_n) = np_n \to \lambda$.

For higher order free cumulants, we show that $k_j(b_n) = np_n + o(1)$. Using the moment-free cumulant relation,

$$\kappa_j(a_{1,n}) = \sum_{\pi \in NC(j)} m_{\pi,n}\mu[\pi, \mathbf{1}_j] \tag{4.6}$$

where $\{m_{\pi,n}\}$ is the multiplicative extension of the moments of $a_{1,n}$.

Note that the moments $m_{k,n}$ of $a_{1,n}$ are equal to p_n for all $k \geq 1$. Hence, $m_{\pi,n} = p_n^{|\pi|}$. Now, since the number of terms in the sum (4.6) is finite, and $np_n \to \lambda$, it is easy to see that

$$\kappa_j(a_{1,n}) = p_n + O(p_n^2),$$

where the contribution p_n comes from the partition $\mathbf{1}_n$ and the rest of the partitions together contribute $O(p_n^2)$. Hence

$$\kappa_j(b_n) = n\kappa_j(a_{1,n}) = np_n + o(1) \to \lambda,$$

and the proof of (a) is complete.

(b) First note that, as discussed in Sections 3.3, the free additive convolution of the free Bernoulli laws with parameters $p_i, 1 \leq i \leq n$, is well defined since all these probability laws on $\mathbb{R}$ are compactly supported. Rigor shall be added to this discussion later in Section 6.5. However, this is also the probability law of $a_{1,n} + \cdots + a_{n,n}$. We have already seen in the proof of Part (a) that all its free cumulants converge to those of the free Poisson law with mean λ. As a consequence, the moments converge to the corresponding moments due to the moment-free cumulant relation. Now, the free Poisson law has compact support and hence we can invoke Lemma 1.2.1 (a) to conclude the proof. ■

Recall Definition 3.10.1 of the compound free Poisson distribution. As in the classical case with compound Poisson (see page 402 of Shorack (2000)[88]), we can realize this distribution as a limit. The next theorem is an extension of Theorem 4.3.1.

Theorem 4.3.2. Suppose ν is a compactly supported probability law on $(0, \infty)$. Let $\{p_n\}$ be a sequence of positive numbers such that $np_n \to \lambda$. Suppose $a_{i,n}, 1 \leq i \leq n$ are self-adjoint free variables with the identical probability law μ_n which is defined as

$$\mu_n(A) = (1 - p_n)\mathbb{I}(0 \in A) + p_n\nu(A), \text{ for all Borel sets } A. \tag{4.7}$$

Then,

(a) the self-adjoint variable $a_{1,n} + \cdots + a_{n,n}$ converges to a compound free Poisson variable with rate λ and jump distribution ν.

(b) The probability law of $a_{1,n} + \cdots + a_{n,n}$ converges weakly to the compound free Poisson law. ◆

Proof. (a) It is enough to show that the j-th free cumulant converges to $\lambda \int_{\mathbb{R}} x^j d\nu(x)$ for every j. Now, using the facts that the mixed free cumulants of all orders vanish, and the variables are identically distributed, we have

$$\kappa_j(\sum_{i=1}^{n} a_{i,n}) = n\kappa_j(a_{1,n}) = n \sum_{\pi \in NC(j)} m_{\pi,n}\mu[\pi, \mathbf{1}_j], \tag{4.8}$$

where μ is the Möbius function and $m_{\pi,n}$ is the multiplicative extension of the moments of $a_{1,n}$.

When π has k blocks of sizes $t_1, \ldots t_k$, then, by using (4.7),

$$m_{\pi,n} = p_n^k \prod_{s=1}^{k} \int x^{t_s} d\nu(x).$$

It is clear that, for $k \geq 2$, the sum of the terms in (4.8) when added over $\pi \neq \mathbf{1}_n$ is of the order $O(p_n^2)$.

On the other hand, $m_{1_n} = p_n \int x^j d\nu(x)$. Thus

$$\begin{aligned}\kappa_j(\sum_{i=1}^{n} a_{i,n}) &= np_n \int_{\mathbb{R}} x^j d\nu(x) + o(1) \\ &\to \lambda \int_{\mathbb{R}} x^j d\nu(x), \text{ as } n \to \infty.\end{aligned}$$

(b) It is enough to note the following two things:

(i) From the results pointed out in the proof of Theorem 4.3.1 (b), there is a compactly supported probability law of $a_{1,n} + \cdots + a_{n,n}$ for each n.

(ii) The limit free cumulants, and hence the limit moments obtained in the proof of Part (a) are associated to a unique probability law. This law has compact support. Hence we can invoke Lemma 1.2.1 (a) to complete the proof. ■

4.4 Sums of triangular arrays

We now wish to move to general triangular arrays. The previous limit theorems can be obtained as special cases.

Theorem 4.4.1. Suppose $(\mathcal{A}_n, \varphi_n)$ is a sequence of $*$-probability spaces and I is an index set. Suppose for each $n \geq 1$, $\{a_{n,1}^{(j)}\}_{j \in I}, \ldots, \{a_{n,n}^{(j)}\}_{j \in I}$ are free and identically distributed variables from $\mathcal{A}_n$. Then the statements (a) and (b) given below are equivalent:

(a) $\{a_{n,1}^{(j)} + \cdots + a_{n,n}^{(j)}\}_{j \in I} \to \{b_j\}_{j \in I}$.

(b) For all $t \geq 1$, and all $\{i(1), ..., i(t)\} \subset I$, $\lim n\varphi_n(a_{n,r}^{(i(1))} \cdots a_{n,r}^{(i(t))})$ exists.

If (b) holds, then the free cumulants of $\{b_j\}_{j \in I}$ are given by

$$\kappa_t(b_{i(1)}, \ldots, b_{i(t)}) = \lim n\varphi_n(a_{n,r}^{(i(1))} \cdots a_{n,r}^{(i(t))}) \text{ for all } t \geq 1, \{i(1), ..., i(t)\} \subset I.$$

◆

Proof. Incidentally, the above limit does not depend on the specific choice of r. We shall provide only a sketch and leave the details of the proof to the reader as an exercise. There are two key steps. One is to remember the identically distributed assumption and freeness assumption. The other is to quickly establish, using the moment-free cumulant relation, that

$$\lim n\varphi_n(a_{n,r}^{(i(1))} \cdots a_{n,r}^{(i(t))}) \text{ exists for all } t \geq 1, \ \{i(1), ..., i(t)\} \subset I$$

if and only if

$$\lim n\kappa_t^{(n)}(a_{n,r}^{(i(1))}, \ldots, a_{n,r}^{(i(t))}) \text{ exists for all } t \geq 1, \{i(1), ..., i(t)\} \subset I$$

and they are equal. Here $\kappa^{(n)}$ is the free cumulant function for φ_n. ■

4.5 Exercises

1. Show that the limit moments in Example 4.1.4 determine a unique probability distribution.
2. Verify the claim on $T_{k,n}$ in Example 4.1.5.
3. Consider the $n \times n$ real symmetric matrix $H_{1,n}$ whose (i, j)-th entry equals 1 if $i + j = n + 1$ and all other entries are 0. Show that $H_{1,n}$ converges in distribution with respect to the state $\varphi_n = \frac{1}{n}$Trace and identify the limit. Extend the result in a suitable way.
4. Prove Lemma 4.1.1.
5. (Extension of the free CLT to non-identical variables) Suppose $\{a_i\}$ are self-adjoint free variables in some $*$-probability space $(\mathcal{A}, \varphi)$ such that $\varphi(a_i) = 0$, $\varphi(a_i^2) = 1$. Suppose further

$$\sup_{1 \leq i < \infty} |\varphi(a_i^k)| < C_k < \infty \text{ for every } k \geq 1. \tag{4.9}$$

Show that

(a) $n^{-1/2}(a_1 + \cdots + a_n) \to s$ where s is a standard semi-circular variable. Hint: For a free cumulant-based proof, equality in (4.5) is not available anymore. One can still get away by establishing an upper bound which is asymptotically negligible, with the help of the moment-free cumulant relation and the following growth bound on the Möbius function:

$$|\mu[\pi, \mathbf{1}_j]| \leq 4^j \text{ for all } j \geq 1.$$

For a proof of the above bound, see Corollary 13.1.5, or Lecture 10 of Nica and Speicher (2006)[74]. As an alternative, one can switch to a more cumbersome moment-based proof.

(b) Suppose further that each a_i has a probability law, say μ_i, $i \geq 1$. Then the probability law $\mu_1^{(n^{1/2})} \boxplus \mu_2^{(n^{1/2})} \boxplus \cdots \boxplus \mu_n^{(n^{1/2})}$ converges weakly to the semi-circular law.

6. Formulate a non-identically distributed version of the free multivariate CLT.

7. Suppose $\{\mu_n, n \geq 1\}$ and μ are probability measures with compact support. Show that the following are equivalent:

   (a) $\mu_n^{\boxplus n}$, the n-fold free additive convolution of μ_n, converges to μ weakly.

   (b) $nm_j(\mu_n) \to \kappa_j(\mu)$ for all $j \geq 1$.

   (c) $n\kappa_j(\mu_n) \to \kappa_j(\mu)$ for all $j \geq 1$.

8. Suppose $(\mathcal{A}_n, \varphi_n)$ is a sequence of $*$-probability spaces and $a_n \in \mathcal{A}_n$ for all $n \geq 1$. Let $\kappa^{(n)}$ denote the free cumulants corresponding to φ_n.

   (a) Show that $\lim n\varphi_n(a_n^j)$ exists for all $j \geq 1$ if and only if $\lim n\kappa_j^{(n)}(a_n)$ exists for all $j \geq 1$. Moreover, if the limits exist then they are equal.

   (b) Formulate and prove an extension of Part (a) when we have k variables $\{a_{n1}, \ldots, a_{nk}\}$ from $\mathcal{A}_n$.

9. Provide a detailed proof of Theorem 4.4.1.

10. Let x_n, y_n be non-commutative self-adjoint variables on the same $*$-probability space $(\mathcal{A}, \varphi)$ such that $x_n \to x$ and $y_n \to c\mathbf{1}_{\mathcal{A}}$ for a constant c. Then show that $(x_n, y_n) \to (x, c\mathbf{1}_{\mathcal{A}})$ jointly.

11. Let $\{x_n\}$ be self-adjoint freely independent variables on a $*$-probability space $(\mathcal{A}, \varphi)$ such that $\sup_n |\varphi(x_n^k)| < \infty$ for all k. Then show that

$$\frac{1}{n}\sum_{i=1}^{n} x_i \to \varphi(x_1)\mathbf{1}_{\mathcal{A}}.$$

12. Suppose $\{x_n\}$ are freely independent identically distributed self-adjoint random variables on a $*$-probability space $(\mathcal{A}, \varphi)$. Suppose $h(x_1, ..., x_m)$ is a self-adjoint polynomial in the m variables $x_1, ..., x_m$ and is symmetric in its arguments. Let U_n be the U-statistics with kernel h defined as

$$U_n = \frac{1}{\binom{n}{m}} \sum_{1 \leq i_1 < i_2 < \cdots < i_m \leq n} h(x_{i_1}, \ldots x_{i_m}).$$

   Let $\theta = \varphi(h(X_1, ..., X_m))$. For $k = 0, 1, ..., m$ define

$$c_k = \kappa_2(h(X_1, ..., X_k, X_{k+1}, ..., X_m), h(X_1, ..., X_k, X_{m+1}, ..., X_{2m-k})).$$

   Then as $n \to \infty$, show that

   (a) $U_n \to \theta\mathbf{1}_{\mathcal{A}}$,

   (b) $\sqrt{n}(U_n - \theta\mathbf{1}_{\mathcal{A}})$ converges in $*$-distribution to a semi-circular variable with variance $m^2 c_1$.

13. Let

$$h(x_1, x_2) := \sum_{k=1}^{K} a_k(f_k(x_1)f_k(x_2) + f_k(x_2)f_k(x_1))$$

where $\{f_k(x)\}$ are self-adjoint polynomials in the variable x, and $\{a_k\}$ are constants. Let $\{x_n\}$ be freely independent self-adjoint identically distributed variables such that

$$\begin{aligned} \varphi(f_k(x_1)) &= 0, \quad \varphi(f_k^2(x_1)) = 1, \quad \text{for } 1 \leq k \leq K, \\ \varphi(f_k(x_1)f_l(x_1)) &= 0 \text{ for } 1 \leq k \neq l \leq K. \end{aligned}$$

Let

$$U_n = \frac{1}{\binom{n}{2}} \sum_{1 \leq i_1 < i_2 \leq n} h(x_{i_1}, x_{i_2}).$$

Show that

$$nU_n \to 2 \sum_{k=1}^{K} a_k(s_k^2 - \mathbf{1}_{\mathcal{A}})$$

where $s_1, ..., s_K$ are free standard semi-circular variables.
[For solutions to Exercises 12 and 13, please refer to Bose and Dey (2020)[23].]

5

Transforms

The three most important transforms in non-commutative probability, namely the $\mathcal{R}$, the S and the Stieltjes transform, are introduced for the single variable case, and their interrelations are given.

The $\mathcal{R}$ transform has the additive property with respect to free additive convolutions, similar to the cumulant generating function in the case of the classical additive convolutions. The S-transform enjoys the multiplicative property with respect to free multiplicative convolutions, like the Mellin transform in the case of the classical product convolutions. The Stieltjes transform has many uses—in particular, in the study of weak convergence and convergence of spectral distribution of non-Hermitian matrices.

The concept of *free infinite divisibility* is described very briefly. The semi-circular and the free Poisson laws are both free infinitely divisible.

We shall make substantial use of the concept of analytic functions. Often the domain where a function is analytic may not be explicitly specified but would be clear from the context. The reader may refer to any standard text for those concepts from complex analysis that are used in this chapter. The uninitiated reader may like to skip the technical details of this chapter, at least at the first reading.

5.1 Stieltjes transform

The Stieltjes transform is a vital tool in the study of probability laws and enjoys properties similar to the characteristic function. The Stieltjes transform of the spectral distribution of a real symmetric matrix is the trace of its resolvent, and this makes it particularly useful. It is also useful in the study of non-Hermitian matrices through a Hermitization technique used for such matrices. Though we shall not be using this transform much, we cover its basic properties due to its importance.

Let

$$\begin{aligned}
\mathbb{C}^+ : &= \{z : z = x + \iota y : x \in \mathbb{R},\ y > 0\}, \\
\mathbb{C}^- : &= \{z : z = x + iy : x \in \mathbb{R},\ y < 0\}, \\
\mathcal{I}z : &= \text{Imaginary part of } z \in \mathbb{C}.
\end{aligned}$$

DOI: 10.1201/9781003144496-5

Definition 5.1.1. (Stieltjes and Cauchy transforms) The *Stieltjes transform* of a random real-valued variable X or its probability law μ on $\mathbb{R}$ is defined as

$$s_X(z) \equiv s_\mu(z) := \int \frac{1}{t-z}\mu(dt), \quad z \in \mathbb{C}^+. \tag{5.1}$$

The *Cauchy transform* is defined as $G_\mu(z) := -s_\mu(z)$. These transforms are also defined for $z \in \mathbb{C}^-$ and $z \in \mathbb{R}$, which are outside the support of μ. ◇

The integral in (5.1) is finite since the integrand is bounded above by $1/|y|$. The smaller domain $\mathbb{C}^+$ will serve our purposes in most situations. We now list some basic properties of the Stieltjes transform. Define, for every $y > 0$, the function $f_{\mu,y}(\cdot)$ on $\mathbb{R}$ as

$$f_{\mu,y}(x) := \frac{1}{\pi}\mathcal{I}s_\mu(x+\iota y).$$

Lemma 5.1.1. The following properties hold for $s_\mu(z)$:

(a) The function s_μ is analytic, and its range is contained in $\mathbb{C}^+$.

(b) $\lim_{y\to\infty} \iota y s_\mu(\iota y) = -1$.

(c) The point masses of μ can be recovered from s_μ by

$$\mu\{t\} = \lim_{y\downarrow 0} y\mathcal{I}s_\mu(t+iy), \ t \in \mathbb{R}.$$

(d) The function $f_{\mu,y}(\cdot) : \mathbb{R} \to \mathbb{R}^+$ is a probability density function. As $y \to 0$, the corresponding sequence of probability laws converges to μ weakly.

(e) For every bounded continuous function $g : \mathbb{R} \to \mathbb{R}$,

$$\int_{\mathbb{R}} g(t)d\mu(t) = \frac{1}{\pi}\lim_{y\downarrow 0}\mathcal{I}\int_{\mathbb{R}} g(x)s_\mu(x+iy)dx.$$

(f) (Inversion formula) For all continuity points a, b of μ, we have

$$\mu(a,\ b) = \frac{1}{\pi}\lim_{y\downarrow 0}\mathcal{I}\int_a^b s_\mu(x+iy)dx.$$

(g) If $s_\mu(\cdot)$ has an almost everywhere-Lebesgue, continuous extension to $\mathbb{C}^+\cup\mathbb{R}$ then μ has a density (with respect to the Lebesgue measure) given by

$$f_\mu(t) := \frac{1}{\pi}\lim_{y\downarrow 0}\mathcal{I}s_\mu(t+iy), \ t \in \mathbb{R}.$$

(h) Suppose μ is a probability law whose support is contained in $[-C,\ C]$, and which has moments $\{m_n(\mu)\}$. Then

$$s_\mu(z) = -\sum_{n=0}^{\infty}\frac{m_n(\mu)}{z^{n+1}} \ \text{ for all } \ z \in \mathbb{C},\ |z| > C. \tag{5.2}$$

Moreover,

$$\lim_{z\in\mathbb{C}^+,\ |z|\to\infty} zs_\mu(z) = -1.$$

♦

Remark 5.1.1. Suppose a is a variable in some $*$-probability space $(\mathcal{A}, \varphi)$ such that for some $C > 0$, $|\varphi(a^n)| \leq C^n$ for all $n \in \mathbb{N}$. Then due to Lemma 5.1.1 (h), it is natural to define the Stieltjes transform of the variable a as

$$s_a(z) := -\sum_{n=0}^{\infty} \frac{\varphi(a^n)}{z^{n+1}} \quad \text{for all} \quad z \in \mathbb{C}, \ |z| > C.$$

●

Proof of Lemma 5.1.1. (a) By noting that $\frac{1}{t-z} = \frac{t-\bar{z}}{|t-z|^2}$, it easily follows that $s_\mu(z) \in \mathbb{C}^+$ if $z \in \mathbb{C}^+$. The analyticity also follows by easy arguments.

(b)

$$\iota y s_\mu(\iota y) = \iota y \int \frac{1}{t - \iota y} \mu(dt) = \int \frac{\iota t y - y^2}{t^2 + y^2} \mu(dt).$$

The absolute value of the integrand is bounded by 1 and converges to -1 as $y \to \infty$. Hence the result follows from the dominated convergence theorem.

(c) Note that

$$y \mathcal{I} s_\mu(x + iy) = \int_{\mathbb{R}} \frac{y^2}{(t-x)^2 + y^2} d\mu(t).$$

The integrand is bounded by 1 and converges to $\mathbb{I}\{x = t\}$ as $y \to 0$. The result then follows by applying DCT.

(d) Let X be a random variable with the probability law μ, and, for every $y > 0$, let Y_y be a scaled Cauchy random variable, independent of X and with the density

$$h_y(t) = \frac{1}{\pi} \frac{y}{t^2 + y^2}, \quad -\infty < t < \infty.$$

Now, the probability law of $X + Y_y$ is absolutely continuous with density

$$\begin{aligned} f_{X+Y_y}(t) &= \frac{1}{\pi} \int_{\mathbb{R}} \frac{y}{(t-s)^2 + y^2} d\mu(s) \\ &= \frac{1}{\pi} \mathcal{I} \int_{\mathbb{R}} \frac{1}{s - (t + \iota y)} d\mu(s) = \frac{1}{\pi} \mathcal{I} s_\mu(t + \iota y) = f_{\mu,y}(t). \end{aligned}$$

This shows that $f_{\mu,y}(\cdot)$ is the probability density function of $X + Y_y$. Note that as $y \to 0$, Y_y converges weakly to the point mass at 0 and hence the claim of weak convergence follows.

(e) Using the notation introduced above, it is easily checked that for any bounded continuous function $g : \mathbb{R} \to \mathbb{R}$,

$$\frac{1}{\pi} \mathcal{I} \int_{\mathbb{R}} g(x) s_\mu(x + iy) dx = \mathrm{E}\, g(X + Y_y).$$

Now the result follows from the weak convergence proved in (d).

(f) Let $\{g_n\}$ and $\{h_n\}$ be bounded continuous functions such that

$$\begin{aligned} g_n(x) &\le I_{[a,\ b]}(x) \le h_n(x), \\ g_n(x) &\to I_{(a,\ b)}(x), \text{ and} \\ h_n(x) &\to I_{[a,\ b]}(x) \text{ for all } x \in \mathbb{R}. \end{aligned} \tag{5.3}$$

Now, using (5.3),

$$\begin{aligned} \int_{\mathbb{R}} g_n(t)d\mu(t) &= \frac{1}{\pi} \lim_{y \downarrow 0} \mathcal{I} \int_{\mathbb{R}} g_n(x) s_\mu(x+iy)dx \text{ (using Part (e))} \\ &\le \frac{1}{\pi} \liminf_{y \downarrow 0} \mathcal{I} \int_a^b s_\mu(x+iy)dx \\ &\le \frac{1}{\pi} \limsup_{y \downarrow 0} \mathcal{I} \int_a^b s_\mu(x+iy)dx \\ &\le \frac{1}{\pi} \lim_{y \downarrow 0} \mathcal{I} \int_{\mathbb{R}} h_n(x) s_\mu(x+iy)dx \\ &= \int_{\mathbb{R}} h_n(t)d\mu(t) \text{ (use Part (a)).} \end{aligned}$$

But, by DCT, since a and b are continuity points of μ, the first and the last expressions both converge to $\mu(a,b)$. The result follows easily now.

(g) In this case,

$$\lim_{y \downarrow 0} f_{\mu,y}(t) = \frac{1}{\pi} \lim_{y \downarrow 0} \mathcal{I} s_\mu(t+\iota y) = f_\mu(t) \text{ for a.e. } t \in \mathbb{R}.$$

Now the result follows from the inversion formula given in (f).

(h) Note that the following series converges uniformly for $t \in [-C,\ C]$:

$$\frac{1}{t-z} = -\sum_{n=0}^{\infty} \frac{t^n}{z^{n+1}},\ z \in \mathbb{C}, |z| > C.$$

Due to this uniform convergence, we can write, for all $z \in \mathbb{C}, |z| > C$,

$$\begin{aligned} \int \frac{1}{t-z} \mu(dt) &= -\sum_{n=0}^{\infty} \int \frac{t^n}{z^{n+1}} \mu(dt), \\ &= -\sum_{n=0}^{\infty} \frac{m_n(\mu)}{z^{n+1}}. \end{aligned}$$

Hence (5.2) is proved. The second part is then a consequence of the DCT. ■

Remark 5.1.2. We note that in Lemma 5.1.1 (h), the series in (5.2) is finite for all $z \in \mathbb{C}$, $|z| > C$. Thus, the Stieltjes transform $s(z)$ is defined irrespective of the value of the imaginary part of z, if $|z| > C$. ●

We now compute the Stieltjes transform of the semi-circular law. We need to recall the notion of the *analytic square root* function. Any complex number $z \neq 0$ will be written in polar coordinates and its *square root function* $r(\cdot)$ is defined as follows:

$$\begin{aligned} z &= re^{\iota\theta}, \text{ where } r > 0 \text{ and } -\pi/2 < \theta \leq 3\pi/2, \\ r(z) &= r^{1/2}e^{\iota\theta/2}. \end{aligned} \tag{5.4}$$

It is known from complex function theory that $r(\cdot)$ *is analytic* on

$$\mathcal{C}_0 := \{z \in \mathbb{C} : \mathcal{R}(z) \neq 0\} \cup \{z \in \mathbb{C} : \mathcal{I}(z) > 0\} \supset \mathbb{C}^+.$$

Note that the specific choice of the square root has made $r(\cdot)$ analytic. We shall sometimes use the notation $\sqrt{z}$ to denote the function $r(z)$ on the domain $\mathcal{C}_0$.

Example 5.1.1. Let s be a semi-circular variable and let $s(\cdot)$ be its Stieltjes transform. Define the open set

$$\mathcal{D} := \{z = x + \iota y \in \mathbb{C} : xy > 0 \text{ or } x^2 - y^2 - 4 \neq 0\}.$$

Note that $r(z^2 - 4)$ is an analytic function on $\mathcal{D}$. We shall show that

$$s(z) = \frac{-z + r(z^2 - 4)}{2}, \ z \in \mathcal{D}. \tag{5.5}$$

First recall that the support of the semi-circular law is $[-2, 2]$. Moreover, the odd moments are all 0, and the even moment of order $2n$ is the n-th Catalan number C_n. Define

$$\mathcal{D}_1 := \{z : z \in \mathbb{C}, \ |z| > 4\}.$$

From Exercises 1 and 6 of Chapter 2, $C_n \leq 4^n$ for all n and $\{C_n\}$ satisfy the equations

$$C_0 = C_1 = 1, \ C_n = C_0C_{n-1} + C_1C_{n-1} + \cdots + C_{n-1}C_0, \ n \geq 2.$$

Using these and Lemma 5.1.1 (h), for all $z \in \mathcal{D}_1$, $s(z)$ satisfies

$$\begin{aligned} -s(z) &= \frac{1}{z} + \frac{1}{z}\sum_{n=1}^{\infty}\sum_{k=1}^{n}\frac{C_{k-1}C_{n-k}}{z^{2n}} \\ &= \frac{1}{z} + \frac{1}{z}\sum_{n=1}^{\infty}\sum_{k=1}^{n}\frac{C_{k-1}}{z^{2k-1}}\frac{C_{n-k}}{z^{2(n-k)+1}} \\ &= \frac{1}{z} + \frac{1}{z}\sum_{k=1}^{\infty}\frac{C_{k-1}}{z^{2k-1}}\sum_{n=k}^{\infty}\frac{C_{n-k}}{z^{2(n-k)+1}} \\ &= \frac{1}{z} - \frac{1}{z}\sum_{k=1}^{\infty}\frac{C_{k-1}}{z^{2(k-1)+1}}s(z) \\ &= \frac{1}{z} + \frac{1}{z}(s(z))^2. \end{aligned}$$

Now we observe the following three things:

(1) The following quadratic equation holds:

$$(s(z))^2 + zs(z) + 1 = 0, \; z \in \mathcal{D}_1. \tag{5.6}$$

(2) $s(\cdot)$ is analytic on $\mathcal{D}_1$.

(3) $zs(z) \to -1$ as $z = \iota y, y \to \infty$.

Let

$$f(z) := \frac{-z + r(z^2 - 4)}{2}, \; z \in \mathcal{D}. \tag{5.7}$$

Then it is easily checked that the properties (1), (2) and (3) for $s(\cdot)$ given above also hold for $f(\cdot)$ on $\mathcal{D}$. Define the non-empty open set

$$\mathcal{D}_2 := \mathcal{D} \cap \mathcal{D}_1.$$

Then, subtracting (5.6) from (5.7), we have for $z \in \mathcal{D}_2$,

$$\begin{aligned} 0 &= (f(z))^2 - (s(z))^2 + z(f(z) - s(z)) \\ &= \big(f(z) - s(z)\big)\big(f(z) + s(z) + z\big). \end{aligned}$$

Define the sets

$$\begin{aligned} \mathcal{C}_1 &:= \{z \in \mathcal{D}_2 : f(z) + s(z) + z = 0\} \\ &= \{z \in \mathcal{D}_2 : zf(z) + zs(z) + z^2 = 0\}, \\ \mathcal{C}_2 &:= \{z \in \mathcal{D}_2 : f(z) - s(z) = 0\} \\ &= \{z \in \mathcal{D}_2 : f(z) = s(z)\}. \end{aligned}$$

The function $f(z) + s(z) + z$ is analytic on $\mathcal{D}_2$. It is not identically zero since by Property (2), for $z = \iota y$, $y > 0$,

$$zs(z) + zf(z) \to -2 \;\text{ as }\; y \to \infty.$$

Hence $\mathcal{C}_1$ is a set of isolated points.

Now if $f(z) - s(z)$ is not identically zero on $\mathcal{D}_2$, then $\mathcal{C}_2$ is also a set of isolated points but that would be a contradiction because $\mathcal{C}_1 \cup \mathcal{C}_2 = \mathcal{D}_2$.

This shows that $s \equiv f$ on $\mathcal{D}_2$. Hence $s(z)$ is given by (5.5) if $z \in \mathcal{D}_2$. Since f is an analytic function on $\mathcal{D}$, $s(z)$ is given by (5.5) on $\mathcal{D}$.

It is left as an exercise to show that the density of the semi-circular law can be derived by using (5.5) and Lemma 5.1.1 (g). ▲

Remark 5.1.3. We had to go through the above elaborate argument since the square root is a multi-valued function and we need to choose an appropriate branch which gives us an analytic function. Henceforth in this book, whenever we write down complex functions which can potentially be multi-valued, it is understood that a specific choice is being made. Moreover, from now on, without any scope for confusion, *in the domain $\mathcal{C}_0$, $\sqrt{z}$ will stand for the analytic function $r(z)$ given in (5.4).* ●

The Stieltjes transform enjoys a "continuity property" similar to the characteristic function. We state this property in the lemma below but omit its proof. The interested reader may consult Bai and Silverstein (2010)[5] for the proof and application of this lemma in random matrices.

Lemma 5.1.2. Suppose $\{\mu_n\}$ is a sequence of probability measures on $\mathbb{R}$ with Stieltjes transforms $\{s_{\mu_n}\}$. If $\lim_{n\to\infty} s_{\mu_n}(z) = s(z)$ for all $z \in \mathbb{C}^+$, then there exists a probability measure μ with Stieltjes transform s if and only if

$$\lim_{v\to\infty} ivs(iv) = -1 \tag{5.8}$$

in which case $\{\mu_n\}$ converges weakly to μ. Conversely, if $\{\mu_n\}$ converges to μ weakly, then $s_{\mu_n}(z) \to s_\mu(z)$ for all $z \in \mathbb{C}^+$. ♦

5.2 $\mathcal{R}$ transform

Definition 5.2.1. (Voiculescu's $\mathcal{R}$-transform) (a) The $\mathcal{R}$-transform of a variable a is defined as the analytic function

$$\mathcal{R}_a(z) := \sum_{n=0}^{\infty} \kappa_{n+1}(a) z^n \quad \text{for all} \quad z \in \mathbb{C} \quad \text{for which the series is finite.} \tag{5.9}$$

(b) The $\mathcal{R}$-transform of a probability law μ is similarly defined as the analytic function $\mathcal{R}_\mu(z) := \sum_{n=0}^{\infty} \kappa_{n+1}(\mu) z^n$ for all $z \in \mathbb{C}$ for which the series is defined.

The radius of convergence of $\mathcal{R}_a$ and $\mathcal{R}_\mu$ shall be denoted by r_a and r_μ respectively. ◇

A word of caution. In the non-commutative probability literature, another closely related function $R_a(z) = \sum_{n=0}^{\infty} \kappa_n(a) z^n$ is also known as the R-transform. In this book, we will use only the $\mathcal{R}$-transform defined above and so there will be no scope for confusion.

If a is a self-adjoint variable whose moments determine a unique probability law μ_a, then $\mathcal{R}_{\mu_a} = \mathcal{R}_a$. Also note that if, for all n, $|\kappa_n(a)| \leq C^n$ and $|\kappa_n(\mu)| \leq C^n$, then r_a and r_μ are at least as large as $1/C$.

All $\mathcal{R}$-transforms that appear subsequently in this book will have positive radii of convergence. The value of the radius shall be described explicitly as and when needed. If it is clear from the context which variable or probability law is being considered, we shall drop the subscript and simply write $\mathcal{R}(z)$ for $\mathcal{R}_a(z)$ and $\mathcal{R}_\mu(z)$.

Example 5.2.1. Consider the probability measure μ where $\mu\{\gamma\} = 1$ for some $\gamma > 0$. Then it is easy to check that

$$\begin{aligned} m_n(\mu) &= \gamma^n \text{ for all } n \geq 1, \\ \kappa_1(\mu) &= \gamma, \\ \kappa_n(\mu) &= 0 \text{ for all } n \geq 2. \end{aligned}$$

Hence $\mathcal{R}_\mu(z) = \gamma$ for all $z \in \mathbb{C}$. ▲

Example 5.2.2. (a) If a is a free Poisson variable with parameter λ, then all its free cumulants are equal to λ. Hence

$$\mathcal{R}_a(z) = \sum_{n=0}^{\infty} \lambda z^n = \frac{\lambda}{1-z} \text{ for all } z \in \mathbb{C}, \ |z| < 1.$$

(b) Suppose b is a free compound Poisson variable with mean λ and jump distribution ν which has compact support. Let X be a random variable with probability law ν and let $C > 0$ be such that the support of ν is contained in $[-C, C]$. Since the n-th order free cumulant of b is $\lambda\,\mathrm{E}(X^n)$, $\mathcal{R}_b(\cdot)$ is given by,

$$\mathcal{R}_b(z) = \lambda \sum_{n=0}^{\infty} \mathrm{E}(X^{n+1}) z^n = \lambda\,\mathrm{E}\left[\frac{X}{1-Xz}\right] \text{ for all } z \in \mathbb{C}, \ |z| < 1/C.$$

▲

Lemma 5.2.1. Suppose $\mathcal{R}_a$ and $\mathcal{R}_b$ are the $\mathcal{R}$-transforms of a and b. Then the following properties hold.

(a) For any scalar t, $\mathcal{R}_{ta}(z) = t\mathcal{R}_a(tz)$, for all z such that $z \in \mathbb{C}$, $|tz| < r_a$.

(b) The variables a and b have identical distribution if and only if they have identical $\mathcal{R}$ transforms. If μ and ν are compactly supported probability laws, then they are identical if and only if $\mathcal{R}_\mu$ and $\mathcal{R}_\nu$ are identical.

(c) If a and b are free, then

$$\mathcal{R}_{a+b}(z) = \mathcal{R}_a(z) + \mathcal{R}_b(z) \text{ for all } z \in \mathbb{C}, \ |z| < \min\{r_a, r_b\}.$$

(d) Suppose μ and ν are two probability laws on $\mathbb{R}$ with compact supports. Then

$$\mathcal{R}_{\mu \boxplus \nu}(z) = \mathcal{R}_\mu(z) + \mathcal{R}_\nu(z) \text{ for all } z \in \mathbb{C}, \ |z| < \min\{r_\mu, r_\nu\}.$$

♦

Proof. (a) Note that $\kappa_n(ta) = t^n \kappa_n(a)$. Hence

$$\mathcal{R}_{ta}(z) = \sum_{n=0} \kappa_{n+1}(a) t^{n+1} z^n = t\mathcal{R}_a(tz).$$

(b) Recall that two power series are equal if and only if all their coefficients are equal. Thus if $\mathcal{R}_a(z) = \mathcal{R}_b(z)$ in some open domain, then their free cumulants agree. Hence their moments agree. The second part follows since moments in this case determine the respective probability laws.

(c) This is really a restatement of Theorem 3.1.1. Recall that free cumulants are multi-linear and all mixed free cumulants of a and b vanish when they are free.

(d) This follows from (c). Note that if the supports of μ and ν are contained in $[-C, C]$, then $r_{\mu \boxplus \nu} \geq 1/C$. ■

Example 5.2.3. If s_γ is a semi-circular variable with variance γ, then all its free cumulants are 0 except that $\kappa_2(s_\gamma) = \gamma$. Hence $\mathcal{R}_{s_\gamma}(z) = \gamma z$.

If s_1 and s_2 are free standard semi-circular variables, then we have seen that $\frac{s_1+s_2}{\sqrt{2}}$ is also a standard semi-circular variable. This can be shown also by using the $\mathcal{R}$-transform as follows:

$$\begin{aligned}
\mathcal{R}_{\frac{s_1+s_2}{\sqrt{2}}}(z) &= \frac{1}{\sqrt{2}}\mathcal{R}_{s_1+s_2}\Big(\frac{z}{\sqrt{2}}\Big) \quad \text{(by Lemma 5.2.1)} \\
&= \frac{1}{\sqrt{2}}\Big[\mathcal{R}_{s_1}\Big(\frac{z}{\sqrt{2}}\Big) + \mathcal{R}_{s_2}\Big(\frac{z}{\sqrt{2}}\Big)\Big] \quad \text{(by freeness)} \\
&= \frac{1}{\sqrt{2}}\Big[\frac{z}{\sqrt{2}} + \frac{z}{\sqrt{2}}\Big] = z.
\end{aligned}$$

The above argument also shows that if μ and ν are semi-circular with variances σ_1^2 and σ_2^2, then $\mu \boxplus \nu$ is again semi-circular with variance $\sigma_1^2 + \sigma_2^2$. ▲

5.3 Interrelation

In Lemma 5.1.1 we have already seen the relation between moments and Stieltjes transforms. Recall that free cumulants are in one-to-one correspondence with moments. This leads to a relation between the $\mathcal{R}$-transform and the Stieltjes transform. This relation will be extreme useful, and, in particular, will come handy when we introduce the multiplicative free convolution. To identify this relation, we shall use two generating functions of the free cumulants and the moments.

Definition 5.3.1. Let a be an element of an NCP.
(a) The *free cumulant generating function* of a is defined as

$$K_a(z) := 1 + \sum_{n=1}^{\infty} k_n(a) z^n, \; z \in \mathbb{C}, \; |z| < r_a.$$

(b) The *generating function for the moments* of a (not to be confused with the moment generating function defined in Chapter 1) is defined as

$$M_a(z) := 1 + \sum_{n=1}^{\infty} m_n(a) z^n, \ z \in \mathbb{C}, \ |z| < m_a,$$

where m_a is the radius of convergence of the series.

For any probability law μ, $K_\mu(z)$ and $M_\mu(z)$ are defined in the obvious way. ◇

Note that if μ is compactly supported, then $K_\mu(z)$ and $M_\mu(z)$ have positive radii of convergence.

Example 5.3.1. Let s be a semi-circular variable and let $\{m_{2n}\}$ be its even moments. Then

$$K_s(z) = 1 + z^2 \text{ for all } z \in \mathbb{C}.$$

Since the support of the semi-circular law is $[-2, 2]$, we have $|m_n| \leq 2^n$. Thus $M_s(z)$ is defined for all $z \in \mathbb{C}$, $|z| < 1/2$. By repeating the arguments given in Example 5.1.1,

$$\begin{aligned} M_s(z) &= 1 + \sum_{n=1}^{\infty} m_{2n}(s) z^{2n} \quad \text{for} \quad z \in \mathbb{C}, \ |z| < 1/2 \\ &= 1 + z^2 (M_s(z))^2. \end{aligned}$$

This eventually yields,

$$M_s(z) = \frac{1 - \sqrt{1 - 4z^2}}{2z^2}, \quad z \in \mathbb{C}, \ |z| < 1/2. \tag{5.10}$$

The details are left as an exercise. ▲

The following lemma collects the relations between $\mathcal{R}$, s, K and M.

Lemma 5.3.1. Suppose a is an element of an NCP. Then

(a)

$$K_a(z) = 1 + z\mathcal{R}_a(z), \ z \in \mathbb{C}, \ |z| < r_a.$$

(b) If a is self-adjoint with probability law μ, then

$$s_\mu(z) = -\frac{1}{z} M_\mu\Big(\frac{1}{z}\Big), \ z \in \mathbb{C}^+, \ |z| > 1/m_a.$$

(c)

$$K_a(zM_a(z)) = M_a(z), \ z \in \mathbb{C}, \ |z| < m_a, \ |zM_a(z)| < r_a, \tag{5.11}$$

$$M_a\Big(\frac{z}{K_a(z)}\Big) = K_a(z), \ z \in \mathbb{C}, |z| < r_a, \ |z/K_a(z)| < m_a. \tag{5.12}$$

Since $M_a(z)$, $K_a(z) \to 1$ as $z \to 0$, the set of all z such that the expressions in (c) are well-defined includes a non-empty ball around the origin.

(d) If a is self-adjoint with probability law μ, then

$$-s_\mu\Big(\mathcal{R}_\mu(z)+\frac{1}{z}\Big) = -s_\mu\Big(\frac{K_\mu(z)}{z}\Big) = z \quad \text{for all} \quad z \in \mathbb{C}^-,\ |z| \text{ small}. \tag{5.13}$$

♦

Proof. The proof of (a) is trivial and (b) follows from Lemma 5.1.1 (h).

(c) Since the expression on the two sides are well-defined on a non-empty ball around 0, the calculations below hold. Recall that

$$m_n = \sum_{\pi \leq \mathbf{1}_n} \kappa_\pi. \tag{5.14}$$

Let $B_1 := \{1 = i_1 < \cdots < i_s\}$ be the first block of π. Then

$$\pi = B_1 \cup \tilde{\pi}_{i_1} \cup \cdots \cup \tilde{\pi}_{i_s},$$

where

$$\tilde{\pi}_{i_j} \in NC\{i_j+1, \ldots i_{j+1}-1\} \cong NC(v_j)$$

with $v_j = i_{j+1} - i_j - 1$ for all $j \leq s$ and $i_{s+1} = n+1$. We rewrite (5.14) as below, where, if $v_j = 0$, then the corresponding block does not contribute.

$$\begin{aligned}
m_n &= \sum_{s=1}^{n} \sum_{B_1:|B_1|=s} \sum_{\substack{\pi \in NC(n) \\ \pi = B_1 \cup \tilde{\pi}_{i_1} \cup \cdots \cup \tilde{\pi}_{i_s}}} \kappa_\pi \\
&= \sum_{s=1}^{n} \sum_{B_1:|B_1|=s} \sum_{\substack{\pi \in NC(n) \\ \pi = B_1 \cup \tilde{\pi}_{i_1} \cup \cdots \cup \tilde{\pi}_{i_s}}} \kappa_s \prod_{j=1}^{s} \big(\kappa_{\tilde{\pi}_{i_j}}\big) \\
&= \sum_{s=1}^{n} \kappa_s \sum_{\substack{v_1,\ldots,v_s \in \{0,1,\ldots,n-s\} \\ v_1+\cdots+v_s = n-s}} \prod_{j=1}^{s} \sum_{\tilde{\pi}_{i_j} \leq \mathbf{1}_{v_j}} \kappa_{\tilde{\pi}_{i_j}} \\
&= \sum_{s=1}^{n} \kappa_s \sum_{\substack{v_1,\ldots,v_s \in \{0,1,\ldots,n-s\} \\ v_1+\cdots+v_s = n-s}} \prod_{j=1}^{s} m_{v_j}.
\end{aligned}$$

As a consequence,

$$\begin{aligned}
M_a(z) &= 1 + \sum_{n=1}^{\infty} m_n z^n \\
&= 1 + \sum_{n=1}^{\infty} \sum_{s=1}^{n} \sum_{\substack{i_1,\ldots,i_s \in \{0,1,\ldots,n-s\} \\ i_1+\cdots+i_s = n-s}} \kappa_s z^s \prod_{j=1}^{s} (m_{i_j} z^{i_j}) \\
&= 1 + \sum_{s=1}^{n} \kappa_s z^s \Big(\sum_{i=1}^{\infty} m_i z^i\Big)^s \\
&= K_a(zM_a(z)),
\end{aligned}$$

proving (5.11). We now establish (5.12) using (5.11). Let

$$t = zM_a(z).$$

Note that when $|z|$ is small, so is t. Thus the steps below are valid for small values of z.

Observe that

$$z = \frac{t}{M_a(z)} = \frac{t}{K_a(t)}.$$

Hence

$$K_a(t) = K_a(zM_a(z)) = M_a(z) = M_a\Big(\frac{t}{K_a(t)}\Big).$$

(d) If μ is degenerate at 0, then (5.13) holds trivially. So suppose this is not the case. Recall that the Stieltjes transform is always defined at least for all $z \in \mathbb{C}$, $\mathcal{I}z \neq 0$, though we have restricted to the domain $\mathbb{C}^+$. In the arguments below, we shall have to be careful to check that the domain condition is not violated. Let

$$H_\mu(z) := \frac{z\mathcal{R}_\mu(z) + 1}{z} = \frac{K_\mu(z)}{z}, \ z \in \mathbb{C}^-, \ |z| < r_\mu. \tag{5.15}$$

Hence we need to show that

$$-s_\mu(H_\mu(z)) = z, \ z \in \mathbb{C}^-, \ |z| \text{ small}. \tag{5.16}$$

As $z \to 0$, $\mathcal{R}_\mu(z) \to \kappa_1(\mu)$ which is real. Also observe that if $z \in \mathbb{C}^-$, then $1/z \in \mathbb{C}^+$, and, as $z \to 0$, the real part of $1/z$ goes to ∞. Thus, for $z \in \mathbb{C}^-$ where $|z|$ is small, $H_\mu(z) \in \mathbb{C}^+$, and hence $s_\mu(H_\mu(z))$ is well-defined.

Now, for $z \in \mathbb{C}^+$, $|z|$ large, all quantities in the display below are well-defined. Moreover, by using Parts (b) and (c),

$$\frac{K_\mu(-s_\mu(z))}{s_\mu(z)} = \frac{K_\mu(\frac{1}{z}M_\mu(\frac{1}{z}))}{s_\mu(z)} = \frac{M_\mu(1/z)}{s_\mu(z)} = -z.$$

In other words (note that for $z \in \mathbb{C}^+$, $-s_\mu(z) \in \mathbb{C}^-$ by Lemma 5.1.1 (a)),

$$H_\mu(-s_\mu(z)) = z, \ z \in \mathbb{C}^+, \ |z| \text{ large}. \tag{5.17}$$

On the other hand, both H_μ and s_μ are analytic in their appropriate domains. Hence the relation (5.17) holds for all $z \in \mathbb{C}^+$, if needed by analytic extension of the function H_μ which was initially defined in (5.15) for $z \in \mathbb{C}^-$, $|z| < r_\mu$.

But this implies that

$$H_\mu(-s_\mu(H_\mu(z))) = H_\mu(z), \tag{5.18}$$

on a non-trivial sub-domain of $D =: \{z : z \in \mathbb{C}^-, \ |z| \text{ small}\}$. It follows that $-s_\mu(H_\mu(z)) = z$ on a non-trivial sub-domain of D and hence the proof is complete. ■

Example 5.3.2. (Free additive convolution of symmetric Bernoulli) Suppose μ is the symmetric Bernoulli probability law which puts mass $1/2$ at ± 1. Then

$$\begin{aligned} s_\mu(z) &= \frac{1}{2}\Big(\frac{1}{-1-z} + \frac{1}{1-z}\Big) \\ &= \frac{z}{1-z^2}. \end{aligned}$$

Let $t(z) = \mathcal{R}_\mu(z) + \frac{1}{z}$. Then Lemma 5.3.1 (d) implies that $z = -\frac{t(z)}{1-(t(z))^2}$. Rearranging, we obtain the quadratic equation

$$z - (t(z))^2 z + t(z) = 0. \tag{5.19}$$

Now we may argue as in Example 5.1.1, remembering that $\mathcal{R}_\mu(0) = \kappa_1(\mu) = 0$, to finally obtain

$$\mathcal{R}_\mu(z) = \frac{\sqrt{1+4z^2}-1}{2z} \quad \text{for small } |z|. \tag{5.20}$$

The details are left as an exercise. This formula can now be used to find the probability law $\mu \boxplus \mu$. Using Lemma 5.2.1 (d),

$$\mathcal{R}_{\mu\boxplus\mu}(z) + \frac{1}{z} = 2\mathcal{R}_\mu(z) + \frac{1}{z} = \frac{\sqrt{1+4z^2}}{z} \quad \text{for small } |z|.$$

Hence, again using Lemma 5.3.1 (d),

$$-s_{\mu\boxplus\mu}\Big(\frac{\sqrt{1+4z^2}}{z}\Big) = z.$$

This, when simplified, yields

$$s_{\mu\boxplus\mu}(z) = -\frac{1}{\sqrt{z^2-4}}, \quad z \in \mathbb{C}^+.$$

This clearly has a continuous extension to $\mathbb{C}^+ \cup \mathbb{R}$ a.e. (only the points $x = \pm 2$ are excluded). Now we can use Lemma 5.1.1 (g) to conclude that the law $\mu \boxplus \mu$ is absolutely continuous with respect to the Lebesgue measure and its density is given by

$$f_{\mu\boxplus\mu}(x) = \begin{cases} \frac{1}{\pi\sqrt{4-x^2}} & \text{if } -2 < x < 2, \\ 0 & \text{otherwise.} \end{cases} \tag{5.21}$$

This calculation can be extended to the n-fold free additive convolution of μ which is related to the *Kesten measures*, see Kesten (1959)[61]. ▲

Example 5.3.3. Suppose μ is the semi-circular law. We already know the formula for its Stieltjes transform. Nevertheless, let us use its free cumulants to derive this formula using Lemma 5.3.1 (d). We know that $K_\mu(z) = 1 + z^2$. It is easy to check directly that the argument $\frac{1+z^2}{z}$ belongs to $\mathbb{C}^+$ whenever $z \in \mathbb{C}^-$ and $|z|$ is small. Hence, by using Lemma 5.3.1 (d),

$$z = -s_\mu\Big(\frac{z\mathcal{R}_\mu(z) + 1}{z}\Big) = -s_\mu\Big(\frac{K_\mu(z)}{z}\Big) = -s_\mu\Big(\frac{1+z^2}{z}\Big). \tag{5.22}$$

This again involves multiple solutions and the steps to rigorously arrive at the correct solution are left as an exercise. Upon solving, we have in the appropriate domain,

$$s_\mu(z) = \frac{-z + \sqrt{z^2 - 4}}{2}, \tag{5.23}$$

where the square root is taken to be as in (5.4). ▲

5.4 S-transform

As we have seen, the $\mathcal{R}$-transform is connected to the free additive convolution. In this section, we define the S-transform which is connected to free multiplicative convolution. Another use of the S-transform is for computing the *Brown measure* of R-diagonal variables—this will be discussed in Chapter 12.

Let m_a be the generating function of the moments defined by a slight modification of $M_a(z)$ as

$$m_a(z) := M_a(z) - 1 = \sum_{n=1}^{\infty} \varphi(a^n) z^n, \ z \in \mathbb{C}, \ |z| < m_a. \tag{5.24}$$

For a probability measure μ, m_μ is defined similarly. Note that if $\varphi(a) \neq 0$, then m_a is invertible as a formal power series. Denote by $m_a^{\langle -1 \rangle}$ its inverse under the composition map. We are now ready to define the S-transform.

Definition 5.4.1. (S-transform) Suppose a is an element of an NCP $(\mathcal{A}, \varphi)$ such that $\varphi(a) \neq 0$. Then the *S-transform* of a is defined as

$$S_a(z) := \frac{1+z}{z} m_a^{\langle -1 \rangle}(z) \ \ \text{(in a non-trivial domain)}.$$

The S-transform of a probability measure with non-zero mean is defined analogously. ◇

Incidentally, the S-transform can be defined even when $\varphi(a) = 0$. See Rao and Speicher (2007)[78]. However, we shall not pursue this here.

The following lemma gives the S-transform for cc^* where c is a circular variable. We shall use it in Chapter 12.

Lemma 5.4.1. Let c and e respectively be a standard circular variable and an elliptic variable in an NCP $(\mathcal{A}, \varphi)$. Then the S transform of cc^* and ee^* are given by

$$S_{cc^*}(z) = S_{ee^*}(z) = \frac{1}{1+z}, \quad z \neq -1.$$

Hence cc^* and $ece*$ have identical distributions. ♦

Proof. From Lemma 3.6.2, it is easy to see that the moments of cc^* are given by

$$\varphi((cc^*)^n) = \varphi(cc^* \cdots cc^*) = C_n, \quad \text{the } n\text{-th Catalan number.}$$

Let s be a standard semi-circular variable with moments $\{m_n(s)\}$. We now proceed with the formal manipulation of the transforms involved. The rigorous justification for all the steps is left as an exercise.

$$\begin{aligned}
m_{cc^*}(z) &= \sum_{n=1}^{\infty} \varphi((cc^*)^n) z^n \\
&= \sum_{n=1}^{\infty} C_n z^n \\
&= \sum_{n=1}^{\infty} m_{2n}(s) z^n \qquad (5.25)\\
&= M_s(\sqrt{z}) - 1 = \frac{1-\sqrt{1-4z}}{2z} - 1 \text{ (from (5.10) of Example 5.3.1)} \\
&= \frac{1-2z-\sqrt{1-4z}}{2z}. \qquad (5.26)
\end{aligned}$$

Let $m_{cc^*}^{\langle -1 \rangle}(z) = t$. Then (5.26) implies

$$\begin{aligned}
1 - 2t - \sqrt{1-4t} &= 2tz \\
\text{or, } 1 - 2t - 2tz &= \sqrt{1-4t} \\
\text{or, } (1 - 2t - 2tz)^2 &= 1 - 4t \\
\text{or, } t &= \frac{z}{(1+z)^2}.
\end{aligned}$$

Hence

$$\begin{aligned}
S_{cc^*}(z) &= \frac{1+z}{z} m_{cc^*}^{\langle -1 \rangle}(z) \\
&= \frac{1+z}{z} \frac{z}{(1+z)^2} = \frac{1}{1+z}, \quad z \neq -1. \qquad (5.27)
\end{aligned}$$

Using the moment structure of an elliptic variable given in (3.3) of Lemma 3.6.1, it is easy to see that $\varphi((ee^*)^n)$ does not depend on ρ and hence must equal $\varphi((cc^*)^n)$ for all $n \geq 1$. Hence ee^* has the same S-transform as that of cc^* and the proof is now complete. ■

Example 5.4.1. The relation between the S and the $\mathcal{R}$ transform is as follows. Define a modified generating function of free cumulants as

$$k_a(z) = \sum_{n=1}^{\infty} \kappa_n(a) z^n = K_a(z) - 1 = z\mathcal{R}_a(z). \tag{5.28}$$

Then

$$S_a(z) = \frac{1}{z} k_a^{\langle -1 \rangle}(z) = \frac{1}{z}[z\mathcal{R}_a(z)]^{\langle -1 \rangle}(z) \text{ (in a non-trivial domain)}. \tag{5.29}$$

We provide a proof through formal manipulations. A rigorous justification is left to the reader as an exercise. To prove this, let

$$F_a(z) := z\mathcal{R}_a(z).$$

Now

$$\begin{aligned}
& S_a(z) = \frac{1}{z}[z\mathcal{R}_a(z)]^{\langle -1 \rangle}(z) \\
\iff \quad & (1+z) m_a^{\langle -1 \rangle}(z) = F_a^{\langle -1 \rangle}(z) \\
\iff \quad & m_a\Big(\frac{F_a^{\langle -1 \rangle}(z)}{(1+z)}\Big) = z \\
\iff \quad & M_a\Big(\frac{F_a^{\langle -1 \rangle}(z)}{(1+z)}\Big) = 1 + z = K_a(t) \text{ (say)} \\
\iff \quad & \frac{F_a^{\langle -1 \rangle}(z)}{(1+z)} = \frac{t}{K_a(t)} = \frac{t}{1+z} \text{ (by Lemma 5.3.1(c))} \\
\iff \quad & F_a(t) = z = K_a(t) - 1 = k_a(t).
\end{aligned}$$

▲

We now wish to show how the S-transform is connected to free multiplicative convolution.

Theorem 5.4.2 (S-transform of product)**.** Suppose a and b are free and $\varphi(a)\varphi(b) \neq 0$. Then in an appropriate non-trivial domain,

$$S_{ab}(z) = S_a(z) S_b(z) = S_{ba}(z). \tag{5.30}$$

◆

Proof. Since the S-transform involves an inverse, it is difficult to work with it directly. The following proof is taken from Anderson, Guionnet and Zeitouni (2009)[1]. It is based on a clever manipulation of appropriate power series by using the moment-free cumulant relation.

We shall be dealing with several functions and their inverses. For convenience, we shall not mention the domains of these functions. The reader can easily check that all these functions have non-trivial domains where they are defined and these domains are compatible with each other and hence simultaneous manipulation of the functions is legitimate.

Define

$$M_1(z) = \sum_{n=0}^{\infty} \varphi\big(b(ab)^n\big) z^n,$$
$$M_2(z) = \sum_{n=0}^{\infty} \varphi\big(a(ba)^n\big) z^n.$$

From Example 3.9.1 of Chapter 3, recall that since a and b are free,

$$\varphi\big((ab)^n\big) = \varphi\big((ba)^n\big) \text{ for every } n \geq 1.$$

Let $ONC(2j+1)$ and $ENC(2j)$ be the collection of all non-crossing partitions of $\{1, 3, \ldots, 2j+1\}$ and $\{2, 4, \ldots, 2j\}$ respectively. Then,

$$\begin{aligned}
\varphi\big((ab)^n\big) &= \sum_{\pi \in NC(2n)} \kappa_\pi(a, b, \ldots, a, b) \text{ (by the moment-free cumulant relation)} \\
&= \sum_{\substack{\pi_1 \in ONC(2n-1), \pi_2 \in ENC(2n) \\ \pi_1 \cup \pi_2 \in NC(2n)}} \kappa_{\pi_1}(a) \kappa_{\pi_2}(b) \text{ (since mixed free cumulants vanish)} \\
&= \varphi\big((ba)^n\big) \text{ (by symmetry of the above expression).} \qquad (5.31)
\end{aligned}$$

This proof is similar to the proof of Lemma 5.3.1 (c). Let $V_1 = \{v_1 < \cdots < v_s\}$ be the first block of π_1. Let $\{W_k, k = 1, \ldots s\}$ be the set of all consecutive integers between the elements of $V_1 \cup \{2n\}$. Note that $\#W_k = v_{k+1} - v_k - 1 =: 2i_k + 1$, say. Here v_{s+1} is taken as $2n + 1$.

For example, if the first block is $\{v_1 = 1, v_2 = 5, v_3 = 11\}$, then $s = 3$,

$$\begin{aligned}
W_1 &= \{2, 3, 4\}, \ \#W_1 = 3 = 5 - 1 - 1 = 2i_1 + 1, i_1 = 1, \\
W_2 &= \{6, 7, 8, 9, 10\}, \#W_2 = 5 = 11 - 5 - 1 = 2i_2 + 1, i_2 = 2, \\
W_3 &= \{12, \ldots, 2n\}, \#W_3 = 2n + 1 - 11 - 1 = 2n - 11 = 2i_s + 1, i_s = n - 6.
\end{aligned}$$

Then separating out the block V_1, (5.31) can be rewritten as

$$\begin{aligned}
\varphi\big((ab)^n\big) &= \sum_{s=1}^{n} \kappa_s(a) \sum_{\substack{i_1 + \cdots + i_s = n-s \\ i_k \geq 0}} \prod_{k=1}^{s} \sum_{\substack{\pi_1 \in ONC(2i_k+1), \pi_2 \in ENC(2i_k) \\ \pi_1 \cup \pi_2 \in NC(2i_k+1)}} \kappa_{\pi_1}(a) \kappa_{\pi_2}(b) \\
&= \sum_{s=1}^{n} \kappa_s(a) \sum_{\substack{i_1 + \cdots + i_s = n-s \\ i_k \geq 0}} \prod_{k=1}^{s} \varphi(b(ab)^{i_k}), \qquad (5.32)
\end{aligned}$$

where the last relation follows after using the moment-free cumulant relation and freeness.

We can use the same arguments on the variable $b(ab)^n$ with a slight deviation that, now we have a product of $2n+1$ arguments and the first and the last elements of the product are both b.

Keeping this in mind, first use the moment-free cumulant relation to write $\varphi\big(b(ab)^n\big)$ in terms of free cumulants, separating out the partitions corresponding to b and a by using freeness. As before, let $V_1 = \{v_1, \ldots, v_s\}$ be the first block in $ONC(2n+1)$. As earlier, let $2i_k + 1 = v_{k+1} - v_k - 1, 1 \le k \le s-1$. In addition, let $2i_0 = 2n + 1 - v_s$. Now, arguing as before, we obtain

$$\varphi\big(b(ab)^n\big) = \sum_{s=0}^{n} \kappa_{s+1}(b) \sum_{\substack{i_0+\cdots+i_s=n-s \\ i_k \ge 0}} \varphi\big((ab)^{i_0}\big) \prod_{k=1}^{s} \varphi\big(a(ba)^{i_k}\big). \tag{5.33}$$

Recall the definition of $k_a(z)$ from (5.28) of Example 5.4.1:

$$k_a(z) := \sum_{n=1}^{\infty} \kappa_n(a) z^n.$$

Multiplying both sides of (5.32) by z^n and adding over n, we get

$$\begin{aligned}
M_{ab}(z) &= 1 + \sum_{n=1}^{\infty} \varphi((ab)^n) z^n \\
&= 1 + \sum_{n=1}^{\infty} \sum_{s=1}^{n} \kappa_s(a) \sum_{\substack{i_1+\cdots+i_s=n-s \\ i_k \ge 0}} \varphi(b(ab)^{i_k}) z^n \quad \text{(using (5.32))} \\
&= 1 + \sum_{s=1}^{\infty} \kappa_s(a) z^s \sum_{\substack{i_1+\cdots+i_s=n-s\ge 0 \\ i_k \ge 0}} \varphi(b(ab)^{i_k}) z^{n-s} \\
&= 1 + \sum_{s=1}^{\infty} \kappa_s(a) z^s (M_1(z))^s \\
&= 1 + k_a\big(zM_1(z)\big) \\
&= 1 + k_b\big(zM_2(z)\big) \quad (\text{since } \ M_{ab}(z) = M_{ba}(z)).
\end{aligned}$$

Similarly, by using (5.33),

$$\begin{aligned}
M_1(z) &= \sum_{s=0}^{\infty} z^s \kappa_{s+1}(b) M_{ab}(z) \big(M_2(z)\big)^s \\
&= \frac{M_{ab}(z)}{zM_2(z)} k_b\big(zM_2(z)\big).
\end{aligned}$$

Hence

$$\begin{aligned}
M_{ab}(z) - 1 &= k_a\big(zM_1(z)\big) \\
&= k_b\big(zM_2(z)\big) \\
&= \frac{zM_1(z)M_2(z)}{M_{ab}(z)}.
\end{aligned}$$

Now note that k_a and k_b are invertible as formal power series, since $\kappa_1(a) = \varphi(a) \neq 0$ and the same is true for b. This yields

$$\begin{aligned} k_a^{\langle -1\rangle}(M_{ab}(z)-1)k_b^{\langle -1\rangle}(M_{ab}(z)-1) &= z^2 M_1(z)M_2(z) \\ &= zM_{ab}(z)(M_{ab}(z)-1). \quad (5.34) \end{aligned}$$

For any variable x, let

$$m_x(z) := M_x(z) - 1. \quad (5.35)$$

Note that the relation (5.11) of Lemma 5.3.1 (c) can be rewritten as

$$m_a(z) = k_a\big(zM_a(z)\big))$$

which implies

$$k_a^{\langle -1\rangle}(z) = (1+z)m_a^{\langle -1\rangle}(z) = zS_a(z). \quad (5.36)$$

Hence (5.34) can be re-expressed as (let $z' = M_{ab}(z) - 1) = m_{ab}(z)$)

$$k_a^{\langle -1\rangle}(z')k_b^{\langle -1\rangle}(z') = z'(1+z')m_{ab}^{\langle -\rangle}(z'). \quad (5.37)$$

Using (5.36), and the definition of the S-transform, (5.37) can be rewritten as

$$\begin{aligned} [z'S_a(z'][z'S_b(z'] &= z'(1+z')m_{ab}^{\langle -1\rangle}(z') \\ &= (z')^2 S_{ab}(z') \end{aligned}$$

and this completes the proof. ■

5.5 Free infinite divisibility

Recall that in classical probability a random variable X is called *infinitely divisible* if, for any $n \in \mathbb{N}$, there exist i.i.d. random variables $X_{n1}, \ldots, X_{nn}$ such that $X_{n1} + \cdots + X_{nn}$ has the same distribution as that of X. Equivalently, X is infinitely divisible if every n-th root of its characteristic function is also a characteristic function. Infinitely divisible distributions are characterized by a representation formula for their characteristic function.

Theorem 5.5.1. (Lévy-Khintchine representation) A probability law μ on $\mathbb{R}$ is infinitely divisible if and only if there are $A \geq 0$, $\gamma \in \mathbb{R}$ and a measure ν on $\mathbb{R}$ which satisfy

$$\nu(\{0\}) = 0, \qquad \int_{\mathbb{R}} (x^2 \wedge 1)d\nu(x) < \infty$$

such that the characteristic function of μ is given by

$$\hat{\mu}(t) = \exp\left[-\frac{1}{2}At^2 + i\gamma t + \int_{\mathbb{R}} (e^{itx} - 1 - itx\mathbb{I}\{|x| \leq 1\})d\nu(x)\right] \text{ for all } t \in \mathbb{R}.$$

◆

The interested reader may consult Sato (1999)[83] and Bose, Dasgupta and Rubin (2002)[22] for further information on the properties of infinitely divisible distributions.

Recall the notion of free additive convolution introduced in Section 3.7. It leads to a natural notion of *free infinite divisibility.*

Definition 5.5.1. (Free infinite divisibility) A probability law μ is said to be *free infinitely divisible* if, for every $n \geq 2$, there exists a probability law μ_n such that $\mu_n^{\boxplus n} = \mu$. ◇

It follows from Example 3.7.1 (a),(c) that the semi-circular law and the free Poisson law are free infinitely divisible.

A Lévy-Khintchine type representation holds for free infinitely divisible probability laws. Instead of the characteristic function, this representation is for the $\mathcal{R}$-transform.

Theorem 5.5.2. (Representation of free infinitely divisible laws) Let μ be a compactly supported probability law on $\mathbb{R}$. Then μ is free infinitely divisible if and only if there exists a compactly supported finite measure ρ on $\mathbb{R}$ such that

$$\mathcal{R}_\mu(z) = \kappa_1 + \sum_{n=1}^{\infty} z^n \int_{\mathbb{R}} x^{n-1} d\rho(x). \tag{5.38}$$

◆

Remark 5.5.1. (a) Since ρ has compact support, its support is contained in $[-C,\ C]$ for some C. Then the series above converges for at least those z with $|z| < 1/C$.

(b) When μ satisfies (5.38), its free cumulants are given by

$$\kappa_1(\mu) = \kappa_1 \;\text{ and }\; \kappa_n(\mu) = \int x^{n-2} d\rho(x) \;\text{ for }\; n \geq 2.$$

●

Before we prove the theorem, let us see some simple examples. Let δ_x denote the probability law which puts full probability at x.

Example 5.5.1. (i) For the semi-circular law μ,

$$\kappa_1 = 0,\ \rho = \delta_0,\ \text{ so that }\ \kappa_2 = 1.$$

(ii) For the free Poisson law μ with parameter λ,

$$\kappa_1 = \lambda \;\text{ and }\; \rho\{1\} = \lambda.$$

(iii) More generally, if μ is compound free Poisson with rate λ and jump distribution ν, then

$$\kappa_1 = \lambda,\ \text{ and }\ d\rho(x) = \lambda x^2 \nu(dx).$$

▲

There is a nice bijection, known as the Bercovici-Pata bijection, between the class of infinitely divisible probability laws and the class of free infinitely divisible probability laws (in a more general sense). The interested reader may consult Bercovici and Pata (2000)[14] and Barndorff-Nielsen and Thorbjørnsen (2004)[10] for further information.

Example 5.5.2. Let μ be free infinitely divisible with mean zero. If $\kappa_4(\mu) = 0$, then μ must be a semi-circular law. To see this, first observe that $\kappa_1(\mu) = m_1(\mu) = 0$. By the representation given in Theorem 5.5.2, we see that $\kappa_n(\mu) = \int_{\mathbb{R}} x^{n-2} d\rho(x)$ for $n \geq 2$. Since $\kappa_4(\mu) = 0$, we get that

$$\int_{\mathbb{R}} x^2 d\rho(x) = 0,$$

which implies that $\rho(\{0\}) = \rho(\mathbb{R})$. In other words, ρ gives full measure to $\{0\}$. This implies that $\rho(\{0\}) = \kappa_2(\mu)$. Hence for any $j > 2$,

$$\kappa_j(\mu) = \int_{\mathbb{R}} x^{j-2} d\rho(x) = 0.$$

Therefore μ must be a semi-circular law.

Incidentally, Arizmendi and Jaramillo (2013)[2] have proved the following interesting limiting version of this: if $\{\mu_n\}$ is a sequence of free infinitely divisible laws such that each has mean 0, variance 1, and the fourth moment converges to 2, then μ_n converges to the standard semi-circular law weakly. ▲

Example 5.5.3. Let μ be a free infinitely divisible law with mean zero and $\kappa_2(\mu) - 2\kappa_3(\mu) + \kappa_4(\mu) = 0$. Then μ is the law of a centered free Poisson variable. As before, since μ is free infinitely divisible, there exists a compactly supported measure ρ on $\mathbb{R}$ such that

$$\kappa_j(\mu) = \int_{\mathbb{R}} x^{j-2} d\rho(x) \text{ for } j \geq 2.$$

Since $\kappa_2(\mu) - 2\kappa_3(\mu) + \kappa_4(\mu) = 0$, we have

$$\int_{\mathbb{R}} (1 - 2x + x^2) d\rho(x) = 0.$$

This implies

$$\int_{\mathbb{R}} (1 - x)^2 d\rho(x) = 0$$

and hence $\rho(\{1\}) = \rho(\mathbb{R})$. So, for all $j \geq 2$, $\kappa_j(\mu) = \int_{\mathbb{R}} x^{j-2} d\rho(x) = \kappa_2(\mu)$. Therefore μ must be the law of a zero mean free Poisson. ▲

Proof of Theorem 5.5.2. First suppose μ is free infinitely divisible and has compact support. Then, for every n, there exists μ_n such that $\mu_n^{\boxplus n} = \mu$. Suppose a_n is a self-adjoint variable with probability law μ_n. Then

$$n\kappa_j(a_n) = \kappa_j(\mu) \quad \text{for all} \quad j \geq 1, n \geq 1. \tag{5.39}$$

On the other hand, for all $j \geq 1$,

$$\begin{aligned}
nm_j(a_n) &= n\kappa_j(a_n) + n\sum_{\pi \neq \mathbf{1}_j} \kappa_\pi(a_n), \\
&= \kappa_j(\mu) + n\sum_{\pi \neq \mathbf{1}_j} \kappa_\pi(a_n) \\
&= \kappa_j(\mu) + n\sum_{\pi \neq \mathbf{1}_j} \kappa_{|V_1|} \cdots \kappa_{|V_k|} \text{ (where } \{V_i\} \text{ are the blocks of } \pi) \\
&= \kappa_j(\mu) + n\sum_{\pi \neq \mathbf{1}_j} O\Big(\frac{1}{n^{|\pi|}}\Big) \text{ (by (5.39))} \\
&= \kappa_j(\mu) + O(n^{-1}) \text{ (since } \pi \neq \mathbf{1}_j) \\
&\to \kappa_j(\mu) \text{ as } n \to \infty.
\end{aligned}$$

This shows that

$$\kappa_j(\mu) = \lim_{n\to\infty} nm_j(a_n) \text{ for all } j \geq 1.$$

However, $\{nm_j(a_n), j \geq 1\}$ is a positive semi-definite sequence for every n. As a consequence, $\{\kappa_{j+2}(\mu), j \geq 0\}$ is also a positive semi-definite sequence.

Now, since μ has compact support, we know that there exists a $C > 0$ such that

$$|m_j(\mu)| \leq C^j \text{ for all } j \geq 1. \tag{5.40}$$

But the moment-free cumulant relation says that

$$\kappa_j(\mu) = \sum_{\pi \in NC(j)} m_\pi(\mu)\ \mu[\pi, \mathbf{1}_j], \tag{5.41}$$

where $\mu[\cdot,\ \cdot]$ is the Möbius function for non-crossing partitions.

By using the properties of the Möbius function, it can be shown that

$$|\mu[\pi, \mathbf{1}_j]| \leq 4^j \text{ for all } j \geq 1.$$

See Corollary 13.1.5 for a detailed proof. Also, recall from Exercise 1 of Chapter 2 that $\#(NC(j)) \leq 4^j$. As a consequence, in view of (5.40) and (5.41), we get that

$$|\kappa_j(\mu)| \leq (16C)^j \text{ for all } j \geq 1. \tag{5.42}$$

Now, since $\{\kappa_{j+2}(\mu), j \geq 0\}$ is a positive semi-definite sequence, by Herglotz's theorem (or Bochner's theorem, see Loéve (1963)[65]), there exists a finite measure, say ρ, with these as its moments. That is,

$$\kappa_{j+2} = \int_{\mathbb{R}} x^j\, d\rho(x) \text{ for all } j \geq 0.$$

Moreover, due to the bound in (5.42), the measure ρ has compact support. This immediately implies that (5.38) holds for all $z \in \mathbb{C}$, $|z| \leq 1/(16C)$.

The proof of the converse is based on ideas from Bercovici and Voiculescu (1992, page 246)[15]. The proof of the following fact is left as an exercise.

Fact: Suppose $\mathcal{R}_n$ is a sequence of $\mathcal{R}$-transforms of compactly supported probability measures $\{\mu_m\}$ such that $\mathcal{R}_n(z) \to \mathcal{R}(z)$ in a neighborhood of 0. Then $\mathcal{R}(z)$ is the $\mathcal{R}$-transform of a compactly supported probability measure μ, and μ_n converges to μ weakly.

We first show that for every $\gamma > 0$ and every $t \in \mathbb{R}$,

$$\mathcal{R}_{\gamma,t}(z) := \frac{\gamma z}{1 - tz} \tag{5.43}$$

is the $\mathcal{R}$-transform of a compactly supported probability measure on $\mathbb{R}$.

For $t = 0$, $\mathcal{R}_{\gamma,0}(z) = \gamma z$ is the $\mathcal{R}$-transform of the semi-circular law with variance γ.

So suppose that $t \neq 0$. We shall obtain the required probability measure as the weak limit of a sequence of probability measures which are constructed via free convolution.

Let $\epsilon_n := \gamma/n$ and consider the two point probability measure

$$\mu_n := (1 - \epsilon_n)\delta_0 + \epsilon_n \delta_t.$$

Then it follows immediately that

$$\begin{aligned} s_{\mu_n}(z) &= \frac{\epsilon_n}{t - z} - \frac{1 - \epsilon_n}{z} \\ &= \frac{z - t(1 - \epsilon_n)}{z(t - z)}, \quad z \in \mathbb{C}^+. \end{aligned}$$

Then, by using Lemma 5.3.1(d), it can be shown that in a neighborhood of 0, which is independent of n,

$$\begin{aligned} \mathcal{R}_{\mu_n}(z) &= \frac{-1 + tz + \sqrt{(1 - tz)^2 + 4t\epsilon_n z}}{2z} \\ &= \frac{t\epsilon_n z}{1 - tz} + O(\epsilon_n^2). \end{aligned} \tag{5.44}$$

We leave the details of the proof as an exercise.

As a consequence,

$$n\mathcal{R}_{\mu_n}(z) \to \mathcal{R}_{\gamma,t}(z), \quad \text{uniformly in a neighborhood of } 0.$$

Since the convergence is uniform, the limit is also going to be an $\mathcal{R}$-transform. Therefore, we can conclude by the quoted fact that, the n-fold additive convolution of μ_n converges to a probability measure with compact support, whose $\mathcal{R}$-transform is $\mathcal{R}_{\gamma,t}(z)$.

Now we show that for every compactly supported finite measure ρ on $\mathbb{R}$,

$$\begin{aligned}\mathcal{R}_\rho(z) &:= \sum_{n=1}^{\infty} z^n \int_{\mathbb{R}} x^{n-1} d\rho(x) \\ &= \int_{\mathbb{R}} \frac{z}{1-xz} d\rho(x)\end{aligned}$$

is the $\mathcal{R}$-transform of a compactly supported probability measure on $\mathbb{R}$.

Since ρ is a compactly supported measure, we can find a compact set K such that for every n, there are real numbers $t_{1n}, \ldots, t_{nn}$ which are inside K, and positive numbers $\gamma_{1n}, \ldots, \gamma_{nn}$ such that the discrete measures ρ_n defined as

$$\begin{aligned}\rho_n\{t_{jn}\} &:= \gamma_{jn},\ 1 \leq j \leq n, \\ \sum_{j=1}^{n} \gamma_{j,n} &= 1,\end{aligned}$$

converge weakly to ρ as $n \to \infty$.

Observe that by (5.43), for each $1 \leq j \leq n$, $n \geq 1$, $\mathcal{R}_{\gamma_{jn}, t_{jn}}(z)$ is the $\mathcal{R}$-transform of compactly supported measure, $\mu_{j,n}$, say, all of whose supports are contained in $[-C,\ C]$ for some $C > 0$. Hence

$$\begin{aligned}\mathcal{R}_{\rho_n}(z) &= \sum_{j=1}^{n} \mathcal{R}_{\gamma_{jn}, t_{jn}}(z)\ (\mathcal{R}\text{-transform of } \mu_{1,n} \boxplus \cdots \boxplus \mu_{n,n}) \\ &= \int_{\mathbb{R}} \frac{z}{1-xz} d\rho_n(x) \\ &\to \int_{\mathbb{R}} \frac{z}{1-xz} d\rho(x) \\ &= \mathcal{R}_\rho(z).\end{aligned} \tag{5.45}$$

The last convergence holds since ρ_n converges weakly to ρ and on the compact support of ρ, for small z, $x \to \dfrac{z}{1-xz}$ is bounded continuous. Morever, this convergence is uniform over z in a non-trivial neighborhood of 0. This implies that $\mathcal{R}_\rho(z)$ is an $\mathcal{R}$-transform. We leave the details as an exercise.

Now we can conclude that the measure μ corresponding to $\mathcal{R}_\rho(\cdot)$ is free infinitely divisible easily. Let ν_n be the compactly supported probability measure with $\mathcal{R}$-transform $\mathcal{R}_{\rho/n}(\cdot)$. Then clearly

$$\mu = \nu_n \boxplus \cdots \boxplus \nu_n\ (n - \text{fold free additive convolution}).$$

Finally, adding a constant κ_1 does not alter the free infinitely divisible property. This completes the proof of the converse. ■

5.6 Exercises

1. Derive the density (2.6) of the standard semi-circular law from the Stieltjes transform formula (5.5) by using Lemma 5.1.1 (g).

2. Suppose $\{\mu_n\}$ are probability measures on $\mathbb{R}$ with densities $\{f_n(\cdot)\}$. Suppose that $f_n \to f$ a.e. with respect to the Lebesgue measure and f is also a density function of a probability measure μ. Show that μ_n converges weakly to μ and $\int_{\mathbb{R}} |f_n(t) - f(t)|dt \to 0$.

3. Suppose μ is a finite measure on $\mathbb{R}$. Then μ has compact support if and only if there exists $C > 0$ such that

$$s_n = \int_{\mathbb{R}} x^n d\mu(x) \text{ is finite and } |s_n| \leq C^n \text{ for all } n \geq 1.$$

In that case, the support of μ is contained in $[-C,\ C]$.

4. Provide the rigorous arguments to establish (5.10) and (5.20) in Examples 5.3.1 and 5.3.2 respectively.

5. Provide the rigorous arguments to establish (5.23) in Example 5.3.3 from (5.22).

6. Suppose the probability law μ puts mass $1/2$ at ± 1.

   (a) Using (5.20) of Example 5.3.2, derive the free cumulants of μ.

   (b) Show that the Stieltjes transform of the 4-fold free convolution $\mu_4 = \mu^{\boxplus 4}$ of μ is given by

$$s_{\mu_4}(z) = \frac{z - 2\sqrt{z^2 - 12}}{z^2 - 16}.$$

   (c) Obtain the density of the n-fold free additive convolution of μ.

   (d) Suppose ν is the probability law which puts mass $1/2$ at 0 and 1. Using the above calculations, write down the density of the n-fold free additive convolution of ν. This probability law is the free binomial probability law with parameter $1/2$.

7. Suppose $\{a_n\} \in \mathcal{A}_n, n \geq 1$, and $a \in \mathcal{A}$ are variables in some NCP. Suppose that $r = \inf\{r_{a_n}\} > 0$. Show that a_n converges to a if and only if $\mathcal{R}_{a_n}(z)$ converges to $\mathcal{R}(z)$ for all $z \in \mathbb{C}$, $|z| < r$.

8. Show that for the modified moment and free cumulants generating functions, $m_a(z)$ and $k_a(z)$, defined, respectively, in (5.24) and (5.28),

$$(1+z)m_a^{\langle -1\rangle}(z) = k_a^{\langle -1\rangle}(z) \text{ and } m_a(z) = k_a(z(1+ma(z))).$$

9. Verify that the formula arrived at in (5.27) through the formal manipulations in the proof of Lemma 5.4.1 can be rigorously justified.

10. If s is a standard semi-circular variable and c is a standard circular variable, show that s^2 and cc^* have the same moments. Use this and Lemma 5.3.1 (d) to find the Stieltjes transform of cc^* and the density of the probability law of cc^*.
11. Justify rigorously Equation (5.29) in Example 5.4.1.
12. Suppose a and s are free, a is self-adjoint and s is standard semi-circular. Show that

$$k_{sas}(z) = m_a(z) \text{ in an appropriate domain.}$$

13. Prove the following statement that was used in the proof of Theorem 5.5.2: Suppose $\mathcal{R}_n$ is a sequence of $\mathcal{R}$-transforms of compactly supported probability measures $\{\mu_n\}$ such that $\mathcal{R}_n(z) \to \mathcal{R}(z)$ in a neighborhood of 0. Then $\mathcal{R}(z)$ is the $\mathcal{R}$-transform of a compactly supported probability measure μ, and μ_n converges to μ weakly.
14. Show that (5.44) holds.
15. Show the uniform convergence claimed for (5.45).

6

C^*-probability space

A C^*-probability space is obtained by imposing additional restrictions on a $*$-algebra and its state. The advantage of working with a C^*-probability space is that the moments of *any* self-adjoint variable in such a space identify a probability law. The price to pay is that all these probability laws have compact support. If a_1 and a_2 are self-adjoint in a C^*-probability space, with probability laws μ_1 and μ_2, then the probability law of $a_1 + a_2$ is $\mu_1 \boxplus \mu_2$ if a_1 and a_2 are free. This provides a more rigorous foundation to free additive convolution introduced initially in Chapter 3. Free multiplicative convolution is a bit more complicated. If a_1 is non-negative (so that $a^{1/2}$ is well-defined) then the probability law of $a_1^{1/2} a_2 a_1^{1/2}$ is the free multiplicative convolution of μ_1 and μ_2 if a_1 and a_2 are free. The reader may skip the details of this chapter at the first reading.

6.1 C^*-probability space

Definition 6.1.1. (C^*-probability space) (a) Any $*$-algebra $\mathcal{A}$ is called a C^*-algebra, if it is a complete normed vector space where the norm $||\cdot||$ satisfies

(i) $||ab|| \leq ||a||\,||b||$, and

(ii) $||a^*a|| = ||a||^2$.

(b) If $(\mathcal{A}, \varphi)$ is a $*$-probability space where $\mathcal{A}$ is a C^*-algebra and φ is tracial and positive, then it is called a *C^*-probability space.* ◇

Thus a C^*-probability space has the extra richness of a norm, an extremely useful ingredient in the subsequent developments. Throughout this chapter $\mathcal{A}$ will denote a C^*-algebra.

Example 6.1.1. Consider the $*$-algebra $\mathcal{M}_n(\mathbb{C})$ of all $n \times n$ matrices with complex entries (random or non-random). Define

$$||A|| = \sup_{x \in \mathbb{R}^n : ||x||=1} |x^\top A^* A x| \text{ for all } A \in \mathcal{M}_n(\mathbb{C}).$$

One can verify that $||\cdot||$ defined above is a norm and satisfies the two conditions in Definition 6.1.1 (a). Further, the state $\varphi_n = \mathrm{E}\,\mathrm{tr}$ is tracial and positive, and hence $(\mathcal{M}_n(\mathbb{C}), \mathrm{E}\,\mathrm{tr})$ is a C^*-probability space. ▲

DOI: 10.1201/9781003144496-6

6.2 Spectrum

We now define the spectrum of any element in $\mathcal{A}$ and develop some of its important properties. These will be useful to us when we discuss free additive and multiplicative convolutions.

An element $a \in \mathcal{A}$ is called *invertible* if it has a multiplicative inverse, that is, there exists $b \in \mathcal{A}$ such that

$$ab = ba = \mathbf{1}_{\mathcal{A}},$$

where $\mathbf{1}_{\mathcal{A}}$ is the multiplicative identity in $\mathcal{A}$. The inverse is always unique.

Definition 6.2.1. (Spectrum) Suppose $\mathcal{A}$ is a C^*-algebra. For any $a \in \mathcal{A}$, its *spectrum* is defined as

$$\mathtt{sp}(a) := \{\lambda \in \mathbb{C} : a - \lambda \mathbf{1}_{\mathcal{A}} \text{ is } \textit{not} \text{ invertible}\}.$$

◇

Example 6.2.1. Consider $A \in \mathcal{M}_n(\mathbb{C})$. Then its spectrum is its set of eigenvalues. Suppose A is Hermitian, that is $A = A^*$. Then its eigenvalues are real and so $\mathtt{sp}(A) \subset \mathbb{R}$. ▲

Lemma 6.2.1. If $a_1, \ldots, a_n \in \mathcal{A}$ are invertible then $a_1 \cdots a_n$ is invertible. The converse holds if $a_1, \ldots, a_n \in \mathcal{A}$ commute. ♦

Proof. The first part is easily established and we omit the proof. For the second part, clearly, it is enough to prove the result for $n = 2$. So, suppose that a_1 and a_2 commute, and $a_1 a_2$ is invertible. Let

$$b = (a_1 a_2)^{-1}.$$

We shall show that $b a_2$ is the inverse of a_1. Observe that

$$a_1 a_2 b = b a_1 a_2 = \mathbf{1}_{\mathcal{A}}. \tag{6.1}$$

But since a_1 and a_2 commute, we have

$$b a_2 a_1 = \mathbf{1}_{\mathcal{A}}. \tag{6.2}$$

From the above relations, $a_2 b$ and $b a_2$ are respectively the right and left inverses of a_1. We must now show that $a_2 b = b a_2$. Using (6.2) and (6.1),

$$a_2 b = b a_2 a_1 a_2 b = b a_2.$$

By a similar argument, a_2 is invertible, and this completes the proof. ■

Theorem 6.2.2. Suppose $a \in \mathcal{A}$. Then
(a) for all $x \in \mathtt{sp}(a)$, $|x| \leq \|a\|$;
(b) $\mathtt{sp}(a)$ is closed and hence is compact;
(c) if a is self-adjoint then $\mathtt{sp}(a) \subset \mathbb{R}$. ◆

Proof. If a is the zero element, then $\mathtt{sp}(a) = \{0\}$, and hence the results are trivially true. So suppose that a is not the zero element.
(a) Consider any element $c \in \mathcal{A}$ such that $0 < \|c\| < 1$. Using property (i) of Definition 6.1.1, it then easily follows that the infinite sum $b := \sum_{n=0}^{\infty} c^n$ converges. Now note that

$$\mathbf{1}_{\mathcal{A}} - c^{n+1} = (\mathbf{1}_{\mathcal{A}} - c)\Big(\sum_{j=0}^{n} c^j\Big) = \Big(\sum_{j=0}^{n} c^j\Big)(\mathbf{1}_{\mathcal{A}} - c).$$

Letting $n \to \infty$, it follows that

$$\mathbf{1}_{\mathcal{A}} = (\mathbf{1}_{\mathcal{A}} - c)b = b(\mathbf{1}_{\mathcal{A}} - c).$$

In other words, $\mathbf{1}_{\mathcal{A}} - c$ is invertible whenever $\|c\| < 1$, and

$$(\mathbf{1}_{\mathcal{A}} - c)^{-1} = \sum_{n=0}^{\infty} c^n = b.$$

Now consider any $\lambda \in \mathbb{C}$ such that $\lambda > \|a\|$. Then

$$(\lambda \mathbf{1}_{\mathcal{A}} - a) = \lambda(\mathbf{1}_{\mathcal{A}} - \lambda^{-1} a)$$

is invertible since $\|\lambda^{-1} a\| < 1$. This proves (a).
(b) We first show that the set of invertible elements is open in $\mathcal{A}$. Let $a \neq 0$ be invertible. For any $b \in \mathcal{A}$, write

$$b = a\big(\mathbf{1}_{\mathcal{A}} - a^{-1}(a - b)\big).$$

From (a), it follows that if $\|a - b\|$ is small enough, then $(\mathbf{1}_{\mathcal{A}} - a^{-1}(a - b))$ is invertible and hence so is b. Thus the set of invertible elements is open.

Now observe that the map $\lambda :\to \lambda \mathbf{1}_{\mathcal{A}} - a$ is continuous with respect to the topology on the C^*-algebra. It immediately follows that $\{\lambda : \lambda \mathbf{1}_{\mathcal{A}} - a \text{ is invertible}\}$ is open.
(c) We first show that if $u \in \mathcal{A}$ is such that $uu^* = u^*u = \mathbf{1}_{\mathcal{A}}$, then

$$\mathtt{sp}(u) \subset \{z \in \mathbb{C} : |z| = 1\}. \tag{6.3}$$

To see this, let $\lambda \in \mathtt{sp}(u)$. Then by (a),

$$|\lambda| \leq \|u\| = \|u^*u\|^{1/2} = 1.$$

Clearly, $\lambda \neq 0$, and

$$u^*(u - \lambda \mathbf{1}_{\mathcal{A}}) = \mathbf{1}_{\mathcal{A}} - \lambda u^* = -\lambda(u^* - \lambda^{-1}\mathbf{1}_{\mathcal{A}}).$$

Since u^* commutes with $u - \lambda\mathbf{1}_{\mathcal{A}}$, which is not invertible, it follows from Lemma 6.2.1 that $u^* - \lambda^{-1}\mathbf{1}_{\mathcal{A}}$ is not invertible, and hence $\lambda^{-1} \in \mathbf{sp}(u^*)$. Using the same arguments with u replaced by u^*, one can show that $|\lambda^{-1}| \leq 1$, which proves (6.3).

Now let $a \in \mathcal{A}$ be self-adjoint. Note that

$$\sum_{n=0}^{\infty} \frac{\|a^n\|}{n!} \leq \sum_{n=0}^{\infty} \frac{\|a\|^n}{n!} = e^{\|a\|} < \infty.$$

So if we define

$$u_n := \sum_{k=0}^{n} \frac{\iota^n}{n!} a^n \quad \text{where} \quad \iota = \sqrt{-1},$$

then it is easy to verify that $\{u_n\}$ is a Cauchy sequence. We denote its limit in $\mathcal{A}$ by

$$u := \sum_{n=0}^{\infty} \frac{\iota^n}{n!} a^n.$$

It is then also easy to check that

$$u^* = \sum_{n=0}^{\infty} \frac{(-\iota)^n}{n!} a^n, \quad \text{and} \quad u^* u = u u^* = \mathbf{1}_{\mathcal{A}}.$$

Let $\lambda \in \mathbf{sp}(a)$. Then

$$\begin{aligned} u - e^{\iota\lambda}\mathbf{1}_{\mathcal{A}} &= \sum_{n=1}^{\infty} \frac{\iota^n}{n!} (a^n - \lambda^n \mathbf{1}_{\mathcal{A}}) \\ &= (a - \lambda\mathbf{1}_{\mathcal{A}}) \sum_{n=1}^{\infty} \frac{\iota^n}{n!} \sum_{j=0}^{n-1} a^j \lambda^{n-j-1} \mathbf{1}_{\mathcal{A}} \\ &= (a - \lambda\mathbf{1}_{\mathcal{A}}) b, \text{ say.} \end{aligned}$$

Clearly, b commutes with $a - \lambda\mathbf{1}_{\mathcal{A}}$. Since $a - \lambda\mathbf{1}_{\mathcal{A}}$ is not invertible, by Lemma 6.2.1, $u - e^{\iota\lambda}\mathbf{1}_{\mathcal{A}}$ is also not invertible. That is, $e^{\iota\lambda} \in \mathbf{sp}(u)$. The inclusion relation in (6.3) shows that

$$|e^{\iota\lambda}| = 1,$$

and hence $\lambda \in \mathbb{R}$. Thus (c) follows, and this completes the proof. ∎

We now proceed to show that $\mathbf{sp}(a)$ is always non-empty. We need a small lemma. Note that by Theorem 6.2.2 (a), $\mathbf{sp}(a)^c$ is non-empty.

Lemma 6.2.3. Suppose $f : \mathcal{A} \to \mathbb{C}$ is a continuous linear function. Then

$$g(z) := f\left((z\mathbf{1}_{\mathcal{A}} - a)^{-1}\right), \; z \in \mathbb{C} \setminus \mathtt{sp}(a),$$

is holomorphic. ♦

Proof. Fix $z_0 \in \mathbb{C} \setminus \mathtt{sp}(a)$. By definition, $z_0\mathbf{1}_{\mathcal{A}} - a$ is invertible, which enables us to write for any $z \in \mathbb{C}$,

$$z\mathbf{1}_{\mathcal{A}} - a = \left(\mathbf{1}_{\mathcal{A}} - (z_0 - z)\,(z_0\mathbf{1}_{\mathcal{A}} - a)^{-1}\,\mathbf{1}_{\mathcal{A}}\right)(z_0\mathbf{1}_{\mathcal{A}} - a)\,.$$

If $|z - z_0| < \|(z_0\mathbf{1}_{\mathcal{A}} - a)^{-1}\|^{-1}$, then the infinite sum

$$\sum_{n=0}^{\infty}(z_0 - z)^n\,(z_0\mathbf{1}_{\mathcal{A}} - a)^{-n-1}$$

converges and equals $(z\mathbf{1}_{\mathcal{A}} - a)^{-1}$. Continuity and linearity of f implies that

$$g(z) = \sum_{n=0}^{\infty}(z_0 - z)^n f\left((z_0\mathbf{1}_{\mathcal{A}} - a)^{-n-1})\right)$$

for all $z \in \mathbb{C}$ with $|z - z_0| < \|(z_0\mathbf{1}_{\mathcal{A}} - a)^{-1}\|^{-1}$. This shows that g is differentiable at z_0. Since this holds for all $z_0 \in \mathtt{sp}(a)^c$, the proof follows. ■

Theorem 6.2.4. For $a \in \mathcal{A}$, $\mathtt{sp}(a)$ is non-empty. ♦

The proof of the theorem uses two facts. The first fact is a consequence of Liouville's theorem in complex analysis. The second fact follows from the Hahn-Banach theorem in functional analysis.

Fact 6.2.1. If $g : \mathbb{C} \to \mathbb{C}$ is an entire function with $\lim_{|z|\to\infty} g(z) = 0$, then g is identically zero on $\mathbb{C}$.

Fact 6.2.2. Let X be a Banach space. Let Y be the collection of all continuous linear functionals from X to $\mathbb{C}$, equipped with the operator norm:

$$\|f\| := \sup_{x \in X : \|x\| \le 1} |f(x)|, \; f \in Y.$$

Then for all $x \in X$,

$$\sup_{f \in Y : \|f\| \le 1} |f(x)| = \|x\|.$$

Proof of Theorem 6.2.4. Suppose $\mathtt{sp}(a) = \emptyset$ for some $a \in \mathcal{A}$. Then a is invertible. By Fact 6.2.2, there exists a continuous linear functional f on $\mathcal{A}$ such that $f(a^{-1}) \neq 0$. By Lemma 6.2.3, $g(z) := f\left((z\mathbf{1}_{\mathcal{A}} - a)^{-1}\right)$ is an entire function. We leave it as an exercise to show that as $|z| \to \infty$, $(z\mathbf{1}_{\mathcal{A}} - a)^{-1} \to 0$. Hence $\lim_{|z|\to\infty} g(z) = 0$. Thus by Fact 6.2.1, g is identically zero. But this is a contradiction. ■

Lemma 6.2.5. For $a \in \mathcal{A}$ and any polynomial p with coefficients in $\mathbb{C}$,

$$\mathtt{sp}(p(a)) = \{p(x) : x \in \mathtt{sp}(a)\}.$$

♦

Proof. We assume without loss of generality that p is not a constant polynomial. Fix $\lambda \in \mathbb{C}$. By the fundamental theorem of algebra, $\lambda - p(x)$ can be factorized into linear functions, that is,

$$\lambda - p(x) = c(\alpha_1 - x) \cdots (\alpha_n - x)$$

for some $c \neq 0, \alpha_1, \ldots, \alpha_n \in \mathbb{C}$. Therefore

$$\lambda \mathbf{1}_{\mathcal{A}} - p(a) = c(\alpha_1 \mathbf{1}_{\mathcal{A}} - a) \cdots (\alpha_n \mathbf{1}_{\mathcal{A}} - a).$$

Lemma 6.2.1 implies that $\lambda \mathbf{1}_{\mathcal{A}} - p(a)$ is not invertible if and only if $\alpha_i \mathbf{1}_{\mathcal{A}} - a$ is not invertible for some i. That is, $\lambda \in \mathtt{sp}(p(a))$ if and only if $\lambda = p(\theta)$ for some $\theta \in \mathtt{sp}(a)$. Hence the proof is complete. ■

Lemma 6.2.6. Suppose $a \in \mathcal{A}$.

(a) The following relation holds:

$$\sup\{|x| : x \in \mathtt{sp}(a)\} = \lim_{n\to\infty} \|a^n\|^{1/n}.$$

(b) If a is self-adjoint, then

$$\|a\| = \sup\{|x| : x \in \mathtt{sp}(a)\}.$$

♦

To prove this, we shall need the following result from functional analysis.

Fact 6.2.3 (Uniform boundedness principle). Let X be a Banach space and Y a normed vector space. Suppose that F is a collection of continuous linear operators from X to Y. If $\sup_{T\in F} \|T(x)\| < \infty$ for all $x \in X$, then, $\sup_{T\in F, \|x\|\leq 1} \|T(x)\| < \infty$.

Proof of Lemma 6.2.6. (a) Let us start by showing that

$$\limsup_{n\to\infty} \|a^n\|^{1/n} \leq \sup\{|x| : x \in \mathtt{sp}(a)\}. \tag{6.4}$$

In view of Theorems 6.2.2 (a) and 6.2.4, there exists $0 < R \leq \infty$ such that

$$R^{-1} = \sup\{|x| : x \in \mathtt{sp}(a)\}.$$

Fix a continuous linear function $f : \mathcal{A} \to \mathbb{C}$. Define

$$g(z) = f\left((\mathbf{1}_{\mathcal{A}} - za)^{-1}\right), \; z \in \mathbb{C}, \; |z| < R,$$

which clearly makes sense because

$$g(z) = \begin{cases} z^{-1}f\left((z^{-1}\mathbf{1}_{\mathcal{A}} - a)^{-1}\right), & \text{if } 0 < |z| < R, \\ f(\mathbf{1}_{\mathcal{A}}), & \text{if } z = 0. \end{cases}$$

The above along with Lemma 6.2.3 shows that g is holomorphic on the set $\{z \in \mathbb{C} : 0 < |z| < R\}$. Since g is continuous at 0, it is holomorphic on $\{z \in \mathbb{C} : |z| < R\}$. This implies that there exist $c_0, c_1, \ldots \in \mathbb{C}$ such that

$$g(z) = \sum_{n=0}^{\infty} c_n z^n, \ |z| < R. \tag{6.5}$$

For $|z| < \|a\|^{-1}$, it is easy to see that

$$(\mathbf{1}_{\mathcal{A}} - za)^{-1} = \sum_{n=0}^{\infty} z^n a^n,$$

and hence

$$g(z) = \sum_{n=0}^{\infty} f(a^n) z^n, \ |z| < \|a\|^{-1}.$$

Comparing coefficients with (6.5), we get $c_n = f(a^n)$ for $n \geq 0$. Thus, for all $|z| < R$,

$$\sum_{n=0}^{\infty} f(a^n) z^n$$

is a convergent sum. Therefore, for every linear functional f on $\mathcal{A}$,

$$\sup_{n \geq 1} |f(a^n z^n)| < \infty. \tag{6.6}$$

Let X be the Banach space of all linear functionals on $\mathcal{A}$. Fix $|z| < R$ and define for $n \in \mathbb{N}$, $T_n : X \to \mathbb{C}$ by $T_n(f) := f(z^n a^n),\ f \in X$. Then $F := \{T_n : n \geq 1\}$ is a collection of continuous linear operators from X to $\mathbb{C}$. By (6.6), for all $f \in X$, $\sup_{T_n \in F} |T_n(f)| < \infty$. Hence the uniform boundedness principle implies that

$$\sup_{T_n \in F, f \in X : \|f\| \leq 1} |T_n(f)| < \infty.$$

Hence using Fact 6.2.2,

$$\sup_{n \geq 1} |z|^n \|a^n\| = \sup_{n \geq 1,\ f \in X : \|f\| \leq 1} |f(z^n a^n)| < \infty.$$

Therefore,

$$\limsup_{n \to \infty} \|a^n\|^{1/n} \leq |z|^{-1}, \ 0 < |z| < R.$$

Letting $|z| \to R$, (6.4) follows. It is now enough to show that

$$\sup\{|x| : x \in \mathsf{sp}(a)\} \leq \liminf \|a^n\|^{1/n}.$$

Observe that if $\lambda \in \mathtt{sp}(a)$, then $\lambda^n \in \mathtt{sp}(a^n)$ and hence

$$|\lambda^n| \leq \sup\{|x| : x \in \mathtt{sp}(a^n)\} \leq \|a^n\|,$$

where the right inequality follows from Theorem 6.2.2. This together with (6.4) completes the proof.

(b) When a is self-adjoint, we have $\|a^2\| = \|a^*a\| = \|a\|^2$. An easy induction argument gives us that $\|a\| = \lim_{n\to\infty} \|a^{2^n}\|^{\frac{1}{2^n}}$. ■

Lemma 6.2.7. The set $\mathcal{C} = \{a : a = a^*\}$ is closed in $\mathcal{A}$. ♦

Proof. Note that for any $b \in \mathcal{A}$,

$$\|b\|^2 = \|b^*b\| \leq \|b^*\|\|b\|.$$

This implies that

$$\|b\| \leq \|b^*\| \leq \|(b^*)^*\| = \|b\|.$$

Now suppose $a_n \in \mathcal{C}$ and $a \in \mathcal{A}$ such that $\|a_n - a\| \to 0$. We have to show that $a = a^*$. The above argument shows that $a_n^* \to a^*$. Now since $a_n^* = a_n \to a$ we have $a = a^*$ and hence $a \in \mathcal{C}$. ■

The proof of the following lemma is left as an exercise.

Lemma 6.2.8. Let $\mathcal{B}$, $\mathcal{B}'$ be two Banach spaces and $\mathcal{B}_0$ be a dense subspace of $\mathcal{B}$. If $T : \mathcal{B}_0 \to \mathcal{B}'$ is linear, and $\sup_{x\in\mathcal{B}_0:\|x\|=1} \|T(x)\| < \infty$, then there exists a unique bounded linear operator $\tilde{T} : \mathcal{B} \to \mathcal{B}'$ such that $T = \tilde{T}$ on $\mathcal{B}_0$. ♦

We need a notation. Let

$$C_{\mathbb{R}}(K) := \{f : f \text{ is a continuous function from } K \text{ to } \mathbb{R}\},\ \ K \subset \mathbb{R} \text{ compact}.$$

Then $C_{\mathbb{R}}(K)$ is a Banach space when equipped with the sup-norm.

The result stated in the next lemma is often referred to as *continuous functional calculus*. It is used to give meaning to $f(a)$ where $f \in C_{\mathbb{R}}(\mathtt{sp}(a))$.

Lemma 6.2.9. (Continuous functional calculus) Fix a self-adjoint $a \in \mathcal{A}$. Then there exists a bounded linear operator $\psi : C_{\mathbb{R}}(\mathtt{sp}(a)) \to \mathcal{A}$ such that

$$\psi(p) = p(a) \text{ for all real polynomials } p. \tag{6.7}$$

♦

Proof. Note that the set of real polynomials, $\mathcal{P}$, is dense in $C_{\mathbb{R}}(\mathtt{sp}(a))$ with respect to the sup-norm. Define $\psi : \mathcal{P} \to \mathcal{A}$ as in (6.7). Observe that for $p \in \mathcal{P}$, $\psi(p)$ is self-adjoint. Hence Lemma 6.2.6 (b) implies that

$$\begin{aligned}
\|\psi(p)\| = \|p(a)\| &= \sup\{|x| : x \in \mathtt{sp}(p(a)\} \\
&= \sup\{|p(x)| : x \in \mathtt{sp}(a)\} \text{ (by Lemma 6.2.5)} \\
&= \|p\|_\infty,
\end{aligned}$$

where $\|\cdot\|_\infty$ is the sup-norm. Thus ψ is continuous on the space of polynomials. Since $\mathcal{P}$ is dense in $C_{\mathbb{R}}(\mathtt{sp}(a))$, an application of Lemma 6.2.8 completes the proof. ■

Definition 6.2.2. For a self-adjoint $a \in \mathcal{A}$ and any continuous function $f : \mathtt{sp}(a) \to \mathbb{R}$, define

$$f(a) = \psi(f),$$

where $\psi : C_{\mathbb{R}}(\mathtt{sp}(a)) \to \mathcal{A}$ is as in Lemma 6.2.9. ◊

Theorem 6.2.10. For a self-adjoint $a \in \mathcal{A}$ and a continuous $f : \mathtt{sp}(a) \to \mathbb{R}$, we have $\mathtt{sp}\,(f(a)) = f\,(\mathtt{sp}(a))\,.$ ♦.

Proof. Lemma 6.2.9 along with the Stone-Weierstrass Theorem implies that, for continuous functions f and g from $\mathtt{sp}(a)$ to $\mathbb{R}$,

$$(fg)(a) = f(a)g(a), \quad \text{and} \quad (f+g)(a) = f(a) + g(a). \tag{6.8}$$

Suppose that $\lambda \notin f\,(\mathtt{sp}(a))$. Theorem 6.2.2 and continuity of f imply that $f\,(\mathtt{sp}(a))$ is a compact set, and therefore bounded away from λ. Hence, $g : \mathtt{sp}(a) \to \mathbb{R}$ defined by

$$g(x) = \frac{1}{f(x) - \lambda}, \; x \in \mathtt{sp}(a),$$

is continuous. Invoking (6.8), we get

$$g(a)\,(f(a) - \lambda) = (f(a) - \lambda)\,g(a) = \mathbf{1}_{\mathcal{A}}.$$

Hence $\lambda \notin \mathtt{sp}\,(f(a))$. In other words,

$$\mathtt{sp}\,(f(a)) \subset f\,(\mathtt{sp}(a))\,.$$

For the reverse inclusion, let $\lambda \in f(\mathtt{sp}(a))$. Then $\lambda = f(x_0)$ for some $x_0 \in \mathtt{sp}(a)$. Suppose if possible, $\lambda \notin \mathtt{sp}(f(a))$. Then $f(a) - \lambda$ being invertible. Let

$$\alpha > \|\,(f(a) - \lambda)^{-1}\,\|.$$

Continuity of f implies the existence of $\varepsilon > 0$ such that

$$|f(x) - \lambda| \le \frac{1}{\alpha}, \; x \in [x_0 - \varepsilon, x_0 + \varepsilon] \cap \mathtt{sp}(a).$$

Let

$$g(x) := \alpha \max\left(0, 1 - \varepsilon^{-1}|x - x_0|\right), \; x \in \mathbb{R}.$$

In other words, g is the continuous piecewise linear function joining the points $(x_0 - \varepsilon, 0)$, (x_0, α), $(x_0 + \varepsilon, 0)$, and vanishes outside $[x_0 - \varepsilon, x_0 + \varepsilon]$. Although g is defined on $\mathbb{R}$, we are interested only in its restriction to $\mathtt{sp}(a)$.

As shown in the proof of Lemma 6.2.9, ψ obtained there is an isometry. Hence it follows that

$$\|g(a)\| = \sup_{x \in \mathbf{sp}(a)} |g(x)| = \alpha.$$

Note that $0 \le g(x) \le g(x_0) = \alpha$ for all $x \in \mathbb{R}$ and $x_0 \in \mathbf{sp}(a)$ together imply the second equality. Using the property of isometry again, we get

$$\|g(a)\,(f(a) - \lambda)\,\| = \sup_{x \in \mathbf{sp}(a)} |g(x)\,(f(x) - \lambda)| \le 1.$$

Therefore,

$$\begin{aligned}
\alpha &= \|g(a)\| \\
&= \|\,(f(a) - \lambda)^{-1}\, g(a)\,(f(a) - \lambda)\,\| \\
&\le \|\,(f(a) - \lambda)^{-1}\,\|\|g(a)\,(f(a) - \lambda)\,\| \\
&< \alpha,
\end{aligned}$$

which is a contradiction. Thus, $\lambda \in \mathbf{sp}(f(a))$, and the proof is complete. ∎

Lemma 6.2.11. (a) If $a \in \mathcal{A}$ is self-adjoint and $\alpha \ge \|a\|$, then

$$\|\alpha \mathbf{1}_{\mathcal{A}} - a\| = \alpha - \inf \mathbf{sp}(a).$$

(b) For self-adjoint $a, b \in \mathcal{A}$,

$$\inf \mathbf{sp}(a + b) \ge \inf \mathbf{sp}(a) + \inf \mathbf{sp}(b).$$

(c) For all $a, b \in \mathcal{A}$,

$$\mathbf{sp}(ab) \cup \{0\} = \mathbf{sp}(ba) \cup \{0\}.$$

♦

Proof. (a) Lemma 6.2.6 (b) implies that

$$\begin{aligned}
\|\alpha \mathbf{1}_{\mathcal{A}} - a\| &= \sup \{|\lambda| : \lambda \in \mathbf{sp}\,(\alpha \mathbf{1}_{\mathcal{A}} - a)\} \\
&= \sup \{|\alpha - \lambda| : \lambda \in \mathbf{sp}\,(a)\} \\
&= \alpha - \inf \mathbf{sp}(a),
\end{aligned}$$

where the last equality follows since $\alpha \ge \|a\|$ implies that $\mathbf{sp}(a) \subset (-\infty, \alpha]$.
(b) Let $\alpha := \|a\|$ and $\beta := \|b\|$. Since $\|a + b\| \le \alpha + \beta$, (a) implies that

$$\begin{aligned}
\alpha + \beta - \inf \mathbf{sp}(a + b) &= \|(\alpha + \beta)\mathbf{1}_{A} - (a + b)\| \\
&\le \|\alpha \mathbf{1}_{A} - a\| + \|\beta \mathbf{1}_{A} - b\| \\
&= \alpha - \inf \mathbf{sp}(a) + \beta - \mathbf{sp}(b),
\end{aligned}$$

where the last equality follows from (a). The proof follows by subtracting $\alpha+\beta$ from both sides.

(c) Let us start by proving the claim that $\mathbf{1}_{\mathcal{A}} - ab$ is invertible if and only if $\mathbf{1}_{\mathcal{A}} - ba$ is. Assuming $\mathbf{1}_{\mathcal{A}} - ab$ is invertible, observe that

$$\begin{aligned}
&(\mathbf{1}_{\mathcal{A}} - ba)\left(b\left(\mathbf{1}_{\mathcal{A}} - ab\right)^{-1} a + \mathbf{1}_{\mathcal{A}}\right)\\
&= b\left(\mathbf{1}_{\mathcal{A}} - ab\right)^{-1} a + \mathbf{1}_{\mathcal{A}} - bab\left(\mathbf{1}_{\mathcal{A}} - ab\right)^{-1} a - ba\\
&= b\left(\mathbf{1}_{\mathcal{A}} - ab\right)\left(\mathbf{1}_{\mathcal{A}} - ab\right)^{-1} a + \mathbf{1}_{\mathcal{A}} - ba\\
&= \mathbf{1}_{\mathcal{A}}.
\end{aligned}$$

A similar algebra shows that

$$\left(b\left(\mathbf{1}_{\mathcal{A}} - ab\right)^{-1} a + \mathbf{1}_{\mathcal{A}}\right)(\mathbf{1}_{\mathcal{A}} - ba) = \mathbf{1}_{\mathcal{A}},$$

and therefore, $\mathbf{1}_{\mathcal{A}} - ba$ is invertible. This proves the "only if" part, and the converse follows by symmetry.

Now suppose that $\lambda \notin \mathtt{sp}(ab) \cup \{0\}$. Write

$$\lambda \mathbf{1}_{\mathcal{A}} - ba = \lambda\left(\mathbf{1}_{\mathcal{A}} - \frac{b}{\lambda} a\right).$$

Since $\mathbf{1}_{\mathcal{A}} - a(b/\lambda)$ is invertible by the choice of λ, the claim proved above implies that so is the right side. Hence, $\lambda \notin \mathtt{sp}(ba)$. In other words,

$$\mathtt{sp}(ab) \cup \{0\} \supset \mathtt{sp}(ba) \cup \{0\}.$$

Symmetry implies the reverse inclusion and thus completes the proof. ■

Definition 6.2.3. (Positive element) An element $a \in \mathcal{A}$ is said to be *positive* if $a = b^*b$ for some $b \in \mathcal{A}$. We write $a \geq 0$. ◇

Lemma 6.2.12. For $a \in \mathcal{A}$, if $a \geq 0$ and $\mathtt{sp}(a) \subset (-\infty, 0]$, then $a = 0$. ♦

Proof. Let $a \geq 0$ and $\mathtt{sp}(a) \subset (-\infty, 0]$. By definition, there exists $b \in \mathcal{A}$ with $a = b^*b$. Lemma 6.2.11 (c) implies that

$$\mathtt{sp}(bb^*) \subset \mathtt{sp}(b^*b) \cup \{0\} \subset (-\infty, 0].$$

Apply Lemma 6.2.11 (b) to $-bb^*$ and $-b^*b$ to get that

$$\sup \mathtt{sp}\left(bb^* + b^*b\right) \leq \sup \mathtt{sp}\left(bb^*\right) + \sup \mathtt{sp}\left(b^*b\right) \leq 0. \tag{6.9}$$

Define

$$x := \frac{1}{2}\left(b + b^*\right), \quad y := \frac{1}{2\iota}(b - b^*), \tag{6.10}$$

where $\iota = \sqrt{-1}$. Clearly x, y are self-adjoint, and

$$x^2 + y^2 = \frac{1}{2}\left(b^*b + bb^*\right).$$

Theorem 6.2.10 implies that

$$\mathtt{sp}(x^2) \subset [0, \infty), \text{ and } \mathtt{sp}(y^2) \subset [0, \infty), \tag{6.11}$$

which in turn implies that $\mathtt{sp}(x^2 + y^2) \subset [0, \infty)$ using Lemma 6.2.11 (b) once again. It follows from (6.9) that $\mathtt{sp}(x^2 + y^2) \subset (-\infty, 0]$ and hence

$$\mathtt{sp}(x^2 + y^2) = \{0\}.$$

Lemma 6.2.6 (b) implies that $\|x^2 + y^2\| = 0$ and hence

$$x^2 = -y^2.$$

Recalling (6.11), it follows by a similar application of Lemma 6.2.6 (b) that $x = y = 0$. Equation (6.10) implies that $b = 0$, from which the desired conclusion follows. ■

Lemma 6.2.13. (a) If $a \in \mathcal{A}$ is self-adjoint, and $\mathtt{sp}(a) \subset [0, \infty)$, then $a \geq 0$.
(b) For any $a \in \mathcal{A}$, $\mathtt{sp}(aa^*) \subseteq [0, \|aa^*\|]$. ♦

Proof. (a) Let $\sqrt{\cdot} : \mathtt{sp}(a) \to \mathbb{R}$ denote the function $x \mapsto \sqrt{x}$, which is defined because $\mathtt{sp}(a) \subset [0, \infty)$. Let

$$b := \sqrt{a}.$$

Recall (6.8) to write

$$a = b \cdot b.$$

Note that $f(a)$ is self-adjoint for any $f \in C_{\mathbb{R}}(\mathtt{sp}(a))$ because, by definition, it is the limit of $p_n(a)$ for some sequence of polynomials $\{p_n\}$. Hence $b = b^*$ and the proof of (a) is complete.
(b) By Theorem 6.2.2 (c) and Lemma 6.2.6 (b), $\mathtt{sp}(a) \subseteq [-\|a\|, \|a\|]$. Thus, all that has to be shown is

$$\mathtt{sp}(a) \subset [0, \infty). \tag{6.12}$$

Let

$$f(t) := \max\{t, 0\}, \quad \text{and} \quad g(t) := \max\{-t, 0\}, \ t \in \mathbb{R}.$$

Denote

$$x = f(a), \quad y = g(a).$$

Clearly, x, y are self-adjoint. Theorem 6.2.10 implies that $\mathtt{sp}(x) \subset [0, \infty)$ and $\mathtt{sp}(y) \subset [0, \infty)$.

Since $fg = 0$ and $f - g$ is identity, (6.8) implies that

$$xy = yx = 0, \text{ and } x - y = a. \tag{6.13}$$

From the hypothesis $a \geq 0$, there exists $b \in \mathcal{A}$ with $a = b^*b$. Let $c := by$ and write

$$c^*c = yb^*by = yay = y(x - y)y = -y^3,$$

where the two rightmost equalities are obtained by (6.13). In other words, $-y^3 \geq 0$. Since $\text{sp}(y) \subset [0, \infty)$, it follows that $\text{sp}(-y^3) \subset (-\infty, 0]$. Lemma 6.2.12 implies that $y^3 = 0$.

Therefore,

$$\{\lambda^3 : \lambda \in \text{sp}(y)\} = \{0\},$$

and hence $\text{sp}(y) = \{0\}$. Lemma 6.2.6 (b) implies that $y = 0$, which in conjunction with (6.13) proves that $x = a$. Recalling that $\text{sp}(x) \subset [0, \infty)$, (6.12) follows, and this completes the proof. ■

Lemma 6.2.14. Suppose $a \in \mathcal{A}$. Then $\|a\|^2 \mathbf{1}_{\mathcal{A}} - a^*a \geq 0$. ◆

Proof. By Lemma 6.2.13 (a), it is enough to show that

$$\text{sp}(\|a\|^2 \mathbf{1}_{\mathcal{A}} - a^*a) \subset [0, \infty).$$

To that end, by using Lemma 6.2.13 (b) in the third step below,

$$\begin{aligned} \lambda \in \text{sp}\big(\|a\|^2 \mathbf{1}_{\mathcal{A}} - a^*a\big) &\Rightarrow (\lambda - \|a\|^2)\mathbf{1}_{\mathcal{A}} + a^*a \text{ is not invertible} \\ &\Rightarrow (\|a\|^2 - \lambda)\mathbf{1}_{\mathcal{A}} - a^*a \text{ is not invertible} \\ &\Rightarrow \|a\|^2 - \lambda \in \text{sp}(a^*a). \end{aligned}$$

But from Lemma 6.2.13 (b), $\text{sp}(a^*a) \subset [0, \|a^*a\|] \subset [0, \|a\|^2]$. As a consequence, $\|a\|^2 - \lambda \leq \|a\|^2$ and hence $\lambda \geq 0$. ■

6.3 Distribution of a self-adjoint element

In Chapter 2, we have seen that the moments of a self-adjoint variable a in a $*$-probability space with a positive state define a unique probability law μ_a if a uniqueness condition is satisfied. We now show that if we have a C^*-probability space, then this happens automatically for all elements in the C^*-algebra.

To show this, we shall make use of the results of the previous section along with the following result. A linear functional γ on $C_{\mathbb{R}}(K)$ is said to be *positive* if $\gamma(f) \geq 0$ for all non-negative $f \in C_{\mathbb{R}}(K)$. The *Riesz representation theorem* provides the following representation for positive γ. The reader may consult Rudin (1987)[81] for a proof.

Theorem 6.3.1. (Riesz representation theorem) Suppose K is a compact subset of $\mathbb{R}$. Then a continuous linear functional γ on $C_{\mathbb{R}}(K)$ is positive and unital if and only if there exists a unique probability law μ_γ on K such that

$$\gamma(f) = \int_K f d\mu_\gamma \quad \text{for all} \quad f \in C_{\mathbb{R}}(K).$$

◆

The next result, known as the Cauchy-Schwarz inequality for C^*-algebras, will also be needed. Its proof is similar to the proof of the usual Cauchy-Schwarz inequality and is left as an exercise.

Lemma 6.3.2 (Cauchy-Schwarz inequality)**.** Suppose φ is a positive linear functional on a C^*-algebra. Then, for all $a, b \in \mathcal{A}$,

$$|\varphi(b^*a)|^2 \leq \varphi(a^*a)\varphi(b^*b).$$

♦

The following lemma facilitates the use of the Riesz representation theorem.

Lemma 6.3.3. Suppose φ is a positive linear functional on a C^*-algebra. Then it is continuous. ♦

Proof. Since φ is linear, it suffices to show that it is bounded. By Lemma 6.3.2,

$$|\varphi(a)|^2 = |\varphi(\mathbf{1}_{\mathcal{A}} \cdot a)|^2 \leq \varphi(a^*a)\varphi(\mathbf{1}_{\mathcal{A}}) \leq \|a\|^2,$$

where the last inequality follows from Lemma 6.2.14. ■

Theorem 6.3.4. For any self-adjoint variable a in a C^*-probability space $(\mathcal{A}, \varphi)$, there exists a unique probability law μ_a supported on a subset of $\mathtt{sp}(a)$ such that

$$\varphi(a^n) = \int_{\mathbb{R}} x^n \mu_a(dx) \ \text{ for all } \ n \geq 1. \tag{6.14}$$

♦

Based on Theorem 6.3.4, self-adjoint variables in a C^*-probability space should be thought of as analogs of random variables with compact support, and φ as the analog of expectation, as discussed in Chapter 2.

Proof. We first show that $\varphi \circ \psi : C_{\mathbb{R}}(\mathtt{sp}(a)) \to \mathbb{R}$ is a continuous, unital, linear, positive functional where ψ is as in Lemma 6.2.9.

The linearity and unital properties follow from definition. The continuity of $\varphi \circ \psi$ follows from Lemma 6.3.3 and Lemma 6.2.9. It remains to show that the range is a subset of $\mathbb{R}$ and it is a positive functional.

Now for any $f \in C_{\mathbb{R}}(\mathtt{sp}(a))$, choose polynomials p_n such that

$$\lim_{n\to\infty} \sup\{|p_n(x) - f(x)| : x \in \mathtt{sp}(a)\} = 0. \tag{6.15}$$

Since a is self-adjoint, for any polynomial p, $\psi(p)$ is also self-adjoint. The proof of this is left as an exercise. On the other hand, for any self-adjoint element $b \in \mathcal{A}$, $\varphi(b) \in \mathbb{R}$. Hence $\varphi \circ \psi(p) \in \mathbb{R}$. Then, by continuity,

$$\varphi \circ \psi(f) = \varphi \circ \psi(\lim_{n\to\infty} p_n) = \lim_{n\to\infty} \varphi \circ \psi(p_n) \in \mathbb{R}.$$

It remains to prove positivity. We need to show that for any $f \in C_{\mathbb{R}}(\mathtt{sp}(a))$, $f \geq 0$ (that is, $f(x) \geq 0$ for all $x \in \mathtt{sp}(a)$), we have $\varphi \circ \psi(f) \geq 0$. Choose p_n as in (6.15). *Moreover, we may assume that p_n are all non-negative.* It then suffices to show that

$$\varphi \circ \psi(p_n) \geq 0 \text{ for all large } n. \tag{6.16}$$

For this, in turn, it is enough to show that, as elements of $\mathcal{A}$, $\psi(p_n) \geq 0$ for all large n. Lemma 6.2.13 implies that this, in turn, is true if $\mathtt{sp}(\psi(p_n)) \subset [0, \infty)$ for large n. But, due to Lemma 6.2.5,

$$\mathtt{sp}(\psi(p_n)) = \{p_n(x) : x \in \mathtt{sp}(a)\}.$$

Now note that f is non-negative and $p_n(\cdot)$ on $\mathtt{sp}(a)$ converges uniformly to $f(\cdot)$ which is non-negative. This implies that $\limsup \mathtt{sp}(\psi(p_n)) \subset [0, \infty)$ for all large n.

Thus $\varphi \circ \psi : C_{\mathbb{R}}(\mathtt{sp}(a)) \to \mathbb{R}$ is a continuous, unital, linear, positive functional.

Now, by Theorem 6.3.1, there exists a probability law μ_a on $\mathtt{sp}(a)$ such that, for all $f \in C_{\mathbb{R}}(\mathtt{sp}(a))$,

$$\varphi \circ \psi(f) = \int_{\mathtt{sp}(a)} f(x)\mu_a(dx).$$

In particular, choosing $f(x) = x^n$,

$$\varphi \circ \psi(f) = \varphi(a^n) = \int_{\mathtt{sp}(a)} x^n \mu_a(dx) \text{ for all } n \geq 1.$$

The uniqueness is a direct consequence of the Riesz representation theorem. This completes the proof. ■

As examples of self-adjoint variables in C^*-probability spaces and their probability laws, we revisit some examples given in the earlier chapters.

Example 6.3.1. The $*$-probability space $(\mathcal{A}, \varphi)$ defined in Example 2.5.2 is a C^*-probability space with the *essential supremum* norm. It is easy to check that for every $f \in \mathcal{A}$, its distribution is $\mu_f = P \circ f^{-1}$, the probability law in the usual sense. ▲

Example 6.3.2. Recall Example 2.5.4. The class of self-adjoint elements of $\mathcal{A}$ is the class of Hermitian $d \times d$ matrices. For a self-adjoint $a \in \mathcal{A}$,

$$\varphi(a^k) = \mathrm{tr}(a^k) = \frac{1}{d}\sum_{i=1}^{d} \lambda_i^k,$$

where $\lambda_1, \ldots, \lambda_d$ are the eigenvalues of a. Thus $\mu_a = \frac{1}{d}\sum_{i=1}^{d} \delta_{\lambda_i}$, where for any x, δ_x denotes the probability law which puts full mass at x. ▲

Example 6.3.3. Let $(\Omega, \mathcal{F}, P)$ be a classical probability space. Let $\mathcal{A}$ be the class of $d \times d$ matrices whose entries are random variables from the class $L^\infty(\Omega)$ of almost surely bounded random variables where random variables are identified if they are almost surely equal. Define $\varphi : \mathcal{A} \to \mathbb{C}$ by

$$\varphi(a) = E\left[\text{tr}(a)\right], \ a \in \mathcal{A}.$$

We leave it as an exercise to show that $(\mathcal{A}, \varphi)$ is a C^*-probability space with the *operator norm* which is defined as

$$||A|| = \sup_{x:||x||=1} ||Ax||.$$

An element $a = ((\alpha_{ij}))_{i,j=1}^d \in \mathcal{A}$ is self-adjoint if α_{ii} is real almost surely and $\overline{\alpha_{ij}} = \alpha_{ji}$ almost surely for all $1 \leq i, j \leq d$. We leave it as an exercise to check that the distribution of a is

$$\frac{1}{d} \sum_{i=1}^{d} P \circ \lambda_i^{-1},$$

where $\lambda_1 \leq \cdots \leq \lambda_d$ are the eigenvalues of a, and hence *random variables* defined on Ω. ▲

6.4 Free product of C^*-probability spaces

Theorem 3.2.1 on the free product of $*$-probability spaces ensured that free variables with specified marginals can always be constructed. It also helped us to define, in a limited way, free additive convolution of probability laws. We now state a similar theorem for the free product of C^*-probability spaces.

We need to introduce one additional property of a state in a C^*-algebra.

Definition 6.4.1. (Faithful state) Let $(\mathcal{A}, \varphi)$ be a C^*-probability space. The state φ is said to be *faithful* if, for all $a \in \mathcal{A}$,

$$\varphi(a^* a) = 0 \text{ implies that } a = 0.$$

◇

The states in Examples 6.3.1–6.3.3 are all faithful states.

The proof of the following theorem shall not be presented in this book. The reader may refer to Nica and Speicher (2006)[74]. See Section 13.3 for the weaker result on the construction of the product of $*$-probability spaces.

Theorem 6.4.1. Let $(\mathcal{A}_i, \varphi_i)_{i \in I}$ be a family of C^*-probability spaces such that φ_i is faithful for each $i \in I$. Then, there exists a C^*-probability space $(\mathcal{A}, \varphi)$, called the free product of $(\mathcal{A}_i, \varphi_i)_{i \in I}$, such that φ is faithful, $\mathcal{A}_i \subset \mathcal{A}, i \in I$ are freely independent in $(\mathcal{A}, \varphi)$ and $\varphi|_{\mathcal{A}_i} = \varphi_i$. ◆

Again, as in Theorem 3.2.1, this result is to be interpreted in the following way:

(a) Identify the identities/unities $\mathbf{1}_{\mathcal{A}_i}$ of $\mathcal{A}_i$ and declare it to be the unity/identity $\mathbf{1}_{\mathcal{A}}$ (say) of $\mathcal{A}$.

(b) For each $i \in I$, there is a norm-preserving unital $*$-homomorphism $W_i : \mathcal{A}_i \to \mathcal{A}$, such that

1. $\varphi \circ W_i = \varphi_i$, $i \in I$,
2. the unital C^*-subalgebras $(W_i(\mathcal{A}_i))_{i \in I}$ form a free family in $(\mathcal{A}, \varphi)$, and
3. $\cup_{i \in I} W_i(\mathcal{A}_i)$ generates $\mathcal{A}$ as a C^*-algebra.

6.5 Free additive and multiplicative convolution

Theorem 6.4.1 provides a rich class of variables with compactly supported distributions. In particular, it can be used to define free additive and multiplicative convolutions of compactly supported probability laws.

Theorem 6.5.1. Let μ, ν be any two compactly supported probability laws on $\mathbb{R}$. Then

(a) there exists a C^*-algebra with two free self-adjoint variables a and b whose probability laws are $\mu_a = \mu$ and $\mu_b = \nu$;

(b) the probability law μ_{a+b} depends only on the probability laws μ and ν. This law is said to be the *free additive convolution* of the laws μ and ν and is written as $\mu \boxplus \nu$. It satisfies

$$\kappa_n(\mu \boxplus \nu) = \kappa_n(\mu) + \kappa_n(\nu) \text{ for all } n \geq 1. \tag{6.17}$$

(c) If, in addition, μ is supported on the non-negative reals, then there exists a unique compactly supported probability law, which we call the *free multiplicative convolution* of μ and ν, written as $\mu \boxtimes \nu$, that satisfies

$$m_n(\mu \boxtimes \nu) = \sum_{\pi \in NC(n)} \kappa_\pi(\mu) \prod_{V \in K(\pi)} m_{|V|}(\nu),\ n \geq 1. \tag{6.18}$$

◆

Proof. (a) One can construct, for example, as in Example 2.5.2, C^*-probability spaces $(\mathcal{A}_1, \varphi_1)$ and $(\mathcal{A}_2, \varphi_2)$, with self-adjoint elements $a \in \mathcal{A}_1$ and $b \in \mathcal{A}_2$, with respective distributions $\mu_a = \mu$ and $\mu_b = \nu$. By Theorem 6.4.1, we may, without loss of generality, assume that a and b are free elements in the same C^*-probability space $(\mathcal{A}, \varphi)$, say.

(b) Since $a+b \in \mathcal{A}$, and is self-adjoint, by Theorem 6.3.4, it has a probability law μ_{a+b}. Since a and b are free, their free cumulants satisfy (6.17). Clearly, the law μ_{a+b} depends only on the laws μ and ν.

(c) In Part (a) of the above construction, let $\mathcal{A}_1$ be the space of L^∞ functions defined on some classical probability space, where a belongs, and has probability law μ with compact support on the non-negative real line. Thus a is an essentially bounded function and is non-negative almost everywhere. Therefore $\sqrt{a}$ is defined. Once again, we denote the copy of $\sqrt{a}$ in $\mathcal{A}$ by $\sqrt{a}$. Then, $\sqrt{a}b\sqrt{a}$ is a self-adjoint element in $\mathcal{A}$. By an appeal to Theorem 6.3.4, it has a probability law $\mu_{\sqrt{a}b\sqrt{a}}$, which we call $\mu \boxtimes \nu$. That is, for a fixed $n \geq 1$,

$$\begin{aligned}
m_n(\mu \boxtimes \nu) &= \varphi\left((\sqrt{a}b\sqrt{a})^n\right) \\
&= \varphi\left((ab)^n\right) \quad (\varphi \text{ is tracial on the relevant sub-algebra}) \\
&= \sum_{\pi \in NC(n)} \kappa_\pi[a, \ldots, a]\varphi_{K(\pi)}[b, \ldots, b] \quad \text{(see Example 3.9.1)} \\
&= \sum_{\pi \in NC(n)} \kappa_\pi(\mu) \prod_{V \in K(\pi)} m_{|V|}(\nu).
\end{aligned}$$

Thus (6.18) follows. ■

Remark 6.5.1. (a) It is immediate that, for any compactly supported probability laws μ and ν,

$$\mu \boxplus \nu = \nu \boxplus \mu.$$

(b) Furthermore, for compactly supported probability laws $\mu_1, \ldots, \mu_n$, their free additive convolution can be defined as the unique compact supported probability law $\mu_1 \boxplus \cdots \boxplus \mu_n$ that satisfies

$$\kappa_j(\mu_1 \boxplus \cdots \boxplus \mu_n) = \sum_{i=1}^{n} \kappa_j(\mu_i), \ j \geq 1.$$

(c) As in Example 3.9.1, one can show that

$$m_n(\mu \boxtimes \nu) = \sum_{\pi \in NC(n)} \kappa_\pi(\nu) \prod_{V \in K(\pi)} m_{|V|}(\mu), \ n \geq 1. \tag{6.19}$$

Therefore $\mu \boxtimes \nu$ and $\nu \boxtimes \mu$ are the same, and both are defined whenever *any one* of μ and ν is supported on the non-negative reals. ●

Example 6.5.1. Let μ be the standard semi-circular law, and let ν be the Bernoulli(p) law, that is, the probability law which puts mass p and $1-p$ on 1 and 0, respectively. Let us identify the probability law $\mu \boxtimes \nu$. Recall that, for all $n \geq 1$ and $\pi \in NC(n)$,

$$\kappa_\pi(\mu) = \begin{cases} 1 & \text{if } n \text{ is even and } \pi \in NC_2(n), \\ 0 & \text{otherwise.} \end{cases}$$

Suppose n is even and $\pi \in NC_2(n)$ so that $|\pi| = n/2$. Since $|\pi|+|K(\pi)| = n+1$, $K(\pi)$ has exactly $n/2+1$ blocks, and hence (6.18) implies that

$$m_n(\mu \boxtimes \nu) = \begin{cases} p^{n/2+1}\#NC_2(n) & \text{if } n \text{ is even,} \\ 0 & \text{otherwise.} \end{cases} \tag{6.20}$$

But these are the moments of $\sqrt{p}XY$, where X and Y are classical independent random variables following Bernoulli(p) and the standard semi-circular laws respectively. Thus $\mu \boxtimes \nu$ is the probability law of $\sqrt{p}XY$. ▲

Example 6.5.2. (Free binomial and free Poisson laws)

(a) In Example 3.3.1 of Chapter 3, we had defined a free binomial variable as a sum $a_1 + \cdots + a_n$ of self-adjoint Bernoulli variables which are freely independent. We can now show that its moments define a valid probability law.

Let μ be the classical Bernoulli law which puts mass p and $1-p$ at 1 and 0 respectively. Since μ is trivially compactly supported, by Theorem 6.5.1 (b), its n-fold free additive convolution $\mu^{\boxplus n}$ is a probability law. This is the free binomial law, and the moments of $a_1 + \cdots + a_n$ are the same as the moments of $\mu^{\boxplus n}$.

(b) Now consider the free binomial probability laws where $p = p_n$, and $np_n \to \lambda$ as $n \to \infty$. In Theorem 4.3.1 (b), we have seen the convergence of this law to the free Poisson law with parameter λ, except that we had not rigorously established the existence of the free binomial law. Now the picture is complete.

(c) In Theorem 4.3.2 we had proved the existence of the free compound Poisson law as a limit law, where the free binomial laws were involved. This picture is also now complete. ▲

6.6 Exercises

1. Prove Lemma 6.2.8.
2. Prove Lemma 6.3.2.
3. In Example 6.3.1, check that the distribution of every f is $P \circ f^{-1}$.
4. Prove that in Example 6.3.3, the distribution of every self-adjoint a is $\frac{1}{d}\sum_{i=1}^{d} P \circ \lambda_i^{-1}$.
5. Show that $(\mathcal{A}, \varphi)$ considered in Example 6.3.3 is a C^*-probability space when we consider the operator norm.
6. Prove equation (6.19).
7. Complete the proofs of the steps leading to (6.20).

8. Let $\mathcal{A}$ be a unital C^*-algebra. Let $\varphi : A \to \mathbb{C}$ be a unital linear functional such that $|\varphi(a)| \leq \|a\|$, for all $a \in \mathcal{A}$. Then show that φ is positive, and hence $(\mathcal{A}, \varphi)$ is a C^*-probability space.

9. Suppose c is a standard circular variable. Find the probability law of cc^*. The probability law of $\sqrt{cc^*}$ is known as the *quarter-circle law*.

10. Suppose e is an elliptic variable. Find the probability law of ee^* and compare it with the probability law of cc^* obtained above.

7

Random matrices

A matrix whose elements are random variables is a *random matrix*. Large dimensional random matrices have found application in various scientific disciplines. In this chapter, we introduce a few important random matrices and the notions of empirical and limiting spectral distribution (LSD). We provide a quick and unified treatment of patterned random matrices of growing dimension, based on the key concepts of *input sequence*, *link function*, *circuits* and *words* and provide a general tightness result.

There are several excellent books on random matrices, written from different theoretical and applied perspectives and at various levels of difficulty. Some of them are referred to at appropriate places in later chapters. Additional books that the reader may find useful and interesting are Girko (2001)[44], Girko (2018)[45], Mehta (2004) [71], Guionnet (2009)[53], Anderson, Guionnet and Zeitouni (2010)[1], Forrester (2010)[41], Baik, Akerman and Di Francesco(2011)[8], Pastur and Shcherbina (2011)[76], Tao (2012)[97] and Livan, Novaes and Vivo (2018)[64].

7.1 Empirical spectral measure

A primary object of interest of a matrix is its spectral distribution. Suppose A_n is any $n \times n$ matrix with eigenvalues $\lambda_1, \lambda_2, ..., \lambda_n$.

Definition 7.1.1. (Empirical spectral measure/distribution) The probability law which puts mass n^{-1} on each eigenvalue λ_i is called the *empirical spectral measure* of A_n. The corresponding distribution function is called the *empirical spectral distribution function* (ESD). We shall refer to both as ESD. ◊

Since there can be complex eigenvalues, the empirical spectral measure of A_n is defined on $\mathbb{C}$. Equivalently, its ESD can be defined on $\mathbb{R}^2$ as

$$F_{A_n}(x, y) := n^{-1} \sum_{i=1}^{n} \mathbb{I}\{\mathcal{R}(\lambda_i) \leq x,\ \mathcal{I}(\lambda_i) \leq y\}.$$

If the eigenvalues are all real, then we may drop the complex argument and define the ESD of A_n on $\mathbb{R}$ by

DOI: 10.1201/9781003144496-7

$$F_{A_n}(x) := n^{-1} \sum_{i=1}^{n} \mathbb{I}\{\lambda_i \leq x\}.$$

If the entries of A_n are random then the empirical spectral measure is a random probability law. In this case, there is yet another probability measure connected to the eigenvalues.

Definition 7.1.2. (Expected empirical spectral distribution) The *expected empirical spectral distribution function* (EESD) of A_n is defined as

$$\mathrm{E}(F_{A_n}(x,y)) := \frac{1}{n} \sum_{i=1}^{n} \mathrm{P}\big[\mathcal{R}(\lambda_i) \leq x,\ \mathcal{I}(\lambda_i) \leq y\}\big].$$

We shall write $\mathrm{E}(F_{A_n})$ in short. The corresponding probability law is called the *expected empirical spectral measure.* We shall refer to both as EESD. ◇

Note that for a non-random matrix, the ESD and the EESD are identical.

7.2 Limiting spectral measure

Our interest is in random matrices whose dimension grows to ∞. In this case, it is natural to inquire if the sequence of empirical spectral measures converges weakly. We have the following definition.

Definition 7.2.1. (Limiting spectral distribution (LSD)) For any sequence of square random matrices $\{A_n\}$, its *limiting spectral distribution (or measure)* is defined as the weak/distributional limit of the sequence $\{\mathrm{E}(F_{A_n})\}$, if it exists, as $n \to \infty$. ◇

Other natural but stronger notions of weak convergence of the spectral distribution are given in Definition 7.2.2.

Suppose the random variable entries of $\{A_n\}$ are defined on the probability space $(\Omega, \mathcal{F}, \mathrm{P})$. Then $\{F_{A_n}(\cdot)\}$ (on $\mathbb{R}$ or on $\mathbb{R}^2$ as the case may be) are random functions (of $\omega \in \Omega$), but we suppress this dependence. Let F be a *non-random* distribution function, either on $\mathbb{R}$ or on $\mathbb{R}^2$ as the case may be. Let

$$C_F := \{t : t \text{ is a continuity point of } F\}.$$

Definition 7.2.2. (a) Say that the ESD of A_n converges to F *almost surely* if, for almost every $\omega \in \Omega$ and for all $t \in C_F$,

$$F_{A_n}(t) \to F(t) \quad \text{as } n \to \infty.$$

(b) Say that the ESD of A_n converges to F *in probability* if for every $t \in C_F$,

$$F_{A_n}(t) \to F(t) \quad \text{in probability.}$$

◇

It is easy to see that (a) $\Rightarrow$ (b) which in turn implies that $\mathrm{E}(F_{A_n})$ converges weakly to F. In all cases, we refer to the limit as the LSD of $\{A_n\}$ and the corresponding probability law as the limiting spectral measure.

LSD of various sequences of random matrices occupy a central place in the literature on random matrix theory (RMT). In this book, we shall mostly deal with the weak convergence of the EESD of real symmetric or Hermitian matrices as given in Definition 7.2.1. The study of the LSD of non-Hermitian matrices is extremely difficult and fewer results are known for random non-Hermitian matrix sequences compared to the real symmetric or Hermitian matrices.

7.3 Moment and trace

Two common tools to establish the LSD of random matrices are the *moment method* and the *method of Stieltjes transform.* The moment method is less sophisticated but shall serve our purpose. The moments of the ESD are connected to the trace of powers of the matrix. This ties up well with non-commutative probability. Consider the ESD F_{A_n} of A_n which is real symmetric. It's h-th moment is given by

$$m_h(F_{A_n}) = \frac{1}{n}\sum_{i=1}^{n} \lambda_i^h = \frac{1}{n}\mathrm{Tr}(A_n^h) = \mathrm{tr}(A_n^h) =: m_h(A_n), \tag{7.1}$$

where Tr denotes the trace of a matrix. This is known as the *trace-moment* formula. This implies that the moments of the EESD, say $m_h(\mathrm{E}(F_{A_n}))$, are given by

$$m_h(\mathrm{E}(F_{A_n})) = \mathrm{E}\left[\frac{1}{n}\mathrm{Tr}(A_n^h)\right] =: \mathrm{E}[m_h(A_n)].$$

Thus Lemma 1.2.1 can be invoked. Consider the following conditions:

(M1) For every $h \geq 1$, $\mathrm{E}[m_h(A_n)] \to m_h$.

(U) The moment sequence $\{m_h\}$ corresponds to a unique probability law.

The following lemma follows easily from Lemma 1.2.1. We omit its proof.

Lemma 7.3.1. Suppose, for a sequence of real symmetric matrices $\{A_n\}$, (M1) and (U) hold. Then the EESD of A_n converges in distribution to F determined by $\{m_h\}$. ♦

The moment $\mathrm{E}[m_h(A_n)]$ involves computing the terms in the following expansion (below $x_{i,j,n}$ denotes the (i,j)-th entry of A_n):

$$\mathrm{E}[\mathrm{tr}(A_n^h)] = \frac{1}{n}\sum_{1\leq i_1,i_2,\ldots,i_h\leq n} \mathrm{E}[x_{i_1,i_2,n}x_{i_2,i_3,n}\cdots x_{i_{h-1},i_h,n}x_{i_h,i_1,n}].$$

Conceptually the method is straightforward but it requires all moments to be finite. In particular cases, the combinatorial arguments involved may become quite unwieldy as h and n increase. However, this approach has been successfully used to obtain the LSD of several important real symmetric random matrices—not only for the EESD but also for the ESD in probability or almost surely by checking additional requirements. The moment conditions are then reduced by resorting to truncation arguments when only some moments are finite. See Bose (2018)[20] for a detailed description of the approach and its application to several random matrices.

There is also an issue of scaling the matrix so that we get a non-trivial LSD. Suppose the entries $\{x_{i,j,n}\}$ of A_n have mean zero and variance 1. Then

$$m_1(A_n) = \frac{1}{n}\operatorname{Tr}(A_n) \quad \text{and} \quad \mathrm{E}[m_1(A_n)] = 0,$$

$$m_2(A_n) = \operatorname{tr}\left({A_n}^2\right) = \frac{1}{n}\sum_{i,j=1}^{n} x_{i,j,n}^2 \quad \text{and} \quad \mathrm{E}[m_2(A_n)] = n.$$

Hence, for stability of the second moment, the appropriate scaled matrix to consider is $n^{-1/2}A_n$.

7.4 Some important matrices

We now describe a few important random matrices that have appeared in the literature. Of these, the Wigner matrix, the IID matrix, the elliptic matrix, the sample covariance matrix and the cross-covariance matrix will have a special place in relation to asymptotic freeness.

Wigner matrix. Wigner (1955)[105] introduced a symmetric random matrix while discussing statistical models for heavy nuclei atoms in physics. Voiculescu (1991)[100] discovered a deep connection of this matrix with free independence.

The symmetric (unscaled) Wigner matrix W_n is defined as

$$W_n := \begin{bmatrix} x_{1,1,n} & x_{1,2,n} & x_{1,3,n} & \cdots & x_{1,n-1,n} & x_{1,n,n} \\ x_{1,2,n} & x_{2,2,n} & x_{2,3,n} & \cdots & x_{2,n-1,n} & x_{2,n,n} \\ & & & \vdots & & \\ x_{1,n,n} & x_{2,n,n} & x_{3,n,n} & \cdots & x_{n-1,n,n} & x_{n,n,n} \end{bmatrix}.$$

Thus, in general, the entries are from a *triangular array*. However, for convenience, we drop the last suffix n and write $W_n = ((x_{i,j}))$. *We adopt this convention for all other matrices to follow.* Figure 7.1 shows the simulated ESD of $n^{-1/2}W_n$ when $\{x_{i,j}\}$ are i.i.d. standard normal or symmetric Bernoulli.

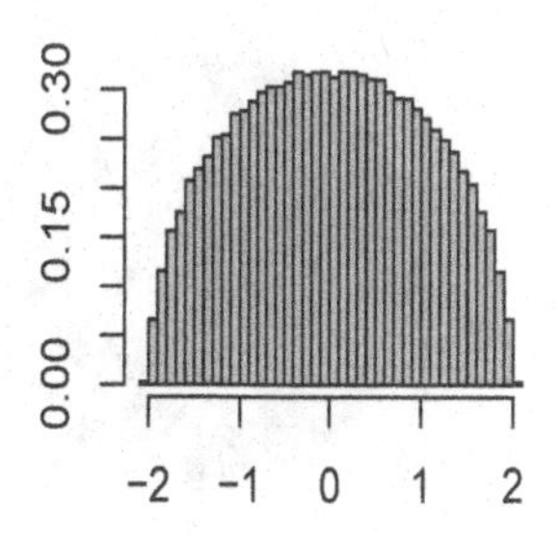

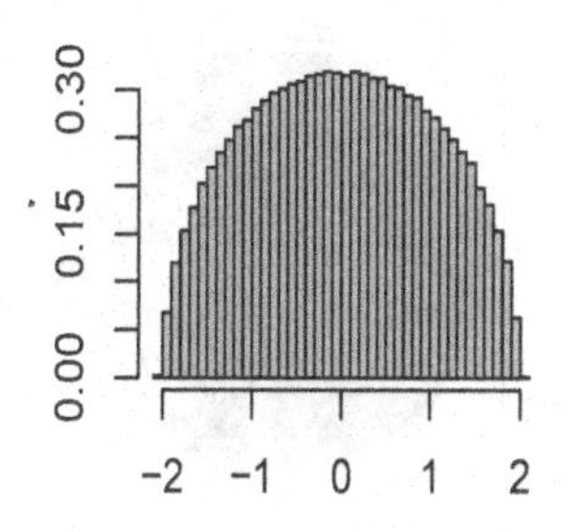

FIGURE 7.1
ESD of $n^{-1/2}W_n$. Left: standard normal $n = 800$; right: standardized symmetric Bernoulli $n = 1000$.

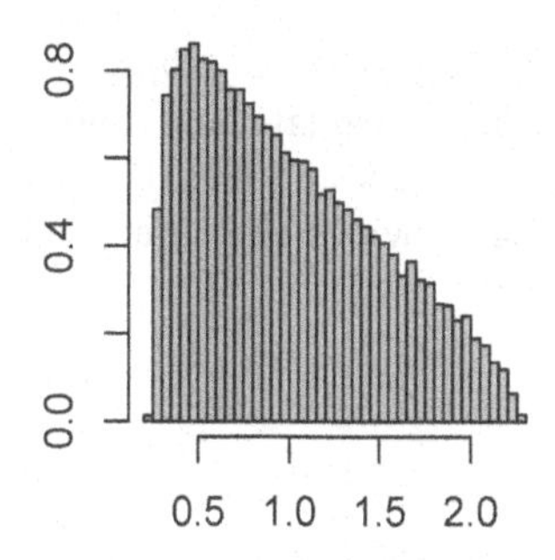

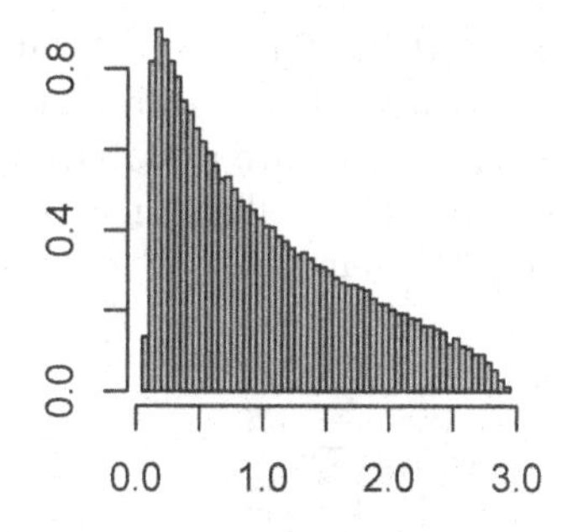

FIGURE 7.2
ESD of the S-matrix. Entries of X are i.i.d. $\mathrm{N}(0, 1)$, $n = 1000$. Left: $p/n = 0.25$; right: $p/n = 0.5$.

Sample covariance matrix. Suppose X_i are p-dimensional (column) vectors, $i = 1, \ldots, n$. Consider the $p \times n$ matrix $X = (X_1 \ \ldots \ X_n)$. The unadjusted covariance matrix is defined as $S := \frac{1}{n}XX^*$. In particular, if $p = n$ and the columns are i.i.d. Gaussian, then it is called a Wishart matrix, after Wishart (1928)[107]. In random matrix theory, it is often referred to as the S-matrix. Figure 7.2 provides eigenvalue distributions for some simulated S-matrices.

IID matrix. The matrix $X_{p \times n}$ where all entries are i.i.d., will be called the *IID matrix.* When $p = n$, X is not symmetric, and the eigenvalues may of course be complex. Figure 7.3 provides the simulated distribution of its eigenvalues.

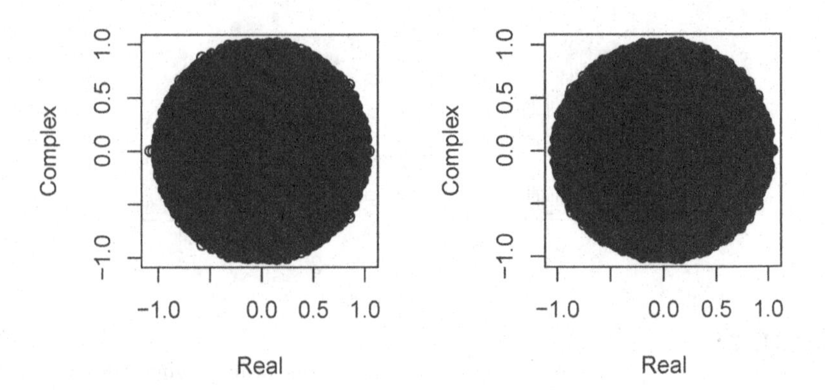

FIGURE 7.3
ESD of the scaled IID matrix. Left: standard normal, $p = n = 800$; right: symmetric Bernoulli, $p = n = 1000$.

Elliptic matrix. The $n \times n$ matrix $E_n = ((x_{i,j}))$ is an *elliptic matrix* if $\{(x_{i,j}, x_{j,i})\}$ $i \neq j$, $i, j \geq 1$ are i.i.d. with common correlation ρ. The diagonal entries $\{x_{i,i}\}$, $i \geq 1$ are also i.i.d. The Wigner matrix and the IID matrix are special cases of the elliptic matrix. Figure 7.4 provides the simulated distribution of its eigenvalues.

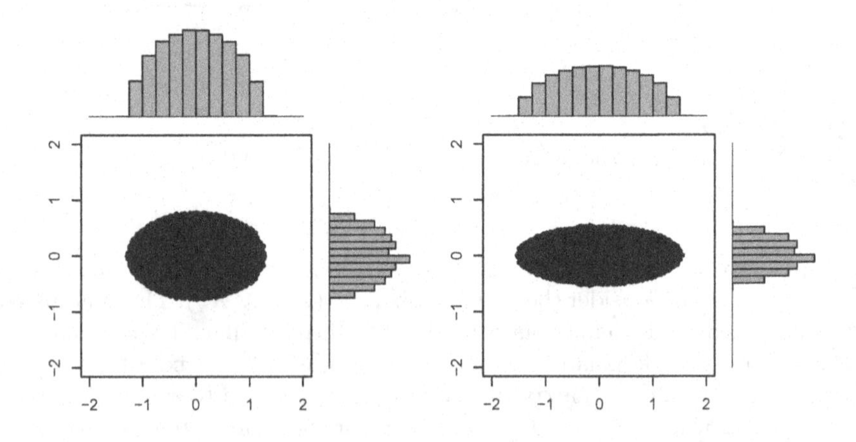

FIGURE 7.4
ESD of $n^{-1/2}E_n$, normal entries with mean 0 and variance 1 and $n = 1000$. Left: $\rho = 0.25$; right: $\rho = 0.50$. Marginals are also depicted.

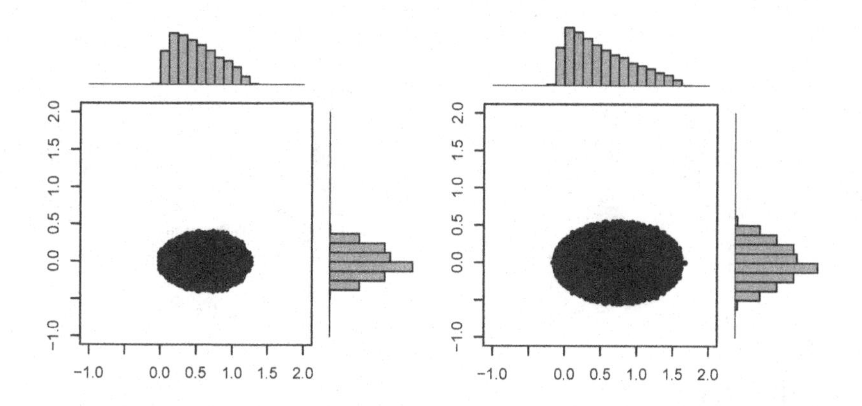

FIGURE 7.5
ESD of $C_n = n^{-1}X_nY_n^*$, normal entries, $\rho = 0.5, n = 1000$. Left: $p = 250$; right: $p = 500$. Marginals are also depicted.

Cross-covariance matrix. Suppose $X_n = ((x_{ij,n}))_{p\times n}$ and $Y_n = ((y_{ij,n}))_{p\times n}$ are random matrices where the entries have mean 0 and variance 1. Then the matrix $C_n := n^{-1}X_nY_n^*$ is called the *cross-covariance matrix*. Note that the sample covariance matrix is a special case when we take $X_n = Y_n$. We consider a special case of this matrix where we assume that $\{(x_{ij,n}, y_{ij,n}) : 1 \leq i \leq p, 1 \leq j \leq n\}$ are independent real-valued bivariate random variables with correlation ρ. Figure 7.5 provides the simulated distribution of its eigenvalues.

Toeplitz matrix. Any matrix of the form $((t_{i-j}))_{1\leq i,j\leq n}$ is called a *Toeplitz* matrix. We shall consider only the symmetric Toeplitz, so that $t_k = t_{-k}$ for all k. The Toeplitz pattern appears in many different places. For example, the autocovariance matrix of a stationary time series has the Toeplitz pattern.

When the dimension of the matrix is ∞ and $\sum_{k=-\infty}^{\infty} |t_k|^2 < \infty$, then it gives rise to the famous Toeplitz linear operator on the space of square summable sequences in the natural way. Under this assumption, one of the famous theorems of Szegő (see Grenander and Szegő (1984)[51]) says that the LSD of the Toeplitz matrix is the distribution of $f(U)$ where the random variable U has the uniform distribution on $[0, 1]$ and the function f is defined as

$$f(x) := \sum_{k=-\infty}^{\infty} t_k e^{2\pi\iota kx}, \; 0 \leq x < 1.$$

Note that f is defined in the L^2 sense and is real valued due to the symmetry condition. Figure 7.6 provides a simulation of the ESD of the symmetric Toeplitz matrices with entries $t_0 = 0$, and $t_k = k^{-1}$ and k^{-2}, $k \geq 1$.

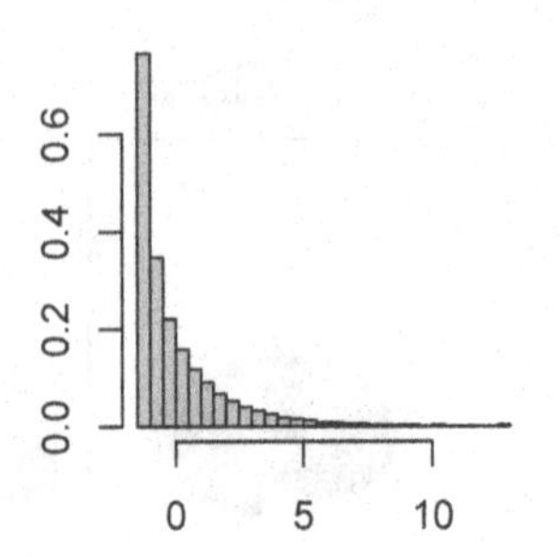

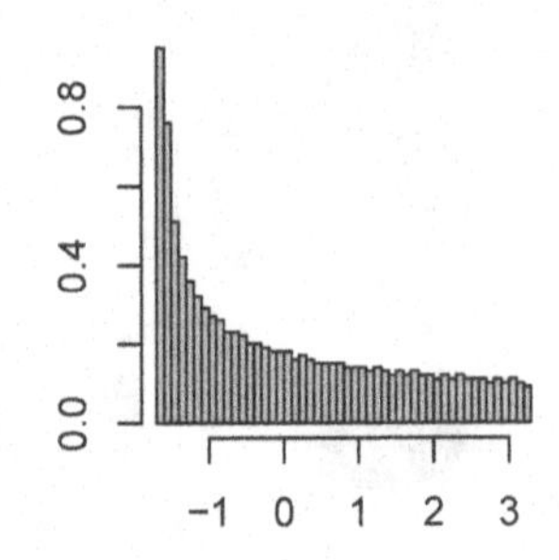

FIGURE 7.6
Non-random Toeplitz matrices. Left: $t_0 = 0$, $t_k = k^{-1}$; right: $t_0 = 0, t_k = k^{-2}$.

The random symmetric Toeplitz matrix is defined as $((x_{|i-j|}))_{1 \leq i,j \leq n}$ where $\{x_{i,n}\}$ are independent random variables with mean 0 and variance 1. Figure 7.7 gives a simulated ESD of $n^{-1/2}T_n$.

$$T_n := \begin{bmatrix} x_{0,n} & x_{1,n} & x_{2,n} & \cdots & x_{n-2,n} & x_{n-1,n} \\ x_{1,n} & x_{0,n} & x_{1,n} & \cdots & x_{n-3,n} & x_{n-2,n} \\ x_{2,n} & x_{1,n} & x_{0,n} & \cdots & x_{n-4,n} & x_{n-3,n} \\ & & & \vdots & & \\ x_{n-2,n} & x_{n-3,n} & x_{n-4,n} & \cdots & x_{0,n} & x_{1,n} \\ x_{n-1,n} & x_{n-2,n} & x_{n-3,n} & \cdots & x_{1,n} & x_{0,n} \end{bmatrix},$$

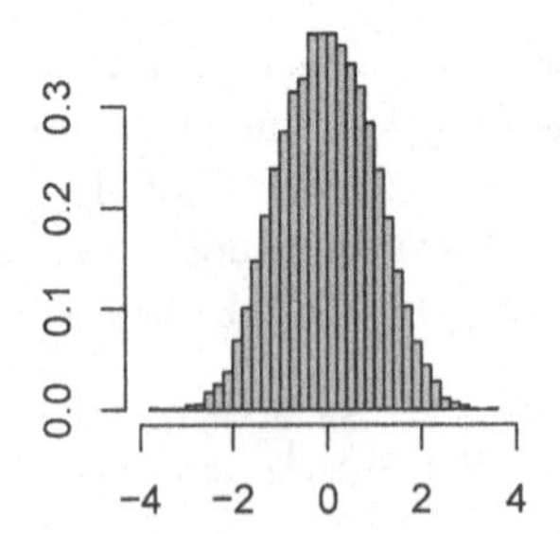

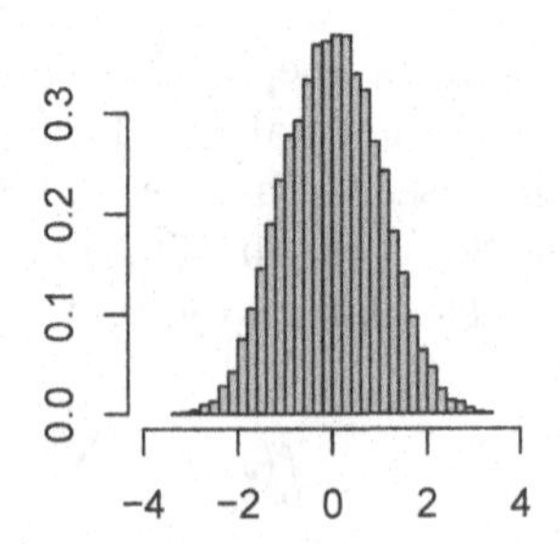

FIGURE 7.7
ESD of $n^{-1/2}T_n$ with independent entries and 25 replications. Left: standard normal, $n = 800$; right: standardized symmetric Bernoulli, $n = 1000$.

Hankel matrix. The symmetric matrix $((t_{i+j-2}))_{1\leq i,j\leq n}$ is called a Hankel matrix. Under the square summability assumption, its infinite dimensional version again gives rise to a linear operator as above. Under this assumption, it also has an LSD. We shall use the following form of the symmetric random Hankel matrix where it is assumed that $\{x_{i,n}\}$ are independent with mean 0 and variance 1. Figure 7.8 provides the simulated ESD of $n^{-1/2}H_n$.

$$H_n := \begin{bmatrix} x_{0,n} & x_{1,n} & x_{2,n} & \cdots & x_{n-3,n} & x_{n-2,n} \\ x_{1,n} & x_{2,n} & x_{3,n} & \cdots & x_{n-2,n} & x_{n-1,n} \\ x_{2,n} & x_{3,n} & x_{4,n} & \cdots & x_{n-1,n} & x_{n,n} \\ & & & \vdots & & \\ x_{n-2,n} & x_{n-1,n} & x_{n,n} & \cdots & x_{2n-4,n} & x_{2n-3,n} \\ x_{n-1,n} & x_{n,n} & x_{n+1,n} & \cdots & x_{2n-3,n} & x_{2n-2,n} \end{bmatrix}.$$

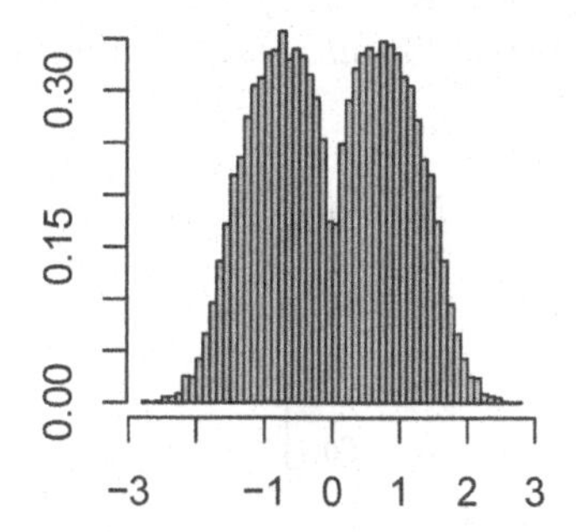

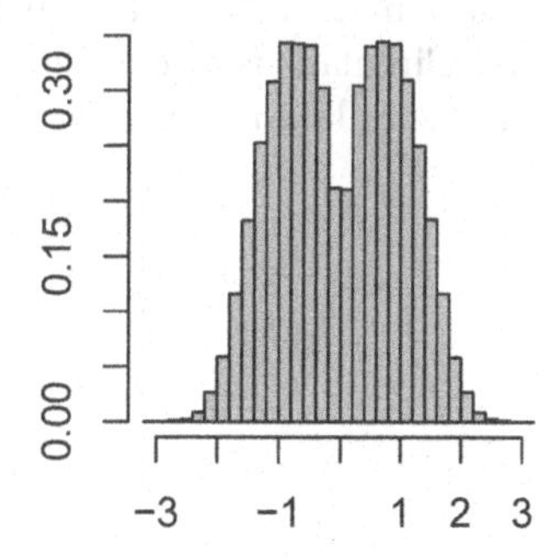

FIGURE 7.8
ESD of $n^{-1/2}H_n$, 25 replications. Left: standard normal, $n = 800$; right: standardized symmetric Bernoulli, $n = 1000$.

Reverse Circulant matrix. This matrix is defined as $((x_{(i+j-2) \bmod n}))$ where $\{x_{i,n}\}$ are independent random variables with mean 0 and variance 1. It is a circulant matrix when we use a left-circular shift instead of the right-circular shift that is used to define the usual circulant matrix. Note that it is a symmetric matrix. Figure 7.9 exhibits the simulated ESD of $n^{-1/2}RC_n$.

$$RC_n := \begin{bmatrix} x_0 & x_1 & x_2 & \cdots & x_{n-2} & x_{n-1} \\ x_1 & x_2 & x_3 & \cdots & x_{n-1} & x_0 \\ x_2 & x_3 & x_4 & \cdots & x_0 & x_1 \\ & & & \vdots & & \\ x_{n-1} & x_0 & x_1 & \cdots & x_{n-3} & x_{n-2} \end{bmatrix}.$$

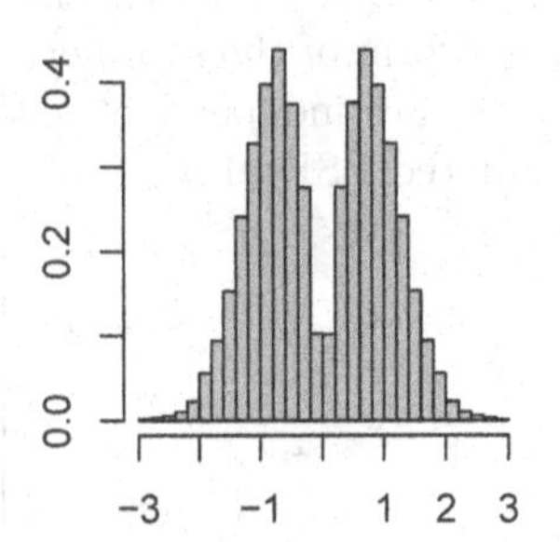

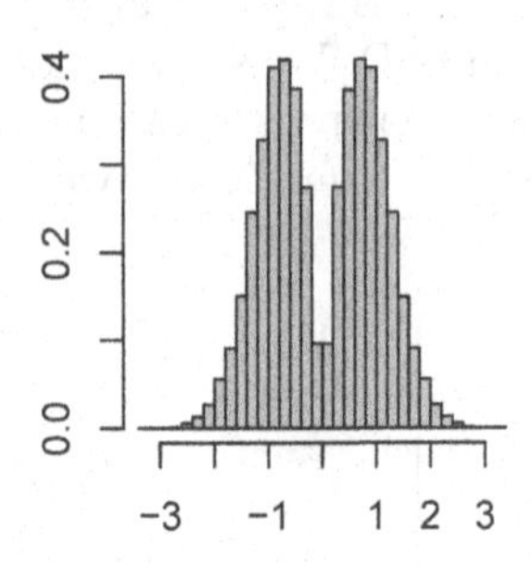

FIGURE 7.9
ESD of $n^{-1/2}RC_n$, 25 replications. Left: standard normal, $n = 800$; right: standardized symmetric Bernoulli, $n = 1000$.

Symmetric Circulant matrix. Consider the usual Circulant matrix. If we impose the restriction of symmetry on this matrix, we obtain the *Symmetric Circulant* matrix SC_n. It can be expressed as $((x_{n/2-|n/2-|i-j||}))$. The Symmetric Circulant is also a *Doubly Symmetric Toeplitz matrix.* Figure 7.10 exhibits the simulated ESD of $n^{-1/2}SC_n$.

$$SC_n := \begin{bmatrix} x_0 & x_1 & x_2 & \dots & x_2 & x_1 \\ x_1 & x_0 & x_1 & \dots & x_3 & x_2 \\ x_2 & x_1 & x_0 & \dots & x_2 & x_3 \\ & & & \vdots & & \\ x_1 & x_2 & x_3 & \dots & x_1 & x_0 \end{bmatrix}.$$

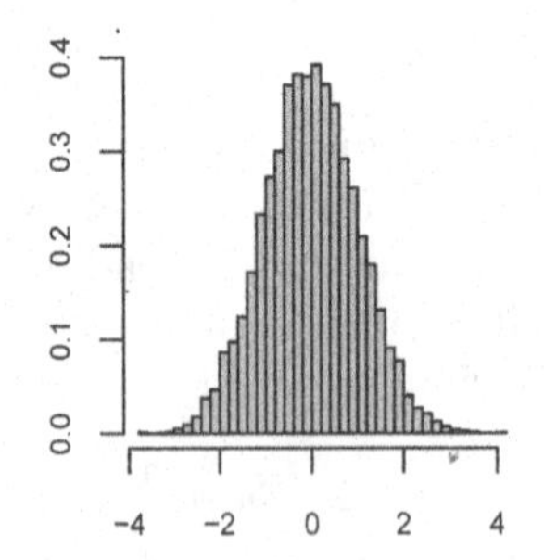

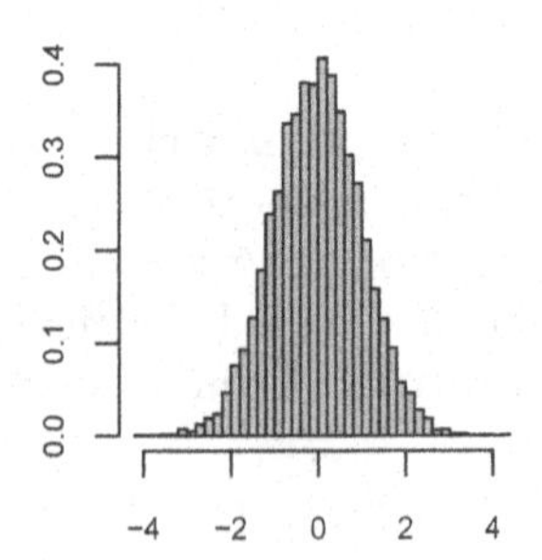

FIGURE 7.10
ESD of $n^{-1/2}SC_n$, 25 replications. Left: standard normal, $n = 800$; right: standardized symmetric Bernoulli, $n = 1000$.

7.5 A unified treatment

The symmetric matrices given above can be treated in a sort of unified way due to some of the common features they have.

Input and link. The above matrices are constructed from a sequence or a bi-sequence of variables $\{x_{i,n}\}$ or $\{x_{i,j,n}\}$, which we call an *input sequence*, as follows: Let $\mathbb{Z}$ be the set of all integers and let $\mathbb{Z}_+$ denote the set of all non-negative integers. Let

$$L_n : \{1, 2, \ldots, n\}^2 \to \mathbb{Z}^d,\ n \geq 1,\ d = 1 \text{ or } 2 \tag{7.2}$$

be a sequence of functions. We shall write $L_n = L$ and call it the *link* function. By abuse of notation, we write $\mathbb{Z}_+^2$ as the common domain of $\{L_n\}$. Let

$$A_n := ((x_{L(i,j)})). \tag{7.3}$$

The S matrix can be written in the form $n^{-1}A_nA_n^*$.

The link functions of the above matrices are as follows (suppressing the subscript n):

(a) Wigner matrix: $L(i,j) = (\max(i,j), \min(i,j))$.

(b) IID and elliptic matrix: $L(i,j) = (i,j)$.

(c) Toepliz matrix: $L(i,j) = |i-j|$.

(d) Hankel matrix: $L(i,j) = i+j-2$.

(e) Reverse Circulant matrix: $L(i,j) = i+j-2 \mod n,\ 1 \leq i,j \leq n$.

(f) Symmetric Circulant matrix: $L(i,j) = n/2 - |n/2 - |i-j||,\ 1 \leq i,j \leq n$.

(g) The Sample covariance matrix does not fall into the above set up. Nevertheless, it can be expressed via two link functions, $L^{(1)}(i,j) = (i,j)$ and $L^{(2)}(i,j) = (j,i)$.

Let us first focus on the symmetric link functions, that is, where $L(i,j) = L(j,i)$.

Repetitions. Define

$$\Delta(L) := \sup_n \sup_{t \in \mathbb{Z}_+^d} \sup_{1 \leq k \leq n} \#\{l : 1 \leq l \leq n,\ L(k,l) = t\}. \tag{7.4}$$

Thus $\Delta(L)$ is the maximum number of times any specific variable is repeated in a row or a column of the matrix. Clearly $\Delta(L) = 1$ for the Wigner and Hankel matrices and is 2 for the Toeplitz matrices. It will be crucial that $\Delta(L) < \infty$ for all the matrices discussed so far. We shall write Δ for $\Delta(L)$ if the link function is clear from the context.

Trace-moment formula. Using (7.1), the h-th moment of $F_{n^{-1/2}A_n}$ is given by the following *trace-moment* formula.

$$\frac{1}{n}\operatorname{Tr}\left(\frac{A_n}{\sqrt{n}}\right)^h = \frac{1}{n^{1+h/2}} \sum_{1\le i_1,\ldots,i_h\le n} x_{L(i_1,i_2)}x_{L(i_2,i_3)}\cdots x_{L(i_{h-1},i_h)}x_{L(i_h,i_1)}. \quad (7.5)$$

Circuit. For fixed h, n, any function $\pi : \{0,1,2,\ldots,h\} \to \{1,2,\ldots,n\}$ with $\pi(0) = \pi(h)$ will be called a *circuit* of *length* h. Condition (M1) of Section 7.3 can be written as

$$\mathrm{E}[m_h(n^{-1/2}A_n)] = \mathrm{E}[\frac{1}{n}\operatorname{Tr}\left(\frac{A_n}{\sqrt{n}}\right)^h] = \frac{1}{n^{1+h/2}} \sum_{\pi:\ \pi \text{ circuit}} \mathrm{E}\,\mathrm{X}_\pi \to m_h, \quad (7.6)$$

where

$$\mathrm{X}_\pi := x_{L(\pi(0),\pi(1))}x_{L(\pi(1),\pi(2))}\cdots x_{L(\pi(h-2),\pi(h-1))}x_{L(\pi(h-1),\pi(h))}. \quad (7.7)$$

A circuit π is *matched* if each value $L(\pi(i-1),\pi(i))$, $1 \le i \le h$, is repeated. If π is non-matched, then $\mathrm{E}(\mathrm{X}_\pi) = 0$. If each value is repeated *exactly twice* (so h is necessarily even), then π is *pair-matched*, and $\mathrm{E}(\mathrm{X}_\pi) = 1$. We shall see that the entire contribution from non-pair matched circuits is negligible. Hence the LSD is determined by the *counts* of different types of pair-matched circuits. This, of course, varies from matrix to matrix and depends on the link function.

Words. To count circuits efficiently, we group them into *equivalence classes* via the following equivalence relation: $\pi_1 \sim \pi_2$ if and only if their L values match at the *same* locations. That is, for all i, j,

$$L(\pi_1(i-1),\pi_1(i)) = L(\pi_1(j-1),\pi_1(j)) \Leftrightarrow L(\pi_2(i-1),\pi_2(i)) = L(\pi_2(j-1),\pi(j)). \quad (7.8)$$

Any equivalence class can be indexed by a *partition* of $\{1,2,\ldots,h\}$. Each *partition block* identifies the positions of the L-matches. We label these partitions by *words* w of length h of letters where the first occurrence of each letter is in alphabetical order.

For example, if $h = 5$, then the partition $\{\{1,4\},\{2,3,5\}\}$ is represented by the word *abbab*. This identifies all circuits π for which $L(\pi(0),\pi(1)) = L(\pi(3),\pi(4))$ and $L(\pi(1),\pi(2)) = L(\pi(2),\pi(3)) = L(\pi(4),\pi(5))$.
The equivalence class corresponding to w will be denoted by

$$\Pi(w) := \{\pi : w[i] = w[j] \Leftrightarrow L(\pi(i-1),\pi(i)) = L(\pi(j-1),\pi(j))\},$$

where $w[i]$ denotes the i-th entry of w.

Let $|w|$ denote the number of distinct letters in w. It is the same as the number of partition blocks in each $\pi \in \Pi(w)$. Clearly

$$|w| = \#\{L(\pi(i-1),\pi(i)) : 1 \le i \le h\}.$$

Note that, for any fixed h, as $n \to \infty$, the number of words remains bounded but the number of circuits in any given $\Pi(w)$ may grow indefinitely. The notions introduced for circuits carry over to words in an obvious way. For instance, a word is *pair-matched* if every letter appears exactly twice. Let

$$\mathcal{P}_2(2k) = \{w : w \text{ is a pair-matched word of length } 2k\}. \tag{7.9}$$

Vertex. Any i (or $\pi(i)$ by abuse of notation) will be called a *vertex*. It is *generating*, if either $i = 0$ or $w[i]$ is the *first* occurrence of a letter. Otherwise, it is called *non-generating*. For example, if $w = ababcab$, then only $\pi(0)$, $\pi(1)$, $\pi(2)$, $\pi(5)$ are generating. Note that, $\pi(0)$ and $\pi(1)$ are always generating. Clearly, for our matrices, a circuit is completely determined, *up to finitely many choices*, by its generating vertices. The number of generating vertices is $|w| + 1$ and hence

$$\#\Pi(w) = \mathrm{O}(n^{|w|+1}).$$

Reduction in counting. We now show that we can restrict attention to only pair-matched words. Define

$$p(w) = \lim_n \frac{1}{n^{1+k}} \#\Pi(w), \;\; w \in \mathcal{P}_2(2k), \tag{7.10}$$

whenever the limit exists.

Note that $n^{-1/2}A_n$ is an element of the $*$-probability space $(\mathcal{M}_n(\mathbb{C}), \mathrm{E}\,\mathrm{tr})$, where tr denotes average trace:

$$\mathrm{tr}(A) = n^{-1}\,\mathrm{Tr}(A), \;\; A \in \mathcal{M}_n(\mathbb{C}).$$

We now state a general result for convergence of random matrices, covering both the weak convergence of the EESD and the algebraic convergence as elements of the above $*$-probability space. As mentioned earlier, we are not going to discuss in any detail the finer probabilistic notions of convergence of random matrices, either of the ESD or of the algebraic convergence as elements of $(\mathcal{M}_n(\mathbb{C}), \mathrm{tr})$. For a brief discussion on the almost sure convergence results for the ESD of patterned random matrices see Section 8.8. We need the following assumption.

Assumption I For every n, the entries of the random matrices come from the collection of random variables $\{x_{i,n}, x_{i,j,n}\}$ which are independent with mean zero and variance 1. Moreover,

$$\sup_{i,j,n} \mathrm{E}(|x_{i,n}|^k + |x_{i,j,n}|^k) \leq B_k < \infty \;\; \text{for all} \;\; k \geq 1.$$

□

Theorem 7.5.1. Let $\{A_n\}$ be a sequence of random matrices with the link function L that satisfies $\Delta(L) < \infty$ and with an input sequence that satisfies

Assumption I. Suppose $p(w)$ exists for all $w \in \mathcal{P}_2(2k)$. Then we have the following:

(a) The EESD of $n^{-1/2}A_n$ converges weakly to a probability law whose odd moments m_{2k+1} are zero, and the even moments are given by

$$m_{2k} = \sum_{w \in \mathcal{P}_2(2k)} p(w), \ k \geq 1.$$

(b) The self-adjoint variable $n^{-1/2}A_n$ converges as an element of $(\mathcal{M}_n(\mathbb{C}), \varphi_n = n^{-1} \operatorname{E} \operatorname{Tr})$ to an element a in the $*$-algebra $\mathcal{A}_a$ generated by a with the state defined by $\varphi(a^k) = m_k$ for all k. ◆

Proof. (a) Fix a positive integer h. Let

N = number of matched but not pair-matched circuits of length h.

We first show that

$$n^{-(1+h/2)} N \to 0 \text{ as } n \to \infty. \tag{7.11}$$

Fix any word w of length h which is matched but not pair-matched. Since the number of such words is bounded, it is enough to show the result for the number of circuits corresponding to w. Either $h = 2k$ or $h = 2k - 1$ for some k. In both cases, $|w| \leq k - 1$. If we fix the generating vertices, then, counting from left to right, as $\Delta < \infty$, the maximum number of choices for any non-generating vertices is bounded above, by, say, D_h. Hence

$$\#\Pi(w) \leq n D_h n^{k-1} = D_h n^{\lfloor (h+1)/2 \rfloor}.$$

Then (7.11) follows immediately.

First suppose that h is odd. Then, in $\operatorname{E}[m_h(n^{-1/2}A_n)]$, there are only terms which are matched but not pair-matched. Since all moments are finite, there is a common upper bound for all the product moments. Now, because of (7.11), $\operatorname{E}[m_h(n^{-1/2}A_n)]$ converges to 0.

Now suppose $h = 2k$. Since the entries are i.i.d mean zero variance 1,

$$\begin{aligned} \lim \operatorname{E}[m_{2k}(n^{-1/2}A_n)] &= \lim \frac{1}{n^{1+k}} \sum_{\pi \text{ circuit of length } 2k} \operatorname{E} X_\pi \\ &= \sum_{\substack{w \text{ matched} \\ \text{of length } 2k}} \lim \frac{1}{n^{1+k}} \sum_{\pi \in \Pi(w)} \operatorname{E} X_\pi, \end{aligned} \tag{7.12}$$

provided the limits exist. By Hölder's inequality and Assumption I, for some constant F_{2k},

$$\Big| \sum_{\pi:\, \pi \in \Pi(w)} \operatorname{E} X_\pi \Big| \leq \#\Pi(w) F_{2k}.$$

Therefore, again by (7.11), those w which are not pair-matched do not contribute to the limit in (7.12). On the other hand, $\mathrm{E}\, X_\pi = 1$ whenever π is pair-matched. Hence

$$\lim E[m_{2k}(n^{-1/2}A_n)] = \sum_{w \in \mathcal{P}_2(2k)} \lim_n \frac{1}{n^{1+k}} \#\Pi(w) = \sum_{w \in \mathcal{P}_2(2k)} p(w). \quad (7.13)$$

Now recall that, once the generating vertices are fixed, the number of choices for each non-generating vertex is upper bounded by Δ. Hence, for each $w \in \mathcal{P}_2(2k)$,

$$p(w) \le \Delta^k.$$

Since $\#(\mathcal{P}_2(2k)) \le \frac{(2k)!}{k!2^k}$, this implies

$$m_{2k} \le \frac{(2k)!}{k!2^k} \Delta^k \quad (7.14)$$

and hence condition (U) of Section 7.3 holds. An application of Lemma 7.3.1 completes the proof of (a).

(b) During the course of the proof of (a), we have really shown the convergence of all moments of $n^{-1/2}A_n$ in the state $\mathrm{E}\,\mathrm{tr}$. ■

Under the conditions of Theorem 7.5.1, the ESD of $n^{-1/2}A_n$ converges to the same limit also almost surely. This claim is established by refining our counting argument and showing that

$$\sum_{n=1}^{\infty} \mathrm{E}\left[m_h(n^{-1/2}A_n) - \mathrm{E}[m_h(n^{-1/2}A_n)]\right]^4 < \infty \;\text{ for every integer }\; h \ge 1.$$

See Bose (2018)[20] for details. In the next chapter, we shall verify that $p(w)$ exists for all $w \in \mathcal{P}_2(2k)$ for specific choices of matrices and identify the limit law in some cases.

7.6 Exercises

1. Show that, in Definition 7.2.2, $(a) \Rightarrow (b)$.
2. Show that if the ESD converges in the sense of Definition 7.2.2 (a), then the EESD converges, and to the same limit.
3. Construct examples where the EESD converges but none of the above convergences hold.
4. Prove Lemma 7.3.1.

5. Consider the following two conditions.

   (M2) For every $h \geq 1$, $\text{Var}[m_h(A_n)] \to 0$.

   (M4) For every $h \geq 1$, $\sum_{n=1}^{\infty} \text{E}[m_h(A_n) - \text{E}(m_h(A_n))]^4 < \infty$.

   Recall also condition (U) from Section 7.3. Then prove the following two statements.

   (a) If (M1), (M2) and (U) hold, then $\{F_{A_n}\}$ converges weakly in probability to F determined by $\{m_h\}$.

   (b) If (M1), (M4) and (U) hold, then the convergence in (a) is almost sure.

6. All the $n \times n$ random matrices considered below have i.i.d. entries with mean 0, variance 1 and finite fourth moment. Show the following by direct calculations:

   (a) If A_n is the Wigner, Hankel, Toeplitz, Symmetric Circulant or Reverse Circulant matrix, then

$$\text{E}\left[n^{-1}\text{Trace}\left(\frac{A_n}{\sqrt{n}}\right)^3\right] \to 0;$$

   (b) If A_n is the Wigner, Hankel or Reverse Circulant matrix, then

$$\text{E}\left[n^{-1}\text{Trace}\left(\frac{A_n}{\sqrt{n}}\right)^4\right] \to 2;$$

   (c) If T_n is the Toeplitz matrix, then

$$\text{E}\left[n^{-1}\text{Trace}\left(\frac{T_n}{\sqrt{n}}\right)^4\right] \to 8/3;$$

   (d) If SC_n is the Symmetric Circulant matrix, then

$$\text{E}\left[n^{-1}\text{Trace}\left(\frac{SC_n}{\sqrt{n}}\right)^4\right] \to 3.$$

7. If A, B and C are Reverse Circulant matrices of the same order, then show that $ABC = CBA$.

8. If A and B are Symmetric Circulant matrices of the same order, then show that $AB = BA$.

8

Convergence of some important matrices

In Theorem 7.5.1, we have seen that the weak convergence of the EESD hinges on the existence of $p(w)$ for every pair-matched word w. The nature of the LSD depends on the link function. In this chapter, we shall use this result to claim the weak convergence of the EESD for several symmetric random matrices.

In particular, we shall show that the EESD of the Wigner matrix, the S-matrix, the random Toeplitz, the random Hankel, the Reverse Circulant and the Symmetric Circulant matrices converge. In the process we will also show that these matrices converge as elements of the $*$-probability spaces $(\mathcal{M}_n(\mathbb{C}), \mathrm{E}\,\mathrm{tr})$ whenever the entries have uniformly bounded moments of all orders and are independent.

We also discuss two non-Hermitian matrices, namely the IID matrix and the elliptic matrix and establish their convergence in $*$-distribution. Even though their ESD also converge weakly almost surely under appropriate assumptions, we do not cover them since the proofs of these results are long and difficult. Appropriate references are provided for the interested reader.

8.1 Wigner matrix: semi-circular law

We shall see that if the entries have uniformly bounded moments of all orders and are independent, then $n^{-1/2}W_n$ converges as an element of $(\mathcal{M}_n(\mathbb{C}), \mathrm{E}\,\mathrm{tr})$ to a semi-circular variable. Moreover, the EESD converges to the semi-circular law.

It turns out that only the following sub-class of words contributes to the limit moments and hence determines the LSD of the Wigner matrix.

Definition 8.1.1. (Catalan word) Any pair-matched word, that is, any $w \in \mathcal{P}_2(2k)$ will be called a *Catalan word* if there is at least one double letter and successively removing double letters leads to the empty word. ◇

For example, *aabb*, *abccba* are Catalan words but *aabbaa* and *abab* are not.

Let $\mathcal{C}(2k)$ denote the set of all Catalan words of length $2k$. Note that $\mathcal{C}(2k) \subset \mathcal{P}_2(2k)$.

DOI: 10.1201/9781003144496-8

Lemma 8.1.1. The set $\mathcal{C}(2k)$ of Catalan words of length $2k$ is in bijection with $NC_2(2k)$. As a consequence, $\#\mathcal{C}(2k) = \frac{(2k)!}{(k+1)!k!}$. ♦

Proof. Recall Lemma 2.3.1 where we showed that $\#NC_2(2k) = C_k = \frac{(2k)!}{(k+1)!k!}$. Thus it is enough to show a bijection. But this is easily done and is essentially a repetition of the proof of Lemma 2.3.1. We sketch a few lines since it will be useful later when we work with the S-matrix.

For any $w \in \mathcal{C}(2k)$, generate a sequence of ± 1 of length $2k$ as follows. For each letter, mark its first and second occurrences by $+1$ and -1 respectively. For example, *abba* and *abccbdda* are represented respectively by $(1, 1, -1, -1)$ and $(1, 1, 1, -1, -1, 1, -1, -1)$. This yields a sequence $\{u_l\}_{1 \le l \le 2k}$ such that:

$$\begin{aligned} u_l &= \pm 1 \text{ for all } l, \\ \sum_{l=1}^{j} u_l &\ge 0 \text{ for all } j \ge 1, \text{ and} \\ \sum_{l=1}^{2k} u_l &= 0. \end{aligned}$$

Conversely, any such sequence yields a Catalan word—the last pair of consecutive $+1, -1$ gives the last pair of identical letters. Then drop this pair and repeat to get the second last pair of letters and continue to get the entire word. This correspondence is a bijection.

But the total number of such sequences was already shown to be equal to the Catalan number C_k in the proof of Lemma 2.3.1. ■

Recall the following assumption from Chapter 7.

Assumption I For every n, $\{x_{i,n}, x_{i,j,n}\}$ are independent with mean zero and variance 1. Moreover,

$$\sup_{i,j,n} \mathrm{E}(|x_{i,j,n}|^k) \le B_k < \infty \text{ for all } k \ge 1.$$

□

The following result deals with a single sequence of Wigner matrices. The joint convergence of several sequences of independent Wigner matrices is covered in Section 9.2.

Theorem 8.1.2. Let W_n be the $n \times n$ Wigner matrix with the entries $\{x_{i,j,n} : 1 \le i \le j,\ j \ge 1, n \ge 1\}$ which satisfy Assumption I. Then:

(a) the EESD of $\{n^{-1/2} W_n\}$ converges to the semi-circular law;

(b) the variable $n^{-1/2} W_n$ in the $*$-probability space $(\mathcal{M}_n(\mathbb{C}), \mathrm{E}\,\mathrm{tr})$ converges to a (standard) semi-circular variable. ♦

Remark 8.1.1. Note that the limit does not depend on the probability law of the variables. This phenomenon is labelled as *universality*. This is true not only for this patterned matrix but also for many other matrix models. In particular, this is true for all the theorems proved in this chapter. ●

Proof. By Theorem 7.5.1 and Lemma 8.1.1, it is enough to show that for $w \in \mathcal{P}_2(2k)$,

$$p(w) = \lim \frac{1}{n^{1+k}} \#\Pi(w) = \begin{cases} 0 & \text{if } w \notin \mathcal{C}(2k), \\ 1 & \text{if } w \in \mathcal{C}(2k). \end{cases} \tag{8.1}$$

Fix a $w \in \mathcal{P}_2(2k)$ and suppose $\pi \in \Pi(w)$. Whenever $w[i] = w[j]$,

$$(\pi(i-1),\ \pi(i)) = \begin{cases} (\pi(j-1),\ \pi(j)) & \text{(constraint (C1)), or} \\ (\pi(j),\ \pi(j-1)) & \text{(constraint (C2)).} \end{cases} \tag{8.2}$$

Since there are k such constraints, there are at most 2^k possible choices of constraints in all. Let λ denote a typical choice. Then

$$\Pi(w) = \cup_\lambda \Pi_\lambda(w) \ \text{ (a disjoint union),} \tag{8.3}$$

where $\Pi_\lambda(w)$ is the subset of Π whose elements obey the constraints corresponding to λ. The union extends over all 2^k possible choices. Some of these sets may be empty.

We now count the number of circuits in $\Pi_\lambda(w)$ for a given λ. For $\pi \in \Pi_\lambda(w)$, consider the graph with vertices $\pi(0), \pi(1), \ldots, \pi(2k)$. By abuse of notation, $\pi(i)$ denotes both a vertex and its numerical value. Edges are drawn according to the following scheme:

(i) Connect $\pi(0)$ and $\pi(2k)$ to ensure that π is a circuit.

(ii) If $w[i] = w[j]$ with a (C1) constraint, then connect $\pi(i-1)$ and $\pi(j-1)$ as well as $\pi(i)$ and $\pi(j)$. Note that $\pi(i-1) = \pi(j-1)$ and $\pi(i) = \pi(j)$.

(iii) If instead the constraint is (C2), then connect $\pi(i-1)$ and $\pi(j)$ as well as $\pi(i)$ and $\pi(j-1)$). Note that $\pi(i-1) = \pi(j)$ and $\pi(i) = \pi(j-1)$.

This graph has $(2k+1)$ edges and may include loops and double edges. For example, if $L(\pi(0), \pi(1)) = L(\pi(3), \pi(4))$ and $L(\pi(1), \pi(2)) = L(\pi(2), \pi(3))$, then we have double edges between $\pi(0)$ and $\pi(4)$ as well as between $\pi(1)$ and $\pi(3)$, and a loop at $\pi(2)$. The numerical values of the vertices in any connected component are the same. The maximum number of vertices whose numerical values can be chosen freely is bounded by the number of *connected components.*

Since $w \in \mathcal{P}_2(2k)$, there are $(k+1)$ generating vertices. All other vertices are connected to one or more of these vertices. Hence the number of connected components is bounded by $(k+1)$.

We now claim that for $w \in \mathcal{P}_2(2k)$, the graph has exactly $(k+1)$ connected components if and only if all constraints are (C2).

To prove this, first suppose the graph has $(k+1)$ connected components. By the pigeon hole principle, there must exist a vertex, say $\pi(i)$, which is connected to itself. This is possible if and only if $w[i] = w[i+1]$ and constraint (C2) is satisfied. But then this implies that $w[i]w[i+1]$ is a double letter.

Remove this double letter and consider the reduced word w' of length $2(k-1)$. We claim that the reduced word still has a double letter. To show this, in the original graph, coalesce the vertices $\pi(i-1)$ and $\pi(i+1)$. Delete the vertex $\pi(i)$ and remove the (C2) constraint edges $(\pi(i-1),\ \pi(i+1))$ and $(\pi(i),\ \pi(i))$ but retain all the other earlier edges. For example, any other edge that might have existed earlier between $\pi(i-1),\ \pi(i+1)$ is now a loop. This gives a new graph with $2k+1-2 = 2(k-1)+1$ vertices and has k connected components. Proceeding as before, there must exist a self-loop implying a double letter yy in w'. Proceeding inductively, after k steps, we are left with just a single vertex with a loop. In other words, $w \in \mathcal{P}_2(2k)$ and all constraints are (C2).

To establish the converse, essentially retrace the steps given above. First identify a double letter (the last new letter is followed by itself). This gives a (C2) constraint. Remove it and proceed inductively. Calesced vertices will fall in the same connected component. We omit the details.

Now observe that if $w \in \mathcal{P}_2(2k)$ and all constraints are (C2), then $w \in \mathcal{C}(2k)$. We denote this choice of constraints by $\lambda = \lambda_0$. Then clearly,

$$\#\Pi_{\lambda_0}(w) \approx n^{k+1}, \tag{8.4}$$

where $a \approx b$ means that $a/b \to 1$. One may be tempted to conclude that the *exact* count of $\Pi_{\lambda_0}(w)$ equals n^{k+1}. However, this is not true—for instance, not all vertices can be chosen to be equal. Such combinations need to be discounted for. It is easy to see that the total number of such choices is negligible compared to n^{k+1}. *The same logic applies to all other matrices we shall come across—the generating vertices can always be freely chosen from asymptotic considerations.*

On the other hand, when $\lambda \neq \lambda_0$, the graph has at most k connected components and hence $\#\Pi_\lambda(w) \leq n^k$. This implies that

$$\frac{1}{n^{k+1}} \# \left(\cup_{\lambda \neq \lambda_0} \Pi_\lambda(w)\right) \to 0. \tag{8.5}$$

The relation (8.1) now follows from (8.4) and (8.5). ■

Remark 8.1.2. (Matrices with complex entries) Random matrices with entries that are complex random variables are of particular interest in many applications, such as wireless communications. In this book we have restricted ourselves to matrices with real entries. However, many of the results in the book—for example, those on the Wigner matrix, the IID matrix, the elliptic matrix and the S-matrix—continue to remain valid, with appropriate changes in their statements, for matrices with such complex entries. In fact, many results are often easier to prove for such matrices.

For instance, consider the self-adjoint Wigner matrix where the diagonal entries are independent real random variables with mean 0 and variance 1, and the off-diagonal entries are complex $((x_{i,j}+\iota y_{i,j}))$ with the condition that $x_{j,i}+\iota y_{j,i}=x_{i,j}-\iota y_{i,j}$, and the x's and y's are real-valued random variables which are i.i.d. with mean zero and variance $1/2$. Further, they satisfy the "all moments uniformly bounded" condition of Assumption I.

In the special case when the entries are complex Gaussian, the matrix is a member of the *Gaussian unitary ensemble* (GUE).

Recall how in the proof of Theorem 8.1.2, first we reduced our calculations to pair-matched words. The proof of this reduction remains the same for a Wigner matrix with complex entries. Then we argued that each pair-match constraint will be either (C1) or (C2). That words with at least one (C1) constraint contribute nothing to the limit is trivial to show now—if there is a (C1) constraint at (i,j), then the contribution of the corresponding pair equals

$$\mathrm{E}\left[x_{i,j}+\iota y_{i,j}\right]^2=\mathrm{E}\left[x_{i,j}^2-y_{i,j}^2+2x_{i,j}y_{i,j}\right]=0.$$

Thus such words are ruled out immediately. Then the rest of the proof is as before leading to the same conclusions as that of Theorem 8.1.2 for the complex Wigner.

As we shall see later, for the elliptic matrix, the IID matrix and the S-matrix, we again reduce to pair-matched words where all contstraints are either (C1) or (C2). Then also words with at least one (C1) constraint can be quickly dropped as above, leaving us with only (C2) constraints and then the rest of the argument is as in the real case. ●

8.2 S-matrix: Marčenko-Pastur law

Let $X=((x_{i,j,n}))$ be the $p\times n$ matrix whose entries satisfy Assumption I. Let $S:=n^{-1}XX^*$ be the $p\times p$ covariance matrix. We first describe the law which turns out to be the LSD of S.

Definition 8.2.1. (*Marčenko-Pastur law*) This law is parametrized by $y>0$. It has a positive mass $1-\frac{1}{y}$ at 0 if $y>1$. Elsewhere it has the density

$$f_{\mathrm{MP}_y}(x):=\begin{cases}\frac{1}{2\pi xy}\sqrt{(b-x)(x-a)} & \text{if } a\le x\le b,\\[2ex] 0 & \text{otherwise,}\end{cases} \tag{8.6}$$

where

$$a:=a(y):=(1-\sqrt{y})^2 \text{ and } b:=b(y):=(1+\sqrt{y})^2.$$

We shall denote this probability law by MP_y. ◇

In Figure 8.1, we have plotted $f_{\mathrm{MP}_y}(\cdot)$ for $y = 0.5, 0.75$ and 1.25. Note that the support of the probability law is compact for any y, and hence its moments determine the law uniquely.

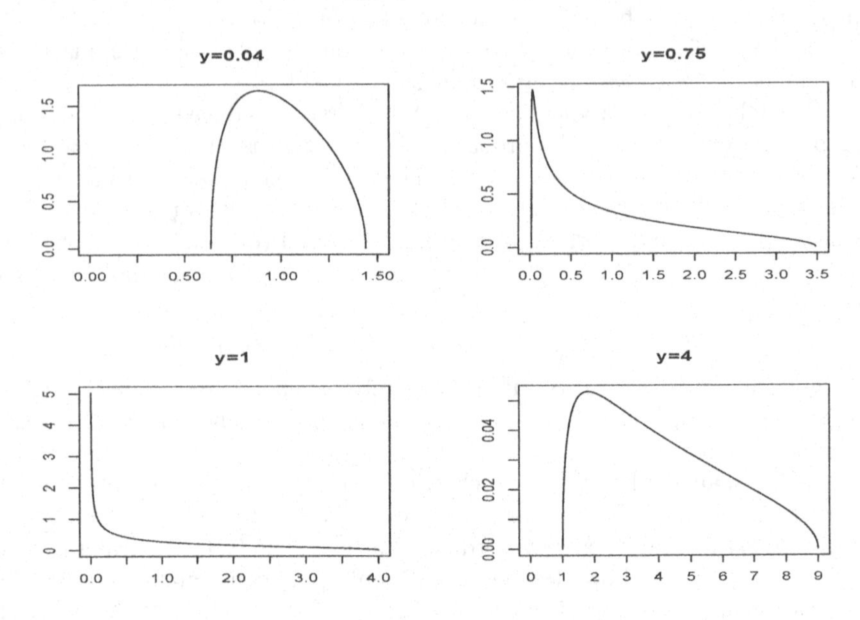

FIGURE 8.1
Density of the Marčenko-Pastur law for various values of y. The point mass of $1 - y^{-1}$ at 0 for $y > 1$ is not shown.

The following lemma gives the moments of MP_y.

Lemma 8.2.1. For every integer $k \geq 1$,

$$m_k(\mathrm{MP}_y) = \sum_{t=0}^{k-1} \frac{1}{t+1} \binom{k}{t} \binom{k-1}{t} y^t. \tag{8.7}$$

♦

Proof. The proof is a straightforward computation.

$$\begin{aligned}
m_k &= \frac{1}{2\pi y} \int_a^b x^{k-1} \sqrt{(b-x)(x-a)}dx \\
&= \frac{1}{2\pi y} \int_{-2\sqrt{y}}^{2\sqrt{y}} (1+y+z)^{k-1} \sqrt{4y - z^2}dz \quad (x = 1+y+z) \\
&= \frac{1}{2\pi y} \sum_{l=0}^{k-1} \binom{k-1}{\ell} (1+y)^{k-1-l} \int_{-2\sqrt{y}}^{2\sqrt{y}} z^l \sqrt{4y - z^2}dz \\
&= \frac{1}{2\pi y} \sum_{2l=0}^{k-1} \binom{k-1}{2l} (1+y)^{k-1-2l} (4y)^{l+1} \int_{-1}^{1} u^{2l} \sqrt{1-u^2}\, du \ (z = 2\sqrt{y}u)
\end{aligned}$$

$$
\begin{aligned}
&= \frac{1}{2\pi y}\sum_{2l=0}^{k-1}\binom{k-1}{2l}(1+y)^{k-1-2\ell}(4y)^{l+1}\int_0^1 w^{l-1/2}\sqrt{1-w}\,dw \ (u=\sqrt{w})\\
&= \sum_{2l=0}^{k-1}\frac{(k-1)!}{l!(l+1)!(k-1-2\ell)!}y^l(1+y)^{k-1-2\ell}\\
&= \sum_{2l=0}^{k-1}\sum_{s=0}^{k-1-2\ell}\frac{(k-1)!}{l!(l+1)!s!(k-1-2\ell-s)!}y^{s+l}\\
&= \sum_{2l=0}^{k-1}\sum_{r=\ell}^{k-1-\ell}\frac{(k-1)!}{l!(l+1)!(r-l)!(k-1-r-l)!}y^r \quad \text{(setting } r=s+l)\\
&= \frac{1}{k}\sum_{r=0}^{k-1}\binom{k}{r}y^r\sum_{\ell=0}^{\min(r,k-1-r)}\binom{s}{l}\binom{k-r}{k-r-\ell-1}\\
&= \frac{1}{k}\sum_{r=0}^{k-1}\binom{k}{r}\binom{k}{r+1}y^r=\sum_{r=0}^{k-1}\frac{1}{r+1}\binom{k}{r}\binom{k-1}{r}y^r.
\end{aligned}
$$

■

We have a natural extension of the concepts introduced in Chapter 7. Due to the appearance of X and its transpose, it is convenient to think in terms of two link functions as a pair of functions given by

$$L^{(1)}, L^{(2)} : \mathbb{N}^2 \to \mathbb{Z}^2,\ L^{(1)}(i,j) := (i,j),\ L^{(2)}(i,j) := (j,i).$$

Then the h-th moment of the ESD of S can be written as

$$m_k(S) = \frac{1}{pn^k}\sum_{\pi} x_{L^{(1)}(\pi(0),\pi(1))}x_{L^{(2)}(\pi(1),\pi(2))}\cdots x_{L^{(2)}(\pi(2k-1),\pi(2k))},$$

where π is a circuit with the following restrictions:

(i) $\pi(0)=\pi(2h)$,
(ii) $1 \le \pi(2i) \le p$ for all $0 \le i \le h$, and
(iii) $1 \le \pi(2i-1) \le n$ for all $1 \le i \le h$.

The concepts of letter, word, vertex and generating vertex remain as earlier. The vertices $\pi(2i), 0 \le i \le h$, and $\pi(2i-1), 1 \le i \le h$, will be called *even vertices* and *odd vertices*, respectively. Now, due to (ii) and (iii), the odd and the even indexed vertices do not have equal roles unless $p = n$.

Recall how the Catalan words of length $2k$ and their count played a fundamental role in the derivation of the Wigner LSD. These words continue to play a crucial role in the LSD of the S-matrix, but we need a more sophisticated counting of certain subsets of these words. In Lemma 2.3.1 we had established bijections between $NC(k)$, $NC_2(2k)$ and $\mathcal{C}_{2k}$. We now wish to refine this bijection. Write any $\pi \in NC_2(2k)$ as

$$\pi = \{(r, s) : r < s, \{r, s\} \text{ is a block of } \pi\}.$$

Note that we have used the same symbol π to denote both, a circuit or a partition, and its usage will be clear from the context. If $(r, s) \in \pi$ then r will be called the *first element* of the block. Note that since $\pi \in NC_2(2k)$, for any $(r, s) \in \pi$, one of r and s is odd and the other is even.

For $0 \leq t \leq k-1$, let

$$A_{t,k} = \{\pi \in NC(k) : \pi \text{ has } t+1 \text{ blocks}\}, \tag{8.8}$$

$$M_{t,k} = \{w \in \mathcal{C}(2k) : w \text{ has } t+1 \text{ } \textit{even} \text{ generating vertices}\}, \tag{8.9}$$

$$NC_{2,t}(2k) = \{\pi \in NC_2(2k) : t+1 \text{ blocks of } \pi \text{ have odd first elements}\}. \tag{8.10}$$

Recall that $NC_2(2k) \cong \mathcal{C}(2k)$. Indeed, any word $w \in \mathcal{C}(2k)$ generates a $\pi \in NC_2(2k)$ and vice versa—the elements of each pair-partition are precisely the positions where the letters of the word match. Thus for each $0 \leq t \leq k-1$,

$$M_{t,k} \cong NC_{2,t}(2k).$$

Note that each $w \in M_{t,k}$ has exactly $k-t$ odd generating vertices. Moreover, using Lemma 2.3.1 and Lemma 8.1.1,

$$\cup_{t=0}^{k-1} M_{t,k} = \mathcal{C}(2k) \cong NC_2(2k) = \cup_{t=0}^{k-1} NC_{2,t}(2k) \cong NC(k) = \cup_{i=1}^{k-1} A_{t,k}.$$

The following Lemma gives the count of $M_{t,k}$ and a bijection between $NC_{2,t}(2k)$ and $A_{t,k}$ for each $0 \leq t \leq k-1$.

Lemma 8.2.2. (a)

$$\#M_{t,k} = \binom{k-1}{t}^2 - \binom{k-1}{t+1}\binom{k-1}{t-1} = \frac{1}{t+1}\binom{k}{t}\binom{k-1}{t} = \frac{1}{k}\binom{k}{t}\binom{k}{t+1}.$$

(b)

$$A_{t,k} \cong NC_{2,t}(2k) \cong M_{t,k}.$$

♦

Proof. (a) Note that $\sum_{t=0}^{k-1} \#M_{t,k} = \#\mathcal{C}(2k)$. In the proof of Lemma 8.1.1, we had set up a bijection between $\mathcal{C}(2k)$ and the set of paths:

$$P =: \{\{u_l\}_{1 \leq l \leq 2k} : u_l = \pm 1,\ S_l := \sum_{j=1}^{l} u_j \geq 0 \text{ for all } l \geq 1, S_{2k} = 0\}$$

and we define $u_0 = 1$. If we go back to that bijection, it is clear that $u_l = +1$, if and only if l is a generating vertex, $1 \leq l \leq 2k-1$. Note that 0 and 1 are always generating vertices and $2k$ is never a generating vertex. In all there are

$k+1$ generating vertices and the number of *even* generating vertices is $t+1$ for some $0 \le t \le k-1$. Now, for every $0 \le t \le k-1$, let P_t be the set of all $\{u_l\} \in P$ which have exactly $(t+1)$ even values l with $u_l = 1$. Then $\{P_t\}$ are disjoint, $P = \cup_{t=0}^{k-1} P_t$ and $\#P_t = \#M_{t,k}$ for all t. We now count the number of paths in P_t.

First we count the number of paths $\{u_l\}$ which satisfy the constraint laid down for P_t except that we ignore the constraint $S_l \ge 0$, $2 \le l \le 2k-2$. Recall that $u_0 = 1$ so 0 is the first even vertex. Choose the remaining t even values of l for the generating even vertices from the $(k-1)$ possible even values (note that $2k$ is not a choice) in $\binom{k-1}{t}$ ways. Now consider the odd generating vertices. The $k-1-t$ odd values of l for the generating odd vertices can be chosen from the $k-1$ possible odd values (note that 1 is always an odd generating vertex, so that is not a free choice) in $\binom{k-1}{k-1-t} = \binom{k-1}{t}$ ways. So the total number of such choices of these paths is $\binom{k-1}{t}^2$.

Now we need to eliminate the paths which have $S_l \le 0$ for at least one l, $2 \le l \le 2k-2$. We count the total number of such paths by a reflection principle. Any such path touches the line $y = -1$ at least once and hence has two consecutive upward movements. Consider the *last* time this occurs so that $u_l = +1$ and $u_{l+1} = +1$. We consider a transformation

$$(u_2, \ldots, u_l = 1, u_{l+1} = 1, \ldots, u_{2k-1}) \mapsto (u_2, \ldots, u_{l-1}, -1, -1, u_{l+2}, \ldots, u_{2k-1}).$$

The resulting sequence is a path from $(1,1)$ to $(2k-1,-3)$. Moreover, this transformation is a bijection from the set of all required paths from $(1,1)$ to $(2k-1,1)$ with exactly t positive steps to the set of all paths from $(1,1)$ to $(2k-1,-3)$ having $u_l = 1$ at $(t-1)$ of the $k-1$ even steps, and $u_l = 1$ at $(k-2-t)$ of the $k-1$ odd steps. The number of all such paths is $\binom{k-1}{t-1}\binom{k-1}{t+1}$. Hence

$$\#M_{t,k} = \#P_t = \binom{k-1}{t}^2 - \binom{k-1}{t+1}\binom{k-1}{t-1} = \frac{1}{t+1}\binom{k}{t}\binom{k-1}{t}.$$

(b) We set up a bijection $f : NC_2(2k) \to NC(k)$ that will also serve as a bijection, for each t, between their subsets, $NC_{2,t}(2k)$ and $A_{t,k}$.

For $\pi \in NC_2(2k)$, collect all its *odd first elements* in the following set:

$$J_\pi = \{2i-1 : 1 \le i \le k, (2i-1, 2j) \in \pi\}.$$

We now give an algorithm to construct the blocks of $f(\pi)$.

Note that $1 \in J_\pi$ always. Let B_1 be the block of $f(\pi)$, to be constructed, containing 1. We build up B_1 by adding elements sequentially as follows. Suppose $(1, 2i_1) \in \pi$. Then i_1 is added to B_1. If $i_1 = 1$ we stop and $B_1 = \{1\}$. If $i_1 > 1$, then $(2i_2, 2i_1 - 1) \in \pi$ for some i_2. Then add i_2 to B_1. If $i_2 = 1$, we stop. Otherwise consider $(2i_3, 2i_2 - 1) \in \pi$ and add i_3 to B_1. We continue this until we reach $2i_k = 2$ for some k. Then $B_1 = \{1 = i_k < i_{k-1} \cdots < i_1\}$ is the block containing 1.

Now we move to the next available element of J_π, say $2j-1$. Suppose it matches with $2j_1$. Then the construction of the next block, say B_2 begins by putting j_1 in it. Then we repeat the method of construction given above until we obtain $2j_1, \ldots, 2j_t = 2j$ for some t. Then $B_2 = \{j_t = j < j_{t-1} < \cdots < j_1\}$. Note that B_2 is disjoint from B_1. We continue this process until all odd first elements are exhausted.

If we carefully follow the steps in the above construction, we can draw the following conclusions:

(i) For all $\pi \in NC_2(2k)$, we have $f(\pi) \in NC(k)$ with blocks $B_1, B_2, \ldots$.

(ii) If $\pi_1 \neq \pi_2$ are elements of $NC_2(2k)$, then $f(\pi_1) \neq f(\pi_2)$.

(iii) For each $2i-1 \in J_\pi$, there is a distinct block of $f(\pi)$. Hence, the number of blocks of $f(\pi)$ is exactly the same as the number of elements of J_π.

The above three observations establish Part (b). ■

Example 8.2.1. Suppose $k=3$. Then $\#NC_2(6) = 5$. We list the five combinations of π, J_π and $f(\pi)$.

1. $\pi = \{(1,2),(3,4),(5,6)\}$, $J_\pi = \{1,3,5\}$, $\#J_\pi = 3$ and $f(\pi) = \{\{1\},\{2\},\{3\}\}$.
2. $\pi = \{(1,4),(2,3),(5,6)\}$, $J_\pi = \{1,5\}$, $\#J_\pi = 2$ and $f(\pi) = \{\{1,2\},\{3\}\}$.
3. $\pi = \{(1,2),(3,6),(4,5)\}$, $J_\pi = \{1,3\}$, $\#J_\pi = 2$ and $f(\pi) = \{\{1\},\{2,3\}\}$.
4. $\pi = \{(1,6),(2,5),(3,4)\}$, $J_\pi = \{1,3\}$, $\#J_\pi = 2$ and $f(\pi) = \{\{1,3\},\{2\}\}$.
5. $\pi = \{(1,6),(2,3),(4,5)\}$, $J_\pi = \{1\}$, $\#J_\pi = 1$ and $f(\pi) = \{\{1,2,3\}\}$. ▲

Remark 8.2.1. The above lemma will come in very handy later. For the moment we point out the following interesting consequences.

(a) *Free Poisson probablity law.* Recall that we defined a free Poisson variable a as a variable for which all free cumulants are equal, say y. We shall now show that the probability law of a exists for $0 < y \leq 1$. Let X and Y be independent random variables where X is distributed as MP_y and Y satisfies $\mathrm{P}(Y=1) = 1 - \mathrm{P}(Y=0) = y$. Then

$$\begin{aligned}
\mathrm{E}(XY)^k &= y\sum_{t=0}^{k-1} \frac{1}{t+1}\binom{k}{t}\binom{k-1}{t} y^t \\
&= \sum_{t=1}^{k} \frac{1}{t}\binom{k}{t-1}\binom{k-1}{t-1} y^t \\
&= \sum_{\pi \in NC(k)} y^{|\pi|}, \text{ by Lemma 8.2.2 (b).}
\end{aligned}$$

But then, by moment-free cumulant relation, these are the moments of the free Poisson variable with mean y. Also note that XY is a bounded random variable, so that its moments determine a unique probability law. This shows that a free Poisson variable a with parameter (mean) y has a probability law μ_a which is the law of XY.

(b) *Semi-circular,* MP_1 *and free Poisson.* Now consider MP_1. Its k-th moment equals $\sum_{t=0}^{k-1} \#M_{t,k} = \#\mathcal{C}(2k)$, which is the $2k$-th moment of the semi-circular law. That is, if X is a semi-circular random variable then X^2 obeys the probability law MP_1. Due to (2.8), this is the same as saying that the law MP_1 is the free Poisson law with rate (mean) 1.

(c) MP_y *is the law of a compound free Poisson variable.* For any $y > 0$, the MP_y law is the law of a compound free Poisson variable whose mean is $1/y$ and jump distribution ν which puts full mass at y. To see this, using the formula for the k-th moment of MP_y, given in Lemma 8.2.1, and Lemma 8.2.2 (a), we have

$$
\begin{aligned}
m_k(\text{MP}_y) &= \sum_{t=0}^{k-1} \frac{1}{k}\binom{k}{t}\binom{k}{t+1} y^t \quad \text{(by Lemma 8.2.1)} \\
&= \sum_{s=0}^{k-1} \frac{1}{k}\binom{k}{k-1-s}\binom{k}{k-s} y^{k-1-s} \quad \text{(using } s = k-1-t) \\
&= \sum_{s=0}^{k-1} \frac{1}{k}\binom{k}{s+1}\binom{k}{s} y^{k-1-s} \\
&= \sum_{s=0}^{k-1} \#M_{s,k} y^{k-1-s} \quad \text{(using Lemma 8.2.2 (a))} \\
&= \sum_{\substack{\pi \in NC(k) \\ \pi = \{V_1, \ldots, V_{s+1}\}}} y^{\sum_{j=1}^{s+1}(|V_j|-1)} \quad \text{(bijection between } M_{s,k} \text{ and } A_{s,k}) \\
&= \sum_{\substack{\pi \in NC(k) \\ \pi = \{V_1, \ldots, V_{s+1}\}}} \prod_{j=1}^{s+1} y^{|V_j|-1}.
\end{aligned}
$$

Now we invoke the moment-free cumulant relation from which it follows that the n-th free cumulant of MP_y is y^{n-1}. Now the result follows once we use Definition 3.10.1 of the compound free Poisson distribution.

(d) *Free cumulants of* MP_y. We have found the free cumulants of MP_y through the above calculations. This is worth recording for future reference:

$$\kappa_n(\text{MP}_y) = y^{n-1}, \; n \geq 1. \tag{8.11}$$

●

The following result is on the convergence of the S-matrix when all its entries are independent. We shall discuss the joint convergence of independent sequences of S-matrices in Section 9.4. Algebraic convergence of single as well as independent copies of pair-correlated cross-covariances matrices will be taken up in Chapter 11.

Theorem 8.2.3. Suppose $\{x_{i,j,n}\}$ satisfies Assumption I and $p \to \infty$ such that $p/n \to y \in (0, \infty)$. Then the following are true.
(a) The EESD of $S = n^{-1}XX^*$ converges to MP_y.
(b) The variable $n^{-1}XX^*$ in the $*$-probability space $(\mathcal{M}_n(\mathbb{C}), \text{E}\,\text{tr})$ converges to a self-adjoint variable whose k-th moment is given by

$$\sum_{t=0}^{k-1} \frac{1}{t+1}\binom{k}{t}\binom{k-1}{t} y^t, \quad k \geq 1.$$

◆

Proof of Theorem 8.2.3. (a). We apply mutatis mutandis, the proof given for the Wigner matrix. It is enough to show that for every $k \geq 1$,

$$\lim_{p\to\infty} \text{E}\, m_k(S) = \sum_{t=0}^{k-1} \#M_{t,k} y^t,$$

where

$$m_k(S) = \frac{1}{pn^k} \sum_{\pi} x_{L^{(1)}(\pi(0),\pi(1))} x_{L^{(2)}(\pi(1),\pi(2))} \cdots x_{L^{(2)}(\pi(2k-1),\pi(2k))}.$$

Since the entries of X are independent and have mean zero, the expectation of a product term in the above sum is 0 if π is not matched. Thus we can focus only on matched circuits and words.

Let $\widetilde{\Pi}(w)$ be the possibly larger class of circuits with range $1 \leq \pi(i) \leq \max(p, n)$, $1 \leq i \leq 2k$. Then for some constant $C > 0$ (which is possible since $p/n \to y, 0 < y < \infty$),

$$\#\Pi(w) \leq C\#\widetilde{\Pi}(w).$$

Hence, by Theorem 7.5.1 (see also 7.11), words for which at least one letter is repeated at least thrice do not contribute to the limit. Thus we can restrict to pair-matched words and circuits, and we need to calculate

$$\lim_n \sum_{w\in\mathcal{P}_2(2k)} \frac{\#\Pi(w)}{n^k p}. \tag{8.12}$$

Clearly, we need *exactly* $(k+1)$ generating vertices (hence k non-generating ones) for a non-zero contribution.

Recall the (C1) and (C2) constraints. Note that for $i \neq j$,

$$L^{(1)}(\pi(i-1), \pi(i)) = L^{(2)}(\pi(j-1), \pi(j)) \tag{8.13}$$

implies a (C2) constraint as defined in (8.2).

On the other hand,

$$L^{(t)}(\pi(i-1), \pi(i)) = L^{(t)}(\pi(j-1), \pi(j)), \; t = 1 \text{ or } 2, \tag{8.14}$$

yields a (C1) constraint as defined in (8.2).

Recall that for the Wigner matrix, in (8.3), we had written $\Pi(w)$ as a disjoint union $\Pi(w) = \bigcup_\lambda \Pi_\lambda(w)$. We do the same here. Also, recall that we have defined $\widetilde{\Pi}(w)$ by extending the range of values of p and n to $\max(p, n)$. We do likewise for $\Pi_\lambda(w)$ and so

$$\widetilde{\Pi}(w) = \cup_\lambda \widetilde{\Pi}_\lambda(w).$$

Now note that unlike the Wigner matrix, for a fixed word w, here $w[i] = w[j]$ implies that *exactly one* of the constraints is satisfied—(C1) if i and j have the same parity and (C2) otherwise. Hence there is a *unique* set of constraints say $\bar{\lambda}$ (depending on w) such that

$$\Pi(w) = \Pi_{\bar{\lambda}}(w) \text{ and } \widetilde{\Pi}(w) = \widetilde{\Pi}_{\bar{\lambda}}(w). \tag{8.15}$$

Now fix a word w. If w is not Catalan then not all constraints are (C2). Let $\bar{\lambda}$ be as in (8.15). Then it follows from the Wigner matrix calculations (see (8.5)) that,

$$n^{-k}p^{-1}\#\Pi_{\bar{\lambda}}(w) \le C[\max(p, n)]^{-(k+1)}\#\widetilde{\Pi}_{\bar{\lambda}}(w) \to 0.$$

Thus non-Catalan words do not contribute in the limit.

Now suppose that w is Catalan. Then all constraints are (C2). Suppose that w is Catalan with $(t + 1)$ even generating vertices (with range p) and $(k - t)$ odd generating vertices (with range n). Again, as in the comments in italics after (8.4), from asymptotic considerations, we can assume free choices of all the generating vertices. This implies that

$$\lim_{n\to\infty} n^{-k}p^{-1}\#\Pi_{\bar{\lambda}}(w) = \lim_{n\to\infty} n^{-k}p^{-1}(p^{t+1}n^{k-t}) = y^t.$$

Hence, using Lemma 8.2.2 (a), $\lim \mathrm{E}[m_k(S)] = \sum_{t=0}^{k-1} M_{t,k}y^t$ and Part (a) is proved.

(b) This follows trivially from the proof of part (a) since we have verified the convergence of all the moments. ■

8.3 IID and elliptic matrices: circular and elliptic variables

Recall the simulations of the eigenvalues of these matrices given in Figures 7.3 and 7.4. It is apparent from these simulations that the ESD and the EESD of the IID and the elliptic matrices converge to the uniform distributions on the disc and on the interior of an ellipse, respectively. However, the proofs of these facts are quite involved and are beyond the scope of this book.

In general, for non-symmetric matrices, it is not easy to show that the EESD or the ESD converges. This is because for such matrices, the ESD and the EESD are probability measures on $\mathbb{C}$. To use the moment method, we would need to consider the mixed moments of the real and complex parts of the eigenvalues. Unfortunately, these cannot be expressed in terms of the traces of the powers of the matrices.

We shall only show that these matrices converge in $*$-distribution as the dimension goes to ∞. Readers interested in the convergence of the ESD are directed to the references given in the notes in Section 8.8.

Since the entries of an elliptic matrix are correlated, we need a modification of Assumption I.

Assumption Ie For every n, the variables $\{x_{i,i,n} : 1 \leq i \leq n\} \cup \{(x_{i,j,n}, x_{j,i,n}) : 1 \leq i < j \leq n\}$ form a collection of independent random variables, and satisfy $\mathrm{E}[x_{i,j,n}] = 0$, $\mathrm{E}[x_{i,j,n}^2] = 1$ for all i, j and $\mathrm{E}[x_{i,j,n}x_{j,i,n}] = \rho$ for $1 \leq i \neq j \leq n$. Moreover,

$$\sup_{i,j,n} \mathrm{E}(|x_{i,j,n}|^k) \leq B_k < \infty \quad \text{for all} \quad k \geq 1.$$

⊡

The following result deals with a single sequence of elliptic matrices. The joint convergence of several sequences of independent elliptic matrices is covered in Section 9.3.

Theorem 8.3.1. (a) Suppose E_n is a sequence of elliptic matrices whose entries satisfy Assumption Ie. Then $n^{-1/2}E_n$, as a variable of $(\mathcal{M}_n(\mathbb{C}), \mathrm{E}\,\mathrm{tr})$, converges in $*$-distribution to an elliptic variable e with parameter ρ.

(b) In particular, if the entries satisfy Assumption Ie with $\rho = 0$, then the scaled IID matrix, $n^{-1/2}C_n$ as an element of $(\mathcal{M}_n(\mathbb{C}), \mathrm{E}\,\mathrm{tr})$, converges in $*$-distribution to a circular variable. ◆

Remark 8.3.1. Recall that the Wigner matrices are a special case of the elliptic matrices where the correlation ρ equals 1. Hence Theorem 8.3.1 implies Theorem 8.1.2. Moreover, Theorem 8.3.1 remains valid for elliptic matrices with complex entries. This will be clear from the proof and the argument is along the lines of what we mentioned in Remark 8.1.2. We omit the details. ●

Note that proofs of all the theorems in this chapter so far have been based on identifying circuits and words that contribute to the limit and then evaluating the contribution of each word.

We now switch gears a bit and introduce a method of keeping track of the counts by using the δ-function, while at the same time drawing from our experience with words. This approach shall come in very handy later, especially when we establish the asymptotic freeness of certain matrices. We recall the definition of the δ-function:

$$\delta_{xy} = \begin{cases} 0 & \text{if } x \neq y, \\ 1 & \text{if } x = y. \end{cases} \tag{8.16}$$

Proof of Theorem 8.3.1. It is enough to prove only (a) since (b) is a special case (with $\rho = 0$). We shall find it convenient to use the notation:

$$a'(r,s) = \delta_{i_r i_s}\delta_{i_{r+1} i_{s+1}}, \quad (8.17)$$
$$b'(r,s) = \delta_{i_r i_{s+1}}\delta_{i_s i_{r+1}}, \quad (8.18)$$
$$\bar{E}_n = n^{-1/2}E_n \quad (8.19)$$
$$x_{ij}^{\epsilon} = \begin{cases} x_{i,j,n} & \text{if } \epsilon = 1, \\ x_{j,i,n} & \text{if } \epsilon = *. \end{cases} \quad (8.20)$$

Let $\epsilon_1, \ldots, \epsilon_t \in \{1, *\}$, $t \geq 1$. Then we have

$$\begin{aligned} \varphi_n\left(\bar{E}_n^{\epsilon_1}\cdots\bar{E}_n^{\epsilon_t}\right) &= \frac{1}{n^{\frac{t}{2}+1}}\,\mathrm{E}[\mathrm{Tr}(A_n^{\epsilon_1}\cdots A_n^{\epsilon_t})] \\ &= \frac{1}{n^{\frac{t}{2}+1}}\sum_{I_t}\mathrm{E}[x_{i_1i_2}^{\epsilon_1}x_{i_2i_3}^{\epsilon_2}\cdots x_{i_ti_1}^{\epsilon_t}], \end{aligned} \quad (8.21)$$

where

$$I_t = \{(i_1, i_2 \ldots i_t) : 1 \leq i_j \leq n, 1 \leq j \leq t\}.$$

As $\mathrm{E}[x_{ij}] = 0$, in the above product, there must be at least two occurrences from one of the sets $\{x_{ij}, x_{ji}\}$ for that term to be non-zero. Most of these combinations are asymptotically negligible once we make the following two observations.

(i) Those combinations where there is at least one match of order three or more are negligible. This follows once we note that all moments in (8.21) are uniformly bounded. Further, the total number of terms with such matches is negligible from the proof of Theorem 7.5.1 (a) (see (7.11)).

(ii) For a given pair-matched combination, if there is at least one constraint which is (C1), then this combination is negligible by the same arguments as used in deriving (8.5) in the proof of Theorem 8.1.2. Thus only those pair-matches where all matches obey the (C2) constraint contribute to the limit moments. So any such pair will be of the form $x_{i_ri_{r+1}}^{\epsilon_r}x_{i_si_{s+1}}^{\epsilon_s}$ where $i_r = i_{s+1}$ and $i_{r+1} = i_s$.

This means that unlike in the proof of Theorem 8.1.2, where each pair had contribution 1 (which had led us to (8.4)), now each pair can contribute 1 or ρ, depending on the type of the pair, (x_{ij}, x_{ij}) or (x_{ij}, x_{ji}), and we need to express these contributions properly.

Observation (i) implies that we may assume that $t = 2k$. Further, only those terms may contribute where the variables $\{x_{ij}\}$ appear *only* in one of the forms $x_{ij}x_{ij}$ or $x_{ij}x_{ji}$, and with no repetition of the paired indices. Hence

$$\varphi_n(\bar{E}_n^{\epsilon_1}\cdots\bar{E}_n^{\epsilon_{2k}}) = \frac{1}{n^{k+1}}\sum_{\pi\in\mathcal{P}_2(2k)}\sum_{I_{2k}(\pi)}\prod_{(r,s)\in\pi}\mathrm{E}[x_{i_ri_{r+1}}^{\epsilon_r}x_{i_si_{s+1}}^{\epsilon_s}] + o(1), \quad (8.22)$$

where $I_{2k}(\pi)$ is the subset of I_{2k} whose elements obey the partition π.

Since the entries of E_n satisfy Assumption Ie, we have

$$\begin{aligned}\mathrm{E}[x^{\epsilon_r}_{i_r i_{r+1}} x^{\epsilon_s}_{i_s i_{s+1}}] &= (a'(r,s) + \rho b'(r,s))\delta_{\epsilon_r \epsilon_s} + (\rho a'(r,s) + b'(r,s))(1-\delta_{\epsilon_r \epsilon_s}) \\ &= ((1-\delta_{\epsilon_r \epsilon_s}) + \rho\delta_{\epsilon_r \epsilon_s})b'(r,s) + (\delta_{\epsilon_r \epsilon_s} + \rho(1-\delta_{\epsilon_r \epsilon_s}))a'(r,s).\end{aligned}$$

Also note that

$$\delta_{\epsilon_r \epsilon_s} + \rho(1-\delta_{\epsilon_r \epsilon_s}) \leq 1 \text{ and } ((1-\delta_{\epsilon_r \epsilon_s}) + \rho\delta_{\epsilon_r \epsilon_s}) = \rho^{\delta_{\epsilon_r \epsilon_s}}.$$

Using these and the two observations (i) and (ii) in (8.22), we get

$$\begin{aligned}\lim_{n\to\infty} \varphi_n\left(\bar{E}_n^{\epsilon_1}\cdots \bar{E}_n^{\epsilon_{2k}}\right) &= \sum_{\pi\in\mathcal{P}_2(2k)} \lim_{n\to\infty} \frac{1}{n^{k+1}} \sum_{I_{2k}(\pi)} \prod_{(r,s)\in\pi} ((1-\delta_{\epsilon_r \epsilon_s}) + \rho\delta_{\epsilon_r \epsilon_s})b'(r,s) \\ &= \sum_{\pi\in\mathcal{P}_2(2k)} \lim_{n\to\infty} \frac{1}{n^{k+1}} \sum_{I_{2k}(\pi)} \prod_{(r,s)\in\pi} \rho^{\delta_{\epsilon_r \epsilon_s}} \delta_{i_r i_{s+1}} \delta_{i_s i_{r+1}}.\end{aligned}$$

Further, using the above observations, the terms on the right side are non-zero only when π is a non-crossing pair-partition. Therefore we get

$$\begin{aligned}\lim_{n\to\infty} \varphi_n\left(\bar{E}_n^{\epsilon_1}\cdots \bar{E}_n^{\epsilon_{2k}}\right) &= \sum_{\pi\in NC_2(2k)} \lim_{n\to\infty} \frac{1}{n^{k+1}} \sum_{I_{2k}(\pi)} \prod_{(r,s)\in\pi} \rho^{\delta_{\epsilon_r \epsilon_s}} \delta_{i_r i_{s+1}} \delta_{i_s i_{r+1}} \\ &= \sum_{\pi\in NC_2(2k)} \rho^{T(\pi)} \\ &= \varphi(e^{\epsilon_1}\cdots e^{\epsilon_{2k}})\end{aligned}$$

where $T(\pi)$ is given by

$$T(\pi) = \#\{(r,s)\in\pi \ : \ \delta_{\epsilon_r \epsilon_s} = 1\}.$$

These limit moments are the moments of an elliptic variable by Lemma 3.6.1 (see (3.3)). Hence the proof of the theorem is complete. ∎

8.4 Toeplitz matrix

In Figure 7.7, we showed the simulated ESD of random symmetric Toeplitz matrices.

Theorem 8.4.1. Suppose $\{x_{i,n}\}$ satisfy Assumption I. Then:

(a) The EESD of $\{n^{-1/2}T_n\}$ converges weakly to a probability law. The limit law is universal, and its $2k$-th moment is a sum of volumes of certain polyhedra within the unit hypercube in $k+1$ dimensions;

(b) The variable $n^{-1/2}T_n$ in the $*$-probability space $(\mathcal{M}_n(\mathbb{C}), \mathrm{E}\,\mathrm{tr})$ converges to a self-adjoint variable whose moments are as described in (a). ◆

Proof. Fix $w \in \mathcal{P}_2(2k)$. Define the *slopes* of indices/vertices as

$$s(i) = \pi(i) - \pi(i-1). \tag{8.23}$$

Now a match of the i-th and j-th vertices occurs if and only if $|s(i)| = |s(j)|$. This gives rise to two possibilities (both may hold simultaneously):

$$\text{either} \quad s(i) + s(j) = 0 \quad \text{or} \quad s(i) - s(j) = 0. \tag{8.24}$$

The following lemma shows that only matches with opposing slopes matter.

Lemma 8.4.2. Let N be the number of circuits with at least one pair of vertices i, j such that $s(i) - s(j) = 0$. Then $n^{-(k+1)}N = O(n^{-1}) \to 0$. ♦

Proof. We can count all possible circuits by counting the total choices for $\pi(0)$ and $s(i), 1 \leq i \leq 2k$.

Note that the matched pairs together yield k linear constraints—if i-th and j-th vertices match then either $s(i) = s(j)$ or $s(i) = -s(j)$. Also note that the circuit condition says that

$$0 = \pi(0) - \pi(2k) = \sum_{i=1}^{2k} s(i). \tag{8.25}$$

This condition would be satsified if and only if all constraints coming from matched pairs are of the form $s(i) = -s(j)$. By assumption, this is not the case and so we have $k+1$ linear constraints amongst $\{\pi(0), s(i), 1 \leq s(i) \leq 2k\}$, that is amongst $\{\pi(0), \pi(i), 1 \leq i \leq 2k\}$. Hence $N = O(n^{2k+1-(k+1)}) = O(n^k)$. This proves the lemma. ■

Using Lemma 8.4.2, it is enough to show that $\lim_{n\to\infty} \frac{1}{n^{1+k}} \#(\Pi^*(w))$ exists, where

$$\Pi^*(w) = \{\pi : w[i] = w[j] \Rightarrow s(i) + s(j) = 0\}.$$

Let

$$v_i = \pi(i)/n \quad \text{and} \quad U_n = \{0, 1/n, 2/n, \ldots, (n-1)/n\}. \tag{8.26}$$

Then $\#(\Pi^*(w))$ equals

$$\#\{(v_0, \ldots, v_{2k}) : v_0 = v_{2k}, v_i \in U_n, v_{i-1} - v_i + v_{j-1} - v_j = 0 \text{ if } w[i] = w[j]\}.$$

Denote the set of generating vertices by

$$S = \{0\} \cup \{\min(i,j) : w[i] = w[j], i \neq j\}. \tag{8.27}$$

Clearly, $\#S = k+1$. If $\{v_i\}$ satisfy the k equations in (8.27); then each v_i is a unique linear combination of $\{v_j\}$ where $j \in S$ and $j \leq i$. Let

$$v_S = \{v_i : i \in S\} \tag{8.28}$$

be the set of all the v_i corresponding to the set of all generating vertices. Then we may write

$$v_i = L_i^T(v_S), \quad \text{for all} \quad i = 0, 1, \ldots, 2k, \tag{8.29}$$

where $L_i^T(\cdot)$ are linear functions of the vertices in v_S and depend on the word w. Clearly,

$$L_i^T(v_S) = v_i \quad \text{if} \quad i \in S, \tag{8.30}$$

and also, adding the k equations $s(i) + s(j) = 0$ implies that $L_{2k}^T(v_S) = v_0$. So

$$\#(\Pi^*(w)) = \#\{v_S : L_i^T(v_S) \in U_n \quad \text{for all} \quad i = 0, 1, \ldots, 2k\}. \tag{8.31}$$

Since $\frac{1}{n^{1+k}}\#(\Pi^*(w))$ is the $(k+1)$ dimensional Riemann sum for the function $\mathbb{I}(0 \le L_i^T(v_S) \le 1 \; \forall \; i \notin S \cup \{2k\})$ over $[0,1]^{k+1}$,

$$\begin{aligned} \lim_{n\to\infty} \frac{1}{n^{1+k}}\#(\Pi^*(w)) &= \underbrace{\int_0^1 \cdots \int_0^1}_{k+1} \mathbb{I}\left(0 \le L_i^T(x) \le 1 \; \forall \; i \notin S \cup \{2k\}\right) \prod_{i \in S} dx_i \\ &= p_T(w), \quad \text{say}. \end{aligned} \tag{8.32}$$

This proves the theorem. ■

The explicit value of the above limits, and hence the moments, can only be obtained by computing the above integrals. There are no known "closed formulae" for these.

8.5 Hankel matrix

In Figure 7.8 we showed simulations of the ESD of a random Hankel matrix.

Theorem 8.5.1. Suppose $\{x_{i,n}\}$ satisfy Assumption I. Then:
(a) The EESD of $\{n^{-1/2}H_n\}$ converges to a probability law which is symmetric about 0. The limit is universal and its $2k$-th moment may be expressed as sums of volumes of certain polyhedra of the unit hypercube in $(k+1)$ dimensions.
(b) The variable $n^{-1/2}H_n$ in the $*$-probability space $(\mathcal{M}_n(\mathbb{C}), \mathrm{E}\,\mathrm{tr})$ converges to a self-adjoint variable whose moments are as described above. ◆

Proof. Let

$$\Pi^*(w) = \{\pi : w[i] = w[j] \Rightarrow \pi(i-1) + \pi(i) = \pi(j-1) + \pi(j)\}. \tag{8.33}$$

It is enough to show that the limit below exists for every $w \in \mathcal{P}_2(2k)$.

$$\lim_{n\to\infty} \frac{1}{n^{1+k}} \#\Pi^*(w) = p_H(w), \quad \text{say}. \tag{8.34}$$

We can proceed as in the proof of the previous theorem to write v_i as a linear combination $L_i^H(v_S)$ of generating vertices for all $i \notin S$. As before, realizing $n^{-(k+1)}\#\Pi^*(w)$ as a Riemann sum, we may conclude that it converges to the expression given below. We emphasize that unlike in the Toeplitz case, we do not automatically have $L_{2k}^H(v_S) = v_{2k}$ for every word w. Hence this restriction appears as an additional constraint. Combining all these observations,

$$p_H(w) = \underbrace{\int_0^1 \cdots \int_0^1}_{k+1} \mathbb{I}(0 \leq L_i^H(x) \leq 1 \ \forall\, i \notin S \cup \{2k\})\mathbb{I}(x_0 = L_{2k}^H(x)) \prod_{i \in S} dx_i. \tag{8.35}$$

Of course, for some words, $L_{2k}^H(x) = x_0$ is a genuine extra linear constraint, and in such cases $p_H(w) = 0$. ■

8.6 Reverse Circulant matrix: symmetrized Rayleigh law

Note that the upper half of the anti-diagonal of this matrix is identical to the corresponding portion of the Hankel matrix. We have shown the simulated ESD of $n^{-1/2}RC_n$ in Figure 7.9. Just as the Catalan words play a crucial role for Wigner matrices, the following words play a crucial role for Reverse Circulant matrices.

Definition 8.6.1. (*Symmetric word*). A pair-matched word is *symmetric* if each letter occurs both in an odd and an even position. We shall denote the set of symmetric words of length $2k$ by $\mathcal{S}_2(2k)$. ◇

For example, $w = aabb$ is a symmetric word while $w = abab$ is not. All Catalan words are symmetric. A simple counting argument leads to the following lemma. We omit its proof.

Lemma 8.6.1. The size of the set $\mathcal{S}_2(2k)$ is $k!$. ♦

Indeed, $\{k!\}$ is the moment sequence of a distribution. Consider a chi-squared random variable with two degrees of freedom, say χ_2^2. Consider a random variable R whose probablity law is the law of the *symmetrized square root* $\sqrt{\chi_2^2/2}$. We may call it a *symmetrized Rayleigh variable.* Its probability law has density and moments

$$f_R(x) = |x|\exp(-x^2), \ -\infty < x < \infty, \tag{8.36}$$

$$m_n(R) = \begin{cases} k! \ \text{if} \ n = 2k, \\ 0 \ \text{if} \ n \ \text{is odd.} \end{cases} \tag{8.37}$$

Theorem 8.6.2. Suppose $\{x_{i,n}\}$ satisfy Assumption I. Then:
(a) The EESD of $\{n^{-1/2}RC_n\}$ converges weakly to the symmetrized Rayleigh law whose $2k$-th moment is $k!$ and the odd moments are 0.
(b) The variable $n^{-1/2}RC_n$ in the $*$-probability space $(\mathcal{M}_n(\mathbb{C}), \mathrm{E}\,\mathrm{tr})$ converges to a self-adjoint variable whose moments are as described in (a). ◆

Proof. The link function is $L : \mathbb{Z}_+^2 \to \mathbb{Z}_+$ where $L(i,j) = (i+j-2)(\mod n)$. The convergence of its EESD may be established by arguments similar but simpler than those given earlier for the Hankel matrix. As in the previous proofs, it is enough to show that

$$\lim_n \varphi_n((n^{-1/2}RC_n)^{2k}) = \sum_{w \in \mathcal{P}_2(2k)} \lim_n \frac{1}{n^{k+1}} \#\Pi(w) = k!.$$

Now, due to Lemma 8.6.1, it is enough to verify the following two statements:
(i) If $w \in \mathcal{P}_2(2k) \cap \mathcal{S}_2^c(2k)$, then $\lim_{n\to\infty} \frac{1}{n^{k+1}} \#\Pi(w) = 0$.
(ii) If $w \in \mathcal{S}_2(2k)$, then for every choice of the generating vertices there is exactly one choice for each non-generating vertex, and hence $\lim_{n\to\infty} \frac{1}{n^{k+1}} \#\Pi(w) = 1$.

Proof of (i). We use the notation from the proof of Theorem 8.4.1. Also let

$$t_i = v_i + v_{i-1}. \tag{8.38}$$

Note that the vertices i, j match if and only if

$$\begin{aligned}(\pi(i-1) + \pi(i) - 2) \mod n &= (\pi(j-1) + \pi(j) - 2) \mod n. \\ \Leftrightarrow t_i - t_j &= 0,\ 1 \text{ or } -1.\end{aligned}$$

Since w is pair-matched, let $\{(i_s, j_s), 1 \le s \le k\}$ be such that $w[i_s] = w[j_s]$, and $j_s, 1 \le s \le k$, is in ascending order with $j_k = 2k$. So $\#\Pi(w)$ equals

$$\sum_{(r_i, 1 \le i \le k) \in \{0, \pm 1\}^k} \#\Big\{(v_0, v_1, \ldots, v_{2k}) : v_0 = v_{2k},\ v_i \in U_n, t_{i_s} - t_{j_s} = r_s\Big\}. \tag{8.39}$$

Let $r = (r_1, \ldots, r_k)$ be an element of $\{0, \pm 1\}^k$ and let $v_S = \{v_i : i \in S\}$. Observe that $v_i = L_i^H(v_S) + a_i^{(r)}, i \notin S$ for some integer $a_i^{(r)}$. Arguing as in the Hankel case, we easily reach the following equality (compare with (8.35)), $\lim_{n\to\infty} \frac{1}{n^{k+1}} \#\Pi(w) =$

$$\sum_{r \in \{0,\pm 1\}^k} \underbrace{\int_0^1 \cdots \int_0^1}_{k+1} \mathbb{I}(0 \le L_i^H(v_S) + a_i^{(r)} \le 1, i \notin S \cup \{2k\}) \\ \times \mathbb{I}(v_0 = L_{2k}^H(v_S) + a_{2k}^{(r)}) dv_S. \tag{8.40}$$

Now assume that $\lim \frac{1}{n^{k+1}}\Pi(w) \neq 0$. Then one of the terms in the above sum must be non-zero. For the integral to be non-zero, we must have

$$v_0 = L_{2k}^H(v_S) + a_{2k}^{(r)}. \tag{8.41}$$

Now $(t_{i_s} - t_{j_s} - r_s) = 0$ for all i_s, j_s, $1 \leq s \leq k$. Hence, trivially, for any choice of $\{\alpha_s\}$,

$$v_{2k} = v_{2k} + \sum_{s=1}^{k} \alpha_s(t_{i_s} - t_{j_s} - r_s). \tag{8.42}$$

Let us choose the integers $\{\alpha_s\}$ as follows: Let $\alpha_k = 1$. Having fixed α_k, α_{k-1}, $\ldots$, α_{s+1}, we choose α_s as follows:

(a) If $j_s + 1 \in \{i_m, j_m\}$ for some $m > s$, then set $\alpha_s = \pm\alpha_m$ accordingly as $j_s + 1$ equals i_m or j_m,

(b) If there is no such m, choose α_s to be any integer.

By this choice of $\{\alpha_s\}$, we ensure that in $v_{2k} + \sum_{s=1}^{k} \alpha_s(t_{i_s} - t_{j_s} - r_s)$, the coefficient of each $v_i, i \notin S$ vanishes. Hence we get

$$\begin{aligned} v_{2k} &= v_{2k} + \sum_{s=1}^{k} \alpha_s(t_{i_s} - t_{j_s} - r_s) \\ &= L(v_S) + a, \quad \text{a linear combination of the vertices } v_S. \end{aligned}$$

However, from (8.41), $v_0 = L_{2k}^H(v_S) + a_{2k}^{(r)}$. Hence, because only generating vertices are left in both the linear combinations,

$$v_{2k} + \sum_{s=1}^{k} \alpha_s(t_{i_s} - t_{j_s} - r_s) - v_0 = 0 \tag{8.43}$$

and thus the coefficient of each v_i in the left side has to be zero, including the constant term.

Now consider the coefficients of $\{t_i\}$ in (8.43). First, since $\alpha_k = 1$, the coefficient of t_{2k} is -1. On the other hand, the coefficient of v_{2k-1} is 0. Hence the coefficient of t_{2k-1} has to be $+1$.

Proceeding to the next step, we know that the coefficient of v_{2k-2} is 0. However, we have just observed that the coefficient of t_{2k-1} is $+1$. Hence the coefficient of t_{2k-2} must be -1. If we continue in this manner, in the expression (8.43) for all odd i, t_i must have coefficient $+1$, and, for all even i, t_i must have coefficient -1.

Now suppose that for some s, i_s and j_s are both odd or are both even. Then, for any choice of α_s, t_{i_s} and t_{j_s} will have opposite signs in the expression (8.43). This contradicts the fact stated in the last paragraph. Hence either i_s is odd and j_s is even, or i_s is even and j_s is odd. Since this happens for all $s, 1 \leq s \leq k$, w must be a symmetric word, proving (i).

Proof of (ii). Let $w \in \mathcal{S}_2(2k)$. First fix the generating vertices. Then we determine the non-generating vertices from left to right. Consider

$$L(\pi(i-1), \pi(i)) = L(\pi(j-1), \pi(j))$$

where $i < j$ and $\pi(i-1), \pi(i)$ and $\pi(j-1)$ have been determined. We rewrite it as

$$\pi(j) = A + dn \text{ for some integer } d \text{ with } A = \pi(i-1) + \pi(i) - \pi(j-1).$$

Clearly, $\pi(j)$ can be determined uniquely from the above equation since $1 \leq \pi(j) \leq n$. Continuing, we obtain the whole circuit uniquely. Hence the first part of (ii) is proved.

As a consequence, for $w \in \mathcal{S}_2(2k)$, only one term in the sum (8.40) will be non-zero and that will be equal to 1. Since there are exactly $k!$ symmetric words, (ii) is proved completely. This completes the proof of the theorem. ■

8.7 Symmetric Circulant: Gaussian law

Recall Figure 7.10 of the simulated ESD of $n^{-1/2}SC_n$ which may convince the reader that the LSD is the standard Gaussian law. We now prove this.

A self-adjoint variable in an NCP is said to be (standard) Gaussian if all its moments agree with the moments of a standard Gaussian random variable.

Theorem 8.7.1. Suppose $\{x_{i,n}\}$ satisfy Assumption I. Then:
(a) The EESD of $\{n^{-1/2}SC_n\}$ converges weakly to the standard Gaussian law.
(b) The variable $n^{-1/2}SC_n$ in the $*$-probability space $(\mathcal{M}_n(\mathbb{C}), \mathrm{E}\,\mathrm{tr})$ converges to a Gaussian variable. ◆

Proof. We only need to show that the even moments of the Symmetric Circulant converge to the corresponding Gaussian moments. Since $\#\mathcal{P}_2(2k) = \frac{(2k)!}{2^k k!}$, it is enough to show that

$$\lim_{n\to\infty} \frac{1}{n^{k+1}} \#\Pi(w) = 1 \text{ for each } w \in \mathcal{P}_2(2k). \tag{8.44}$$

Note that the link function of the Symmetric Circulant matrix is

$$L(i,j) = n/2 - |n/2 - |i-j||.$$

Let the slopes be defined as before by

$$s(l) = \pi(l) - \pi(l-1). \tag{8.45}$$

Vertices i, j, $i < j$, match if and only if

$$\begin{aligned} |n/2 - |s_i|| &= |n/2 - |s_j|| \\ \Leftrightarrow s(i) - s(j) &= 0, \pm n, \quad \text{or} \quad s(i) + s(j) = 0, \pm n. \end{aligned}$$

Then there are six possibilities in all. We first show that the first three possibilities do not contribute asymptotically.

Lemma 8.7.2. Fix a pair-matched word w with $|w| = k$. Let N be the number of pair-matched circuits of w which have at least one pair $i < j$ such that $s(i) - s(j) = 0, \pm n$. Then, as $n \to \infty$, $N = O(n^k)$ and $n^{-(k+1)}N \to 0$. ♦

Proof. Let $(i_1, j_1), (i_2, j_2), \ldots, (i_k, j_k)$ denote the pair-partition corresponding to the word w, i.e. $w[i_l] = w[j_l], 1 \leq l \leq k$. Suppose, without loss of generality that $s(i_k) - s(j_k) = 0, \pm n$. Clearly, a circuit π becomes completely specified if we know $\pi(0)$ and all the $\{s(i)\}$.

As already observed, if we fix some value for $s(i_l)$, then there are at most six options for $s(j_l)$. We may choose the values of $\pi(0), s(i_1), s(i_2), \ldots, s(i_{k-1})$ in $O(n^k)$ ways, and then we may choose values of $s(j_1), s(j_2), \ldots, s(j_{k-1})$ in $O(6^k)$ ways. For any such choice, from the sum restriction $\sum_{i=1}^{2k} s(i) = \pi(2k) - \pi(0) = 0$, we know $s(i_k) + s(j_k)$. On the other hand, by hypothesis, $s(i_k) - s(j_k) = 0, \pm n$. Hence the pair $(s(i_k), s(j_k))$ has only six possibilities. Thus there are at most $O(n^k)$ circuits with the given restrictions and the proof of the lemma is complete. ■

Now we continue with the proof of Theorem 8.7.1. Due to the above lemma, it now remains to show that with

$$\Pi'(w) = \{\pi : \pi \text{ is a circuit }, w[i] = w[j] \Rightarrow s_i + s_j = 0, \pm n\}, \tag{8.46}$$

$$\lim_{n \to \infty} \frac{1}{n^{k+1}} \#\Pi'(w) = 1. \tag{8.47}$$

Suppose

$$s_i + s_j = 0, \pm n \text{ for some } \ i < j.$$

If we know the circuit up to position $(j-1)$, then $\pi(j)$ has to take one of the values $A - n, A, A + n$, where

$$A = \pi(j-1) - \pi(i) + \pi(i-1).$$

Noting that $-(n-2) \leq A \leq (2n-1)$, exactly one of these three values will fall between 1 and n, and be a valid choice for $\pi(j)$. Thus if we first choose the generating vertices arbitrarily, then the non-generating vertices are determined from left to right uniquely so that $s(i) + s(j) = 0, \pm n$. This automatically yields $\pi(0) = \pi(2k)$ as follows: We have

$$\pi(2k) - \pi(0) = \sum_{i=1}^{2k} s_i = dn \text{ for some } d \in \mathbb{Z}. \tag{8.48}$$

But since $|\pi(2k) - \pi(0)| \leq n - 1$, we must have $d = 0$. Thus

$$\#\Pi'(w) = n^{k+1}$$

and hence

$$\lim_{n\to\infty} \frac{1}{n^{k+1}} \#\Pi'(w) = 1,$$

proving the theorem. ■

8.8 Almost sure convergence of the ESD

For all the random matrices that we have discussed so far, their ESD converge weakly almost surely under Assumption I. But the details of the proof of this fact are beyond the scope of this book. However, let us mention the line of arguments in brief for the symmetric matrices.

Denote any of these symmetric matrices by A_n. Recall that we have verified the conditions (M1) and (U) for A_n (see proof of Theorem 7.5.1). Now consider the following additional condition:

(M4) For every $h \geq 1$, $\sum_{n=1}^{\infty} \mathrm{E}[m_h(A_n) - \mathrm{E}(m_h(A_n))]^4 < \infty$.

It is easy to prove that if (M1), (M4) and (U) hold, then the ESD of A_n converges weakly almost surely. The condition (M4) is indeed true for all the symmetric matrices that we have discussed. For the convergence of the ESD, the moment condition in Assumption I may also be reduced by appropriate truncation arguments.

Below we offer some notes on the history of these matrices as well as sources where the results on the convergence of the ESD may be found.

(a) (Wigner matrix). If the entries in W_n are i.i.d. with mean 0 and variance 1, then the ESD of $n^{-1/2}W_n$ converges weakly to the semi-circular law almost surely. See Bose (2018)[20] for the details.

Historically, Wigner (1958)[106] assumed the entries $\{x_{i,j}, i \leq j\}$ to be i.i.d. real Gaussian and established the convergence of the EESD to the semi-circular law. Improvements and extensions can be found in Grenander (1963, pages 179 and 209)[49], Arnold (1967)[3] and Bai (1999)[4]. In an interesting article, Schenker and Schulz-Baldes (2005)[84] investigate how many independent matrix elements in the Wigner matrix are needed so that the semi-circular limit is retained. See Bose, Saha, Sen and Sen (2021)[27] for a collection of results on the LSD of several variations of the Wigner matrix.

(b) (S-matrix). Suppose the entries of Z are i.i.d. with mean 0, variance 1 and finite fourth moment. Suppose $p \to \infty$ such that $p/n \to y \neq 0$. Then the ESD of $S = n^{-1}ZZ^*$ converges weakly almost surely to the Marčenko-Pastur law MP_y. See Bose (2018)[20] for the details.

The first success in finding the LSD of S was due to Marčenko and Pastur (1967)[69]. Several authors have worked on this matrix over the years. Some of them are Grenander and Silverstein (1977)[50], Wachter (1978)[103], Jonsson (1982)[60], Yin (1986)[108], Bai and Yin (1988)[6] and Bai (1999)[4]. Yin and Krishnaiah (1985)[109] investigated the LSD of the S-matrix when the underlying probability law is isotropic. For further developments on the S-matrix, see Bai and Zhou (2008)[7].

When $p \to \infty$ and $p/n \to 0$, an interesting thing happens. Let I_p be the $p\times p$ identity matrix. Suppose that $p \to \infty$ and $p/n \to 0$. Then $\sqrt{np^{-1}}(n^{-1}S - I_p)$ converges to a semi-circular variable and its ESD converges to the the standard semi-circular law. See Bose (2018)[20] for a detailed proof.

(c) (IID matrix). The ESD of $\{n^{-1/2}C_n\}$ converges almost surely to the uniform probability law on the unit disc in the complex plane when the entries are from one i.i.d. sequence with mean 0 and variance 1. There were many weaker versions of this result over a long period of time. The proof of this result is quite long and difficult and is beyond the scope of this book. The interested reader may consult Bordenave and Chafai (2012)[19] for details on this and related matrices.

(d) (Elliptic matrix). Under suitable conditions, the ESD of $n^{-1/2}E_n$ where E_n is an elliptic matrix with $\rho \neq 1$, converges weakly almost surely to the uniform distribution on the interior of an ellipse whose axes are parallel to the x- and y-axes and the lengths of the minor and major axes of the ellipse depend only on the value of the correlation ρ. See Sompolinsk, Sommers, Crisanti and Stein (1988)[90], Götze, Naumov and Tikhomirov (2015)[47] and Nguyen and O'Rourke (2015)[73].

(e) (Toeplitz and Hankel matrices). The ESD converges weakly almost surely if the entries are from one i.i.d. sequence with mean 0 and variance 1. This was proved by Bryc, Dembo and Jiang (2006)[29]. See also Hammond and Miller (2005)[56]. A detailed proof using the approach discussed above may be found in Bose (2018)[20].

The convergence of the ESD is not known for non-symmetric random Toeplitz and Hankel matrices, even though simulation evidence supports the convergence. The convergence of the ESD of $n^{-1}AA^*$ when $A_{p\times n}$ is a non-symmetric Toeplitz or Hankel matrix is known when $p/n \to y \neq 0$. When $p/n \to 0$, the ESD of $\sqrt{np^{-1}}(n^{-1}AA^* - I_p)$ converges to the same limit as that of the symmetric Toeplitz matrix. See Bose (2018)[20] for the details.

(f) (Reverse Circulant and other circulant matrices). The Reverse Circulant and Symmetric Circulant are special cases of the class of k-Circulant matrices, which are treated in detail in Bose and Saha (2018)[26].

8.9 Exercises

1. Consider the covariance matrix S where the entries satisfy Assumption I. Let I_p be the $p \times p$ identity matrix. Suppose that $p \to \infty$ and $p/n \to 0$. Then show that the $p \times p$ matrix $\sqrt{np^{-1}}(n^{-1}S - I_p)$ converges to a semi-circular variable and its EESD converges to the the standard semi-circular law. Hint: Show that each Catalan word contributes 1 to the limit and other words do not contribute. For more details, see Theorem 3.3.1 of Bose (2018)[20]. This result will be used in the proof of Theorem 11.5.2.

2. Using the moments of the Marčenko-Pastur law show that its Stieltjes transform $s(z)$ satisfies the quadratic equation
$$yz(s(z))^2 + (y + z - 1)s(z) + 1 = 0, \quad \forall z \in \mathbb{C}^+. \tag{8.49}$$
The only valid solution of the above equation is given by
$$s(z) = \frac{1 - y - z + i\sqrt{((1 + \sqrt{y})^2 - z)(z - (1 - \sqrt{y})^2)}}{2yz}, \quad z \in \mathcal{D} \tag{8.50}$$
where $\mathcal{D}$ is an appropriate domain in $\mathbb{C}$. Identify the probability law from the Stieltjes transform.

3. Suppose (X, Y) is uniformly distributed over the unit disc. Find the marginal distribution of X.

4. Suppose (X, Y) is uniformly distributed on the interior of an ellipse centered at 0 with major and minor axis parallel to the two coordinate axes. Find the marginal distribution of X.

5. Suppose $\{E_n\}$ is a sequence of elliptic matrices whose entries satisfy Assumption Ie except that now we assume that $\mathrm{E}[x_{i.j,n}x_{j,i,n}] = \rho_{|i-j|}$. Show that under appropriate assumption on $\{\rho_i\}$, $n^{-1/2}E_n$ converges in distribution.

6. Prove Lemma 8.6.1.

7. Suppose $\{F_n\}$ is a sequence of circulant matrices (not necessarily symmetric) with entries that satisfy Assumption I. Show that as elements of $(\mathcal{M}_n(\mathbb{C})), n^{-1}\,\mathrm{E}\,\mathrm{Tr})$, $n^{-1/2}F_n$ converges to a complex Gaussian variable. [A variable is called *complex Gaussian* if its $*$-moments agree with those of the random variable $X + \iota Y$ where X and Y are independent Gaussian random variables].

9

Joint convergence I: single pattern

In Chapter 8, we have studied the convergence of patterned random matrices, one at a time. Joint convergence of random matrices has a significant place in statistics, operator algebras and wireless communications. We shall now focus on the joint convergence of independent copies of these matrices.

If a sequence of patterned random matrices converges, then their independent copies converge jointly. In particular, independent copies of any of the matrices—namely, Wigner, elliptic, S, Toeplitz, Hankel, Reverse Circulant and Symmetric Circulant—converge. One interesting consequence is that any symmetric matrix polynomial formed from copies of any of these patterned matrices has an LSD. Further, the first three matrices exhibit asymptotic freeness.

9.1 Unified treatment: extension

We extend the ideas of Section 7.5 to accommodate multiple independent matrices with the same pattern. Let $X_{i,n}$, $1 \leq i \leq m$, be symmetric, independent patterned random matrices of order n with a common link function L. We shall refer to $1 \leq i \leq m$ as m distinct *indices*. The (j,k)-th entry of the matrix $X_{i,n}$ will be denoted by $X_{i,n}(L(j,k))$. Suppressing the dependence on n, we shall simply write X_i for $X_{i,n}$. We view $\{\frac{1}{\sqrt{n}}X_i\}_{1\leq i\leq m}$ as elements of $(\mathcal{M}_n(\mathbb{C}), n^{-1}\,\mathrm{E}\,\mathrm{Tr})$.

A typical monomial q in these matrices is of the form

$$q(\{X_i\}_{1\leq i\leq m}) = X_{t_1}\cdots X_{t_k},$$

where $t_j \in \{1,2\ldots,m\}$ for all $1 \leq j \leq k$. We know that $\{\frac{1}{\sqrt{n}}X_i\}_{1\leq i\leq m}$ converges if for *all such* q, the following quantity converges:

$$\begin{aligned}\varphi_n(q) &= \frac{1}{n^{1+k/2}}\,\mathrm{E}[\mathrm{Tr}(X_{t_1}\cdots X_{t_k})] \qquad (9.1)\\ &= \frac{1}{n^{1+k/2}}\sum_{j_1,\ldots,j_k}\mathrm{E}[X_{t_1}(L(j_1,j_2))X_{t_2}(L(j_2,j_3))\cdots X_{t_k}(L(j_k,j_1))].\end{aligned}$$

All developments below are with respect to one fixed monomial q at a time.

DOI: 10.1201/9781003144496-9

Recall the notions of circuit, L-values, matches and pair-matches from Chapter 7. Leaving aside the scaling factor, a typical element in (9.1) can now be written as

$$\mathrm{E}\Big[\prod_{j=1}^{k} X_{t_j}(L(\pi(j-1),\pi(j)))\Big]. \tag{9.2}$$

A circuit π is said to be *index-matched*, if the value of $(t_j, L(\pi(j-1),\pi(j))$ for every j is repeated at least twice in the above product. When a specific monomial q is under consideration, the index-matches are with reference to q. If a circuit is not index-matched, then the expectation in (9.2) is zero. Hence we need to consider only index-matched circuits. Let

$$H = \{\pi : \pi \ \text{ is an index-matched circuit}\}.$$

Define an equivalence relation on H by demanding that $\pi_1 \sim_I \pi_2$, if, for all i, j such that $t_i = t_j$,

$$X_{t_i}(L(\pi_1(i-1),\pi_1(i))) = X_{t_j}(L(\pi_1(j-1),\pi_1(j)))$$

$$\Longleftrightarrow$$

$$X_{t_i}(L(\pi_2(i-1),\pi_2(i))) = X_{t_j}(L(\pi_2(j-1),\pi_2(j))).$$

Any equivalence class induces a partition of $\{1, 2, \ldots, k\}$ where each block of the partition also has an index attached to it. Any such class can be expressed as an *indexed word* w where letters appear in alphabetic order of their first occurrence and with a subscript to distinguish the index. For example consider the indexed word $a_2b_1a_2b_1c_1d_3c_1d_3$. The four corresponding partition blocks are $\{1,3\}, \{2,4\}, \{5,7\}$ and $\{6,8\}$ with the indices $2, 1, 1$ and 3 respectively. A typical (indexed) letter in an indexed word would be referred to as $w_{t_i}[i]$. The class of index-matched words is always with reference to a fixed q. For instance, the above word corresponds to the monomial $q = X_2X_1X_2X_1X_1X_3X_1X_3$. An indexed word is pair-matched if every indexed letter appears exactly two times. The above word is an example of an indexed pair-matched word.

The class of indexed circuits corresponding to an index-matched word w and the class of indexed pair-matched words are denoted, respectively, by

$$\begin{aligned}\Pi_q(w) &= \{\pi : w_{t_i}[i] = w_{t_j}[j] \text{ iff} \\ &\qquad X_{t_i}(L(\pi(i-1),\pi(i))) = X_{t_j}(L(\pi(j-1),\pi(j)))\} \\ \mathcal{P}_2(k,q) &= \{\text{all indexed pair-matched words } w \text{ of length } k \ (k \text{ even})\}. \end{aligned} \tag{9.3}$$

Note that we already have the corresponding *non-indexed* versions of the above notions from Chapter 7. If we drop all the indices from an indexed word, then we obtain a *non-indexed* word. For any monomial q, dropping the index amounts to dealing with only one matrix, that is, with the marginal distribution.

Recall that $w[i]$ denotes the i-th entry of a non-indexed word w. Also recall

$$\begin{aligned}\Pi(w) &= \{\pi : w[i] = w[j] \Leftrightarrow L(\pi(i-1), \pi(i)) = L(\pi(j-1), \pi(j))\} \quad (9.4)\\ \mathcal{P}_2(k) &= \{\text{all pair-matched words } w \text{ of length } k \ (k \text{ even})\}. \quad (9.5)\end{aligned}$$

For any indexed word $w \in \mathcal{P}_2(k, q)$, consider the non-indexed word w' obtained by dropping the indices. Then $w' \in \mathcal{P}_2(k)$. Since we are dealing with one fixed monomial at a time, this yields a bijective mapping, say

$$\psi_q : \mathcal{P}_2(k, q) \to \mathcal{P}_2(k). \quad (9.6)$$

We now extend Assumption I of Chapter 7 in a natural way. Recall the quantity Δ introduced in Chapter 7.

Assumption II. The matrices, $\{X_{i,n}\}_{1 \le i \le m}$, have a common link function L which satisfies $\Delta < \infty$. The input sequence of each matrix is independent with mean zero and variance 1, and they are also independent across i. Further, they satisfy Assumption I. ◻

Theorem 9.1.1. Suppose $\{X_i\}_{1 \le i \le m}$ satisfy Assumption II. Consider any monomial $q := X_{t_1} X_{t_2} \cdots X_{t_k}$. Assume that, whenever k is even,

$$p_q(w) = \lim_{n \to \infty} \frac{1}{n^{k/2+1}} \# \, \Pi(\psi_q(w)) \text{ exists for all } w \in \mathcal{P}_2(k, q). \quad (9.7)$$

(a) Then

$$\lim_{n \to \infty} \varphi_n(X_{t_1} X_{t_2} \cdots X_{t_k}) = \begin{cases} \sum_{w \in \mathcal{P}_2(k,q)} p_q(w) & \text{if } k \text{ is even,} \\ 0 & \text{otherwise.} \end{cases} \quad (9.8)$$

and

$$\sum_{w \in \mathcal{P}_2(k,q)} p_q(w) \begin{cases} \le \dfrac{k! \Delta^{k/2}}{(k/2)! 2^{k/2}} & \text{if each index appears an even number of times} \\ = 0 & \text{otherwise.} \end{cases}$$

(In the first case, k is necessarily even).

(b) If (9.7) holds for every monomial q (of even order), then $\{\frac{1}{\sqrt{n}} X_i\}_{1 \le i \le m}$ as variables of $(\mathcal{M}_n(\mathbb{C}), n^{-1} \operatorname{E} \operatorname{Tr})$ converge to $\{a_i\}_{1 \le i \le m}$ which are elements of the polynomial algebra $\mathbb{C}\langle a_1, a_2, \ldots, a_m \rangle$ in non-commutative indeterminates with the state φ defined on it by

$$\varphi(a_{t_1} \cdots a_{t_k}) := \lim_{n \to \infty} \varphi_n(X_{t_1} \cdots X_{t_k}) \quad (9.9)$$

for every $k \ge 1$ and every $t_1, \ldots t_k$, and $\lim_{n \to \infty} \varphi_n(q)$ is as in (9.8).

(c) In particular, the conclusion in (b) holds for Wigner, Toeplitz, Hankel, Reverse Circulant and Symmetric Circulant matrices. ◆

Note that when k is even, $\frac{k!}{(k/2)!2^{k/2}}$ is the number of non-indexed pair-matched words of length k.

Remark 9.1.1. Suppose (9.7) holds for all monomials (of even order). Let $q = X_{t_1} \cdots X_{t_k}$ be any symmetric matrix. Then an application of Theorem 9.1.1 (a) implies that the EESD of $n^{-k/2}q$ converges under Assumption II. ●

Proof of Theorem 9.1.1. (a) Fix a monomial $q = q(\{X_i\}_{1\leq i\leq m}) = X_{t_1} \cdots X_{t_k}$. Since ψ_q is a bijection,

$$\Pi_q(w) = \Pi(\psi_q(w)) \text{ for } w \in \mathcal{P}_2(k,q).$$

Hence using (9.7),

$$\begin{aligned}\lim_{n\to\infty} \frac{1}{n^{k/2+1}} \#\Pi_q(w) &= \lim_{n\to\infty} \frac{1}{n^{k/2+1}} \#\Pi(\psi_q(w)) \\ &= p(\psi_q(w)) \\ &= p_q(w).\end{aligned}$$

For simplicity, write

$$\begin{aligned}\boldsymbol{j} &= (j_1, \ldots, j_k), \\ \mathbb{T}_{\boldsymbol{j}} &= \mathrm{E}[X_{t_1}(L(j_1,j_2))X_{t_2}(L(j_2,j_3)) \cdots X_{t_k}(L(j_k,j_1))].\end{aligned}$$

Then

$$\varphi_n(q) = \frac{1}{n^{k/2+1}} \sum_{j_1,\ldots,j_k} \mathbb{T}_{\boldsymbol{j}}. \tag{9.10}$$

In the monomial, if any index appears once, then by independence and the mean zero condition, $\mathbb{T}_{\boldsymbol{j}} = 0$ for every $\boldsymbol{j}$. Hence $\varphi_n(q) = 0$.

So henceforth assume that each index that appears in the monomial appears at least twice. Further, if the circuit in $\mathbb{T}_{\boldsymbol{j}}$ is not index-matched, then $\mathbb{T}_{\boldsymbol{j}} = 0$.

Now form the following matrix M whose (i,j)-th element is given by

$$M(L(i,j)) = |X_{t_1}(L(i,j))| + |X_{t_2}(L(i,j))| + \cdots + |X_{t_k}(L(i,j))|.$$

Observe that

$$|\mathbb{T}_{\boldsymbol{j}}| \leq \mathrm{E}[M(L(j_1,j_2)) \cdots M(L(j_k,j_1))].$$

From equation (7.11) in the proof of Theorem 7.5.1, we conclude that the total contribution of all circuits which have at least one match of order three or more is zero in the limit.

As a consequence of the above discussion, if k is odd, then $\varphi_n(q) \to 0$.

So now assume that k is even. In that case, we need to consider only circuits which are pair-matched. Further, pair-matches must occur within the same index. Hence, if $\boldsymbol{j}$ corresponds to any such circuit, then by independence and the mean zero and variance one conditions, $\mathbb{T}_{\boldsymbol{j}} = 1$.

Using all the facts established so far,

$$\lim_{n\to\infty} \varphi_n(q) = \lim_{n\to\infty} \frac{1}{n^{k/2+1}} \sum_{\substack{\pi \text{ pair-matched} \\ \text{within indices}}} \mathrm{E}[X_{t_1}(L(\pi(0),\pi(1)))\cdots \times X_{t_k}(L(\pi(k-1),\pi(k)))]$$

$$= \lim_{n\to\infty} \frac{1}{n^{k/2+1}} \sum_{w\in\mathcal{P}_2(k,q),\ \pi\in\Pi(\psi_q(w))} \mathrm{E}[X_{t_1}(L(\pi(0),\pi(1)))\cdots \times X_{t_k}(L(\pi(k-1),\pi(k)))]$$

$$= \sum_{w\in\mathcal{P}_2(k,q)} p(\psi_q(w)).$$

The last claim in Part (a) follows since

$$\sum_{w\in\mathcal{P}_2(2k,q)} p_q(w) = \sum_{w\in\mathcal{P}_2(2k,q)} p(\psi_q(w)) \leq \sum_{w\in\mathcal{P}_2(k)} p(w) \leq \frac{(2k)!\Delta^k}{k!2^k}.$$

The last inequality was established in (7.14) in the proof of Theorem 7.5.1.

(b) The φ defined in (9.9) for monomials, can be extended to all polynomials linearly. It is clearly unital, tracial and positive since φ_n is so.

(c) This follows from (b) since for these matrices $p(w)$ exists for all $w \in \mathcal{P}(k)$. Recall that if $w \notin \mathcal{P}_2(2k)$ for some k, then $p(w) = 0$. All this follows from the proof of Theorem 7.5.1 in Chapter 7. ■

Almost sure joint convergence. The above theorem can be upgraded to an almost sure result as follows. Define

$$\widetilde{\mu_n}(q) = \frac{1}{n^{1+k/2}} \mathrm{Tr}[X_{t_1}\cdots X_{t_k}] = \frac{1}{n^{1+k/2}} \sum_{j_1,\ldots,j_k} X_{t_1}(L(j_1,j_2))X_{t_2}(L(j_2,j_3))\cdots X_{t_k}(L(j_k,j_1)).$$

Almost sure convergence of $\{\frac{1}{\sqrt{n}}X_i\}_{1\leq i\leq m}$ as elements of $(\mathcal{A}_n, \tilde{\phi}_n = \frac{1}{n}\mathrm{Tr})$ would hold if we can show that for every q,

$$\mathrm{E}\left[|\widetilde{\mu_n}(q) - \varphi_n(q)|^4\right] = \mathrm{O}(n^{-2}).$$

This is tedious and involves careful bookkeeping of terms and their orders. Then we can use the already proved convergence of $\varphi_n(q)$ and the

Borel–Cantelli lemma. In particular, this would imply that the almost sure weak limit of the ESD exists for any symmetric polynomial. The LSD is uniquely determined by the limit moments.

In the upcoming sections, we shall verify the condition (9.7) for different patterned matrices and identify structures in the limits whenever we can.

Example 9.1.1. In Theorem 9.2.1, we will verify (9.7) for independent Wigner matrices and hence claim their joint convergence. In Example 9.2.1 we shall identify the joint limit in terms of free variables. As a particular example, suppose that W_1, W_2 are two $n \times n$ independent Wigner matrices whose entries satisfy Assumption II. Then it follows that the matrix $n^{-3/2}W_1W_2W_1$ converges in the algebraic sense and its EESD converges weakly. Figure 9.1 provides the simulated eigenvalue distribution of this matrix. ▲

Example 9.1.2. Suppose W_1 and W_2 are $n \times n$ independent Wigner matrices whose entries satisfy Assumption II. Then the matrix $n^{-1}(W_1^2+W_2^2)$ converges in distribution and its EESD converges weakly. In Example 9.2.1, we will identify the limit in terms of free variables. Figure 9.2 provides the simulated ESD of the of this matrix.

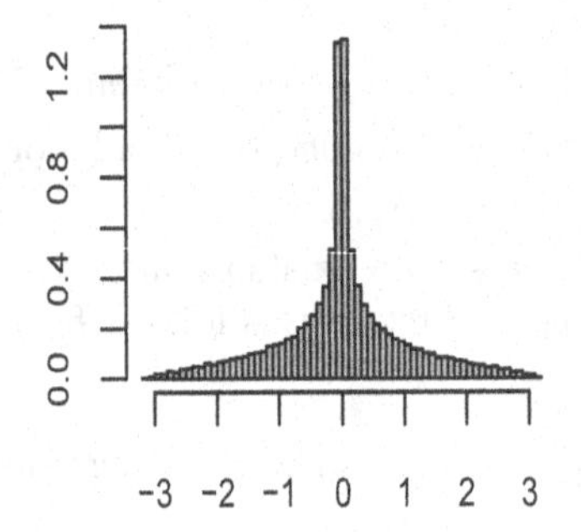

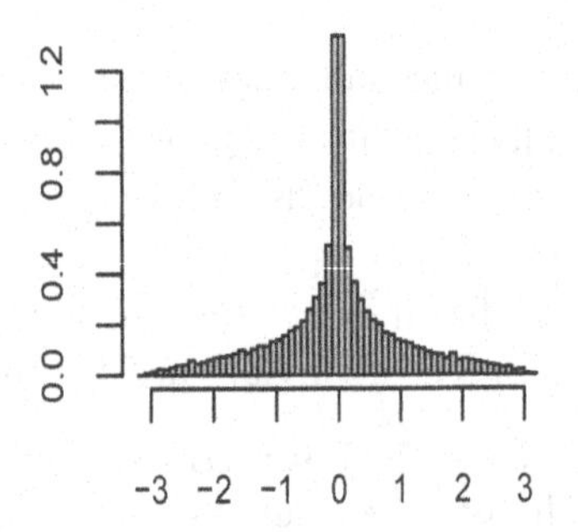

FIGURE 9.1
ESD of $n^{-3/2}W_1W_2W_1$, $n = 400$. Left, standard normal; right, standardized symmetric Bernoulli.

▲

9.2 Wigner matrices: asymptotic freeness

The convergence of the ESD of the Wigner matrix to the semi-circular law has been shown in Section 8.1.

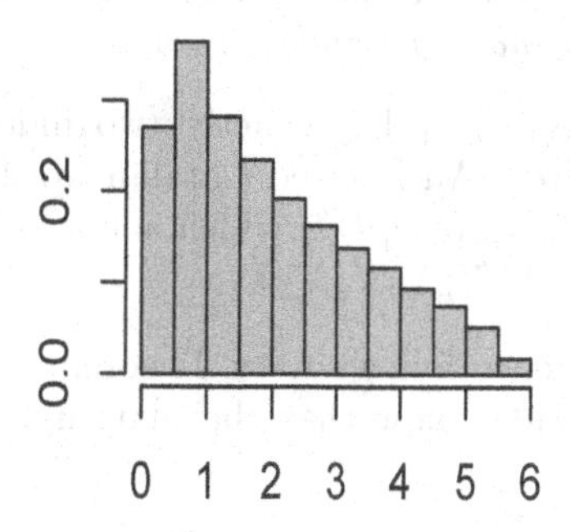

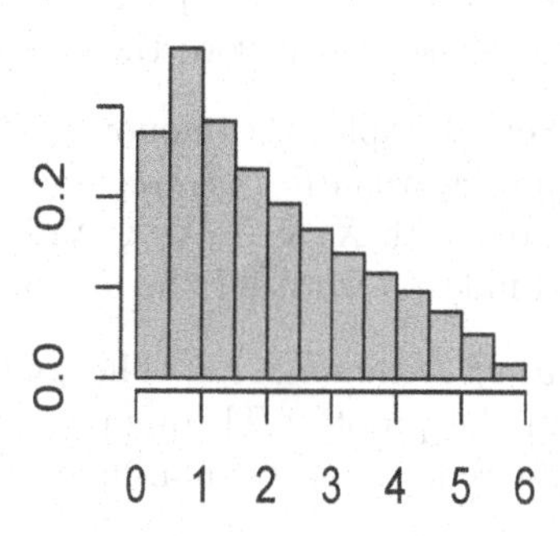

FIGURE 9.2
ESD of $n^{-1}(W_1^2 + W_2^2)$, $n = 400$. Left, standard normal; right, standardized symmetric Bernoulli.

The joint convergence of Wigner matrices was first studied in Voiculescu (1991)[100]. Independent Wigner matrices whose entries satisfy Assumption II become asymptotically free. This is a fundamental result in random matrix theory. We now state and prove this result.

Theorem 9.2.1. Let $\{W_n^{(i)}\}_{1\leq i\leq m}$ be m independent sequences of $n \times n$ Wigner matrices which satisfy Assumption II. Then $\{n^{-1/2}W_n^{(i)}\}_{1\leq i\leq m}$, as elements of $(\mathcal{M}_n(\mathbb{C}), \mathrm{E}\,\mathrm{tr})$, converge jointly to free semi-circular variables. ♦

Remark 9.2.1. (a) Recall that Wigner matrices are a special case of the elliptic matrices where the correlation is 1. In the next section, we shall see that independent elliptic matrices are also asymptotically free. The above theorem then follows as a special case. Nevertheless, we give a separate proof for Wigner matrices.

(b) The theorem is valid for Wigner matrices with complex random variables as entries—in particular, for the GUE. This will be clear from the proof and the argument is along the lines of what we mentioned in Remark 8.1.2. We omit the details.

(c) Due to our discussion in the previous section on the almost sure convergence, the almost surely free version of the theorem is also valid. ●

Note that from Theorem 9.1.1, $\lim \varphi_n(q) = 0$ if k is odd or if any index appears an odd number of times in the monomial q. Henceforth we thus assume that the order of the monomial is even and that each index appears an even number of times.

We extend the definition of Catalan words given in Definition 8.1.1 to move from one Wigner matrix to several Wigner matrices.

Definition 9.2.1. (Indexed Catalan word) Fix a monomial q of length $k \geq 2$. If for a $w \in \mathcal{P}_2(k, q)$, sequentially deleting all double letters of the same index leads to the empty word then we call w an *indexed Catalan word.* ◇

For example, the monomial $X_1X_2X_2X_1X_1X_1$ has exactly two indexed Catalan words $a_1b_2b_2a_1c_1c_1$ and $a_1b_2b_2c_1c_1a_1$. An indexed Catalan word associated with $X_1X_2X_2X_1X_1X_1X_2X_2$ is $a_1b_2b_2a_1c_1c_1d_2d_2$ which is not even a valid indexed word for the monomial $X_1X_1X_1X_1X_2X_2X_2X_2$.

Proof of Theorem 9.2.1. The joint convergence follows from Theorem 9.1.1. From the proof of Theorem 8.1.2, we already know that the marginals are semi-circular, and for non-indexed words,

$$p(w) = \begin{cases} 1 & \text{if } w \text{ is a Catalan word,} \\ 0 & \text{otherwise.} \end{cases} \tag{9.11}$$

It remains to show that $\{a_i\}_{1 \leq i \leq m}$ are free.

Fix any monomial $q = x_{t_1}x_{t_2}\cdots x_{t_{2k}}$ where each index appears an even number of times. Let w be an indexed Catalan word. It remains Catalan when we ignore the indices. Hence from above, $p_q(w) = p(\psi_q(w)) = 1$. Likewise, if w is not indexed Catalan then the word $\psi_q(w)$ cannot be Catalan and hence $p_q(w) = p(\psi_q(w)) = 0$.

Hence if $\mathcal{C}(2k, q)$ denotes the set of indexed Catalan words corresponding to a monomial q of length $2k$, then from the above discussion

$$\lim_{n\to\infty} \varphi_n(q) = \#(\mathcal{C}(2k, q)).$$

From Lemma 8.1.1, it is known that the set of Catalan words of length $2k$, $\mathcal{C}_{2k}$, is in bijection with the set of non-crossing pair-partitions, $NC_2(2k)$.

Note that the elements of any block of a pair-partition must share the same index. We denote a typical block of π as $(r, s), r < s$. Recalling the δ-function defined in (8.16), then we have

$$\#(\mathcal{C}(2k, q)) = \sum_{\pi \in NC_2(2k)} \prod_{(r,s)\in\pi} \delta_{t_r t_s}.$$

This holds for every monomial q. Thus for every choice of k and $\{t_i\}$,

$$\varphi(a_{t_1}\cdots a_{t_{2k}}) = \sum_{\pi \in NC_2(2k)} \prod_{r,s\in\pi} \delta_{t_r t_s}.$$

Now, if we invoke the moment-free cumulant relation, we can identify the free cumulants of $\{a_i\}$: all mixed free cumulants of $\{a_i\}_{1\leq i\leq m}$ are zero, the second order free cumulant of each a_i equals 1, and all other free cumulants are 0. Therefore, $\{a_i\}_{1\leq i\leq m}$ are free standard semi-circular variables. ■

Example 9.2.1. A consequence of the above theorem is that we can identify the LSD of any *symmetric* matrix polynomial of independent Wigner matrices in terms of free semi-circular variables.

For instance, the LSD of $n^{-1}(W_1^2 + W_2^2)$ is the probability law of $s_1^2 + s_2^2$ where $s_i, i = 1, 2$, are standard free semi-circular variables. This law is the free Poisson law with parameter 2. We already provided a simulation in Figure 9.2. Similarly, the LSD of $n^{-3/2}W_1W_2W_1$ is the same as the probability law of $s_1s_2s_1$ where $s_i, i = 1, 2$, are standard free semi-circular variables. We provided a simulation of this law in Figure 9.1. ▲

9.3 Elliptic matrices: asymptotic freeness

In Theorem 8.3.1 of Chapter 8, we proved that $n^{-1/2}E_n$ converges to an elliptic variable. We now explore the joint convergence of independent copies of these matrices. In Figures 9.3 and 9.4 we provide simulation for the eigenvalues for two matrix polynomials in independent copies of elliptic matrices.

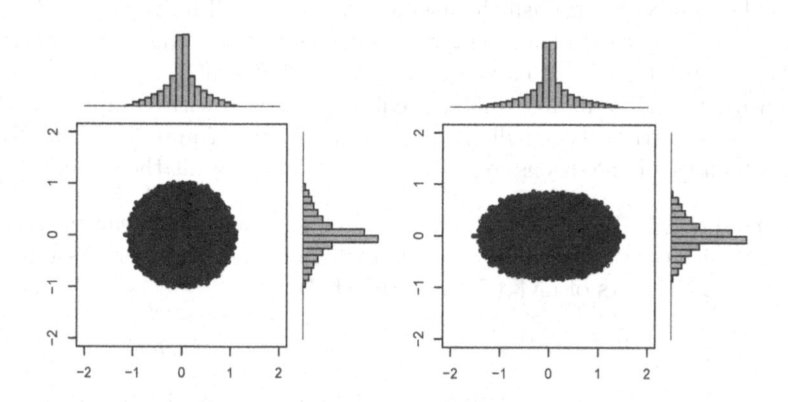

FIGURE 9.3
ESD of $n^{-3/2}E_1E_2E_1$, bivariate normal, $n = 1000$. Left: $\rho = 0.25$; right $\rho = 0.5$

Recall Assumption Ie from Chapter 8.

Assumption Ie For every n, $\{x_{i,i,n} : 1 \leq i \leq n\} \cup \{(x_{i,j,n}, x_{j,i,n}) : 1 \leq i < j \leq n\}$ is a collection of independent random variables which satisfy $\mathrm{E}[x_{i,j,n}] = 0$, $\mathrm{E}[x_{i,j,n}^2] = 1$ for all i, j, and $\mathrm{E}[x_{i.j,n}x_{j,i,n}] = \rho$ for $1 \leq i \neq j \leq n$. Moreover,

$$\sup_{i,j,n} \mathrm{E}(|x_{i,j,n}|^k) \leq B_k < \infty \quad \text{for all} \quad k \geq 1.$$

⊡

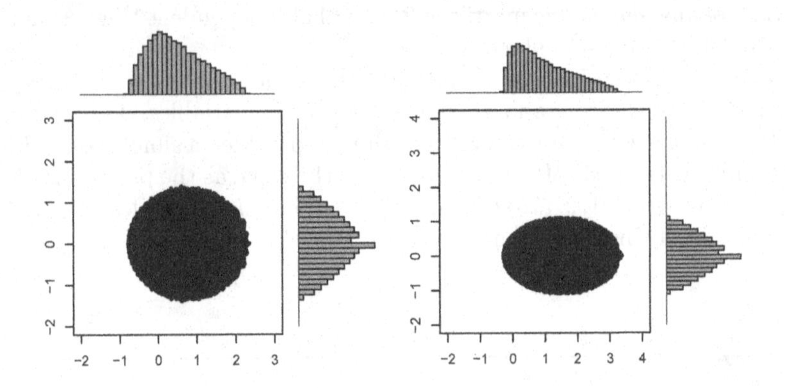

FIGURE 9.4
ESD of $n^{-3/2}(E_1^2 + E_2^2)$, bivariate normal, $n = 1000$. Left: $\rho = 0.25$; right $\rho = 0.5 n = 1000$.

We now prove that independent elliptic matrices $\{E_i\}$ whose entries satisfy Assumption Ie, with possibly different $\{\rho_i\}$, are asymptotically free. In particular, this will establish the asymptotic freeness of independent IID matrices ($\rho_i = 0$ for all i). The asymptotic freeness result of the previous section for Wigner matrices is also a special case ($\rho_i = 1$ for all i).

Theorem 9.3.1 remains valid for elliptic matrices with complex random variables as entries. This will be clear from the proof and the argument is along the lines of the discussion in Remark 8.1.2. We omit the details.

Theorem 9.3.1. Let $E_n^{(1)}, E_n^{(2)}, \ldots, E_n^{(m)}$ be m independent elliptic random matrices with correlations $\rho_i, 1 \leq i \leq m$ and whose entries satisfy Assumption Ie. Then as elements of $(\mathcal{M}_n(\mathbb{C}), \varphi_n = n^{-1} \mathrm{E}\,\mathrm{Tr})$,

$$(n^{-1/2} E_n^{(1)}, \ldots, n^{-1/2} E_n^{(m)}) \to (e_1, \ldots, e_m),$$

where $e_1, \ldots, e_m$ are free and elliptic with parameters ρ_i, $1 \leq i \leq m$. ◆

Proof. We essentially extend the approach used in the proof of Theorem 8.3.1 which was for a single matrix. Let $\tau_1, \ldots, \tau_t \in \{1, 2, \ldots, m\}$ and $\epsilon_1, \ldots, \epsilon_t \in \{1, *\}$. We borrow the notation from that proof. Let

$$\bar{E}_n^{(\tau_j)} = n^{-1/2} E_n^{(\tau_j)}, \ 1 \leq j \leq m.$$

Now

$$\varphi_n(\bar{E}_n^{(\tau_1)\epsilon_1} \bar{E}_n^{(\tau_2)\epsilon_2} \cdots \bar{E}_n^{(\tau_t)\epsilon_t}) = n^{-(1+t/2)} \sum_{(i_1, i_2, \ldots, i_{2k}) \in I_{2k}} \mathrm{E}[x_{i_1 i_2}^{(\tau_1)\epsilon_1} x_{i_2 i_3}^{(\tau_2)\epsilon_2} \cdots x_{i_t i_1}^{(\tau_t)\epsilon_t}].$$

Since $\{x_{ij}\}$ have mean zero and all their moments are finite, as in the proof of Theorem 8.3.1, if t is odd then

$$\lim_{n\to\infty} \varphi_n(\bar{E}_n^{(\tau_1)\epsilon_1} \bar{E}_n^{(\tau_2)\epsilon_2} \cdots \bar{E}_n^{(\tau_t)\epsilon_t}) = 0.$$

So we assume $t = 2k$. Arguing similarly to the steps used in the proof of Theorem 8.3.1, we conclude that while calculating moments by the moment-trace formula, to begin with, only those combinations where only pair-matches appear can potentially contribute to the limit moments. Hence,

$$\sum_{I_{2k}} \mathrm{E}[x_{i_1 i_2}^{(\tau_1)\epsilon_1} x_{i_2 i_3}^{(\tau_2)\epsilon_2} \cdots x_{i_{2k} i_1}^{(\tau_{2k})\epsilon_{2k}}] = \sum_{\pi \in \mathcal{P}_2(2k)} \sum_{I_{2k}(\pi)} \prod_{(r,s)\in\pi} \mathrm{E}[x_{i_r i_{r+1}}^{(\tau_r)\epsilon_r} x_{i_s i_{s+1}}^{(\tau_s)\epsilon_s}] + o(n^{k+1}) \tag{9.12}$$

where $I_{2k}(\pi) \subset I_{2k}$ consists of elements which obey the partition π. Define

$$\begin{aligned} a'(r,s) &= \delta_{i_r i_s} \delta_{i_{r+1} i_{s+1}}, \\ b'(r,s) &= \delta_{i_r i_{s+1}} \delta_{i_s i_{r+1}}, \\ a_\tau(r,s) &= (\rho_\tau + (1-\rho_\tau)\delta_{\epsilon_r \epsilon_s}) a'(r,s), \\ b_\tau(r,s) &= (1 - (1-\rho_\tau)\delta_{\epsilon_r \epsilon_s}) b'(r,s). \end{aligned}$$

Then

$$\mathrm{E}[x_{i_r i_{r+1}}^{(\tau_r)\epsilon_r} x_{i_s i_{s+1}}^{(\tau_s)\epsilon_s}] = \big(a_{\tau_{rs}}(r,s) + b_{\tau_{rs}}(r,s)\big)\delta_{\tau_r \tau_s},$$

where $\tau_{rs} = \tau_r = \tau_s$. Now, it is also clear from the arguments given in the proof of Theorem 8.3.1 that amongst the pair-matches in equation (9.12), only those terms that obey the (C2) constraint survive. Thus, those terms with only the b-factors survive and we obtain

$$\begin{aligned} \lim_{n\to\infty} \varphi_n(\bar{E}_n^{(\tau_1)\epsilon_1} \cdots \bar{E}_n^{(\tau_{2k})\epsilon_{2k}}) &= \lim_{n\to\infty} \frac{1}{n^{k+1}} \sum_{\pi\in\mathcal{P}_2(2k)} \sum_{I_{2k}(\pi)} \prod_{(r,s)\in\pi} b_{\tau_{rs}}(r,s)\delta_{\tau_r \tau_s} \\ &= \sum_{\pi\in NC_2(2k)} \rho_1^{T_1(\pi)} \cdots \rho_m^{T_m(\pi)} \prod_{(r,s)\in\pi} \delta_{\tau_r \tau_s}, \end{aligned} \tag{9.13}$$

where

$$T_\tau(\pi) := \#\{(r,s)\in\pi \; : \; \delta_{\epsilon_r \epsilon_s} = 1, \tau_r = \tau_s = \tau\}.$$

On the other hand, for any variables $e_1, \ldots, e_m$, by the moment-free cumulant relation (2.17) of Chapter 2,

$$\varphi(e_{\tau_1}^{\epsilon_1} e_{\tau_2}^{\epsilon_2} \cdots e_{\tau_{2k}}^{\epsilon_{2k}}) = \sum_{\pi\in NC(2k)} \prod_{V\in\pi} \kappa_{|V|}[e_{\tau_1}^{\epsilon_1}, \ldots, e_{\tau_{2k}}^{\epsilon_{2k}}]. \tag{9.14}$$

Now if $\{e_i\}$ are free elliptic with parameters $\{\rho_i\}$, then we know that

$$\begin{aligned} \kappa(e_i, e_i) &= \kappa(e_i^*, e_i^*) = \rho_i \text{ for } i = 1, \ldots, m \\ \kappa(e_i, e_i^*) &= \kappa(e_i^*, e_i) = 1 \text{ for } i = 1, \ldots, m \end{aligned}$$

and all other free cumulants are zero. Therefore equation (9.14) reduces to

$$\varphi(e_{\tau_1}^{\epsilon_1} e_{\tau_2}^{\epsilon_2} \cdots e_{\tau_{2k}}^{\epsilon_{2k}}) = \sum_{\pi \in NC_2(2k)} \rho_1^{T_1(\pi)} \cdots \rho_m^{T_m(\pi)} \prod_{(r,s)\in\pi} \delta_{\tau_r \tau_s}. \tag{9.15}$$

Using the freeness criteria given in Lemma 3.7 (b), and comparing (9.13) and (9.15), we conclude that $n^{-1/2}E_n^{(1)}, \ldots, n^{-1/2}E_n^{(m)}$ converge in $*$-distribution to free $e_1, \ldots, e_m$ which are elliptic with parameters $\rho_1, \ldots \rho_m$ respectively. ■

9.4 S-matrices in elliptic models: asymptotic freeness

Recall the Marčenko-Pastur law MP_y, $y > 0$, introduced in (8.6). It has a positive mass $1 - \frac{1}{y}$ at 0 if $y > 1$. Elsewhere it has the density

$$f_{MPy}(x) := \begin{cases} \frac{1}{2\pi xy}\sqrt{(b-x)(x-a)} & \text{if } a \le x \le b, \\ 0 & \text{otherwise} \end{cases} \tag{9.16}$$

where $a := a(y) := (1 - \sqrt{y})^2$ and $b := b(y) := (1 + \sqrt{y})^2$.

In Section 8.2 we have seen that the EESD of the sample covariance matrix, $S = n^{-1}XX^*$, converges to MP_y when $p/n \to y$ if the entries of X satisfy Assumption I. We now wish to study the joint convergence of independent copies of the sample covariance matrix. Figures 9.5 and 9.6 provide two simulations. Theorem 9.4.1 claims the asymptotic freeness of independent S-matrices whose entries satisfy Assumption Ie.

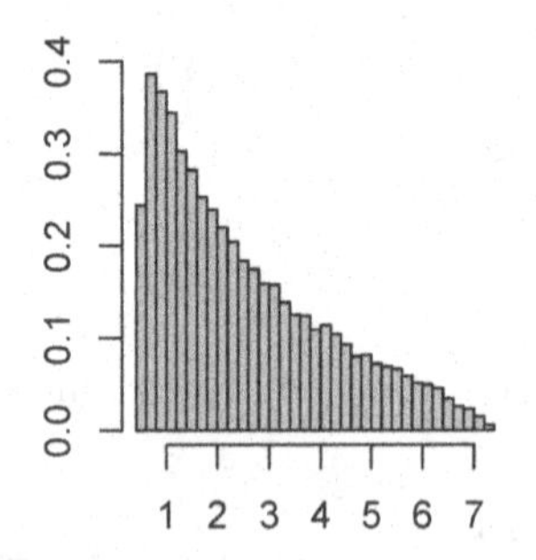

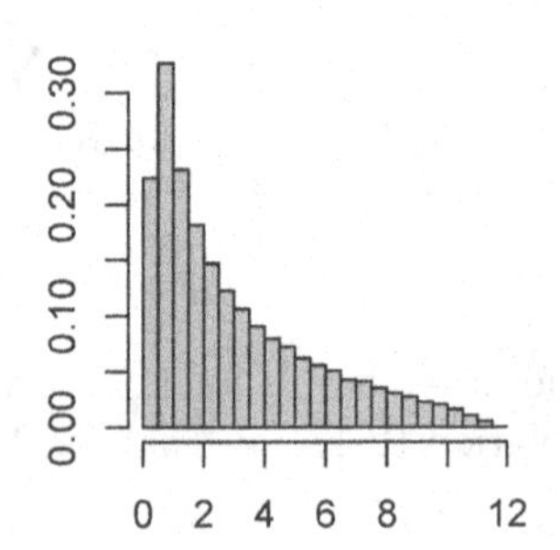

FIGURE 9.5
ESD of $S_1^2 + S_2^2$, $n = 1000$. Left: $p = 250$; Right: $p = 500$.

Theorem 9.4.1. Let $X_{p\times n}$ be an elliptic rectangular random matrix whose entries satisfy Assumption Ie. Let $S = \frac{1}{n}X_{p\times n}X_{n\times p}^*$ and suppose $\frac{p}{n} \to y > 0$ as $p \to \infty$. Then as $p \to \infty$, the following hold:

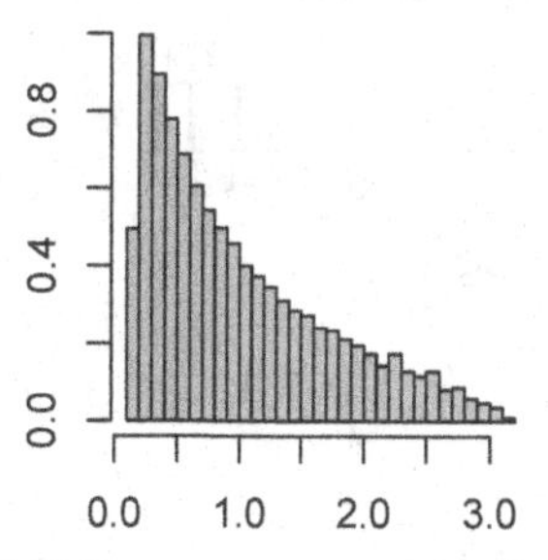

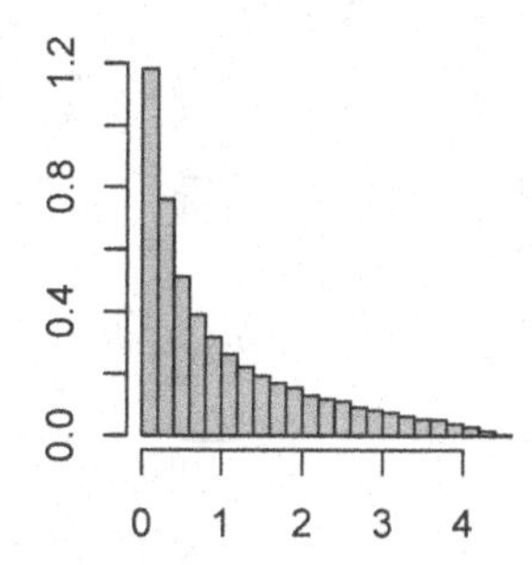

FIGURE 9.6
ESD of S_1S_2, $n = 1000$. Left: $p = 250$; Right: $p = 500$.

(a) As an element of $(\mathcal{M}_p(\mathbb{C}), p^{-1}\,\mathrm{E}\,\mathrm{Tr})$, S converges.

(b) The EESD of S converges to the Marčenko-Pastur law with parameter y.

(c) Let $S^{(\tau)}, 1 \leq \tau \leq m$, be independent S-matrices with entries $\{x_{i,j,n}^{(\tau)}\}$ that satisfy Assumption Ie with possibly different correlation parameters ρ_τ. If $\frac{p}{n} \to y > 0$ as $p \to \infty$, then $S^{(1)}, \ldots, S^{(m)}$ are asymptotically free.

In particular, S-matrices with i.i.d. entries in X, which satisfy Assumption I, are asymptotically free. ◆

Remark 9.4.1. (MP_y variables and their moments) By abuse of notation, a variable in some NCP is said to be an MP_y variable if its moments agree with the moments of the MP_y law. Before we prove Theorem 9.4.1 let us see what the *joint moments of free self-adjoint* MP_y *variables* would look like. Recall from (8.11) that the free cumulants of MP_y are given by

$$\kappa_n(\mathrm{MP}_y) = y^{n-1}, \; n \geq 1. \tag{9.17}$$

Let $S_1, S_2, \ldots S_m$ be free $MP_{y_1}, MP_{y_2}, \ldots, MP_{y_m}$ variables, respectively, in an NCP $(\mathcal{A}, \varphi)$. Let $\tau_i \in \{1, 2 \ldots, m\}$, $1 \leq i \leq k$. Let

$$J_i = \{t : 1 \leq t \leq k, \; \tau_t = i\}, \; 1 \leq i \leq m.$$

Note that

$$J_1 \cup J_2 \ldots \cup J_m = \{1, 2, \ldots, k\}.$$

Let

$$T_i = \# J_i = \sum_{t=1}^{k} \delta_{\tau_t, i}, \; 1 \leq i \leq m.$$

Let $A_{t_1,\ldots,t_m,k}$ be defined as

$$\{\pi \in NC(k) : \pi = \cup_{i=1}^m \pi_i, \pi_i \in NC(J_i) \; \text{ has } \; t_i + 1 \; \text{ blocks}, \; 1 \leq i \leq m\}.$$

Then by freeness of $\{S_i\}$ and (9.17),

$$\begin{aligned}
\varphi(S_{\tau_1}S_{\tau_2}\cdots S_{\tau_k}) &= \sum_{t_1=0}^{T_1-1}\cdots\sum_{t_m=0}^{T_m-1}\sum_{\substack{\pi\in A_{t_1,t_2,\ldots,t_m,k}\\ \pi_i=\{V_{ij}:\ 1\le j\le t_i+1\}\\ 1\le i\le m}}\prod_{i=1}^{m}\prod_{j=1}^{t_i+1} y_i^{|V_{ij}|-1} \\
&= \sum_{t_1=0}^{T_1-1}\cdots\sum_{t_m=0}^{T_m-1}\left(\prod_{i=1}^{m} y_i^{T_i-t_i-1}\right)(\#A_{t_1,t_2,\ldots,t_m,k}) \\
&= \sum_{\substack{0\le t_i\le T_i-1\\ 1\le i\le m-1}}\left(\prod_{i=1}^{m} y_i^{t_i}\right)(\#A_{T_1-t_1-1,T_2-t_2-1,\ldots,T_m-t_m-1,k}).
\end{aligned}$$

The moments of MP_y are a special case: if $\tau_1=\tau_2=\cdots=\tau_k=1$, then

$$\begin{aligned}
\varphi(S_1^k) &= \sum_{t=0}^{k-1} y^t(\#A_{k-t-1,k}) = \sum_{t=0}^{k-1}\frac{1}{k}\binom{k}{k-t-1}\binom{k}{k-t}y^t \\
&= \sum_{t=0}^{k-1}\frac{1}{k}\binom{k}{t+1}\binom{k}{t}y^t = \sum_{t=0}^{k-1}\frac{1}{t+1}\binom{k-1}{t}\binom{k}{t}y^t.
\end{aligned}$$

•

Proof of Theorem 9.4.1. We have

$$\frac{1}{p}\mathrm{E}[\mathrm{Tr}(S^k)] = \frac{1}{pn^k}\sum_{I'_{2k}}\mathrm{E}[x_{i_1i_2}^{\epsilon_1}x_{i_2i_3}^{\epsilon_2}\cdots x_{i_{2k}i_1}^{\epsilon_{2k}}], \tag{9.18}$$

where for $t=1,\ldots,k$,

$$\begin{aligned}
I'_{2k} &= \{(i_1,\ldots,i_{2k})\ :\ 1\le i_{2t}\le n,\ 1\le i_{2t-1}\le p,\ t=1,\ldots,k\}, \\
\epsilon_{2t} &= *,\ \text{ and }\ \epsilon_{2t-1}=1.
\end{aligned}$$

We shall use the two crucial observations (i) and (ii) made in the proof of Theorem 8.3.1. Using the first observation, only pair-matches contribute and hence we have

$$\frac{1}{pn^k}\sum_{I'_{2k}}\mathrm{E}[x_{i_1i_2}^{\epsilon_1}x_{i_2i_3}^{\epsilon_2}\cdots x_{i_{2k}i_1}^{\epsilon_{2k}}] = \frac{1}{pn^k}\sum_{\pi\in\mathcal{P}_2(2k)}\sum_{I'_{2k}(\pi)}\prod_{(r,s)\in\pi}\mathrm{E}[x_{i_ri_{r+1}}^{\epsilon_r}x_{i_si_{s+1}}^{\epsilon_s}]+o(1),$$

where $I'_{2k}(\pi)\subset I'_{2k}$ consists of elements which obey the partition π.

Using the properties of the random variables that appear in the matrices, we have

$$
\begin{aligned}
\mathrm{E}[x_{i_r i_{r+1}}^{\epsilon_r} x_{i_s i_{s+1}}^{\epsilon_s}] =& \big[\delta_{i_r i_s}\delta_{i_{r+1} i_{s+1}} + \rho\delta_{i_r i_{s+1}}\delta_{i_{r+1} i_s}\delta_{\{i_r, i_{r+1}\le \min\{p,n\}\}}\big]\delta_{\epsilon_r \epsilon_s} \\
&+ \big[\delta_{i_r i_{s+1}}\delta_{i_{r+1} i_s} + \rho\delta_{i_r i_s}\delta_{i_{r+1} i_{s+1}}\delta_{\{i_r, i_{r+1}\le \min\{p,n\}\}}\big](1-\delta_{\epsilon_r \epsilon_s}) \\
=& \big[\delta_{\epsilon_r \epsilon_s} + \rho(1-\delta_{\epsilon_r \epsilon_s})\delta_{\{i_r, i_{r+1}\le \min\{p,n\}\}}\big]\delta_{i_r i_s}\delta_{i_{r+1} i_{s+1}} \\
&+ \big[\rho\delta_{\epsilon_r \epsilon_s}\delta_{\{i_r, i_{r+1}\le \min\{p,n\}\}} + (1-\delta_{\epsilon_r \epsilon_s})\big]\delta_{i_r i_{s+1}}\delta_{i_{r+1} i_s} \\
=& g(r,s) + f(r,s) \text{ say.}
\end{aligned}
$$

Note that $|g(r,s)| \le \delta_{i_r i_s}\delta_{i_{r+1} i_{s+1}}$ and $|f(r,s)| \le \delta_{i_r i_{s+1}}\delta_{i_{r+1} i_s}$. Also note that $p/n \to y > 0$.

Now, using the second crucial observation (ii) made in the proof of Theorem 8.3.1, only pair-matched non-crossing words (partitions) where all matches are (C2) contribute. Note that $g(r,s) = 0$ whenever there is a (C1) match and hence only the terms that involve $f(r,s)$ survive. Moreover, if all matches are (C2), then

$$f(r,s) = \delta_{i_r i_{s+1}}\delta_{i_{r+1} i_s} = 1 \quad \text{for all } (r,s) \in \pi.$$

Hence

$$
\begin{aligned}
\lim_{p\to\infty} \frac{1}{pn^k} \sum_{I'_{2k}} \mathrm{E}[x_{i_1 i_2}^{\epsilon_1} x_{i_2 i_3}^{\epsilon_2} \cdots x_{i_{2k} i_1}^{\epsilon_{2k}}] &= \lim_{p\to\infty} \frac{1}{pn^k} \sum_{\pi\in\mathcal{P}_2(2k)} \sum_{I'_{2k}(\pi)} \prod_{(r,s)\in\pi} f(r,s) \\
&= \sum_{\pi\in NC_2(2k)} \lim_{n\to\infty} \frac{1}{pn^k} \sum_{I'_{2k}(\pi)} \prod_{(r,s)\in\pi} 1. \qquad (9.19)
\end{aligned}
$$

The above expression is free of ρ and thus would be the same if the entries were i.i.d. As a cross-check, note that the above count and limit are exactly what we calculated during the course of the proof of Theorem 8.2.3 (see (8.12) there). This completes the proofs of Parts (a) and (b). We prove Part (c) after a remark.

Remark 9.4.2. The proof of Theorem 8.2.3 used generating and non-generating vertices while in the above proof we have introduced and used the δ function. The link between the two approaches is elucidated below with an example.

We know that any $w \in \mathcal{C}(2k)$ is identified with a partition $\pi_w \in NC_2(2k)$. For example,

$$w = abbccddaee \Rightarrow \pi_w = \{\{1,8\},\{2,3\},\{4,5\},\{6,7\},\{9,10\}\} \in NC_2(10).$$

When using the generating vertices, we distinguished between the even and odd generating vertices. Suppose the total number of even generating vertices is $t+1$ for some $0 \le t \le k-1$. Then the number of odd generating vertices is $k-t$. Observe that $1 \le k-t \le k$. The total contribution then equals $T = p^{t+1}n^{k-t}$ and $T/(pn^k) \to y^t$. For the word w, we have $t+1 = 4$ even generating vertices, namely $\pi(0), \pi(2), \pi(4), \pi(6)$, and 2 odd generating vertices, namely $\pi(1), \pi(9)$. The limit contribution is y^3.

On the other hand, for any $\pi \in NC(2k)$, there are k blocks. Let $t+1$ be the number of blocks which have an *odd first element.* Then $0 \leq t \leq k-1$ and the limit contribution equals y^{k-t-1}. For example, of the five blocks of π_w, there are $t+1 = 2$ blocks, namely $\{1, 8\}, \{9, 10\}$ that start with an odd element and hence the contribution of π_w equals $y^{k-1-t} = y^3$.

The δ function is handy while dealing with non-crossing pair-partitions and freeness, and shall be used extensively. ●

(c) This proof proceeds along the arguments given above and those given in the proof of Theorem 9.3.1.

For notational simplicity we deal with only the case $m = 2$. For $\tau_1, \tau_2, \ldots, \tau_k \in \{1, 2\}$, we have

$$\frac{1}{p} \mathrm{E}\,\mathrm{Tr}[S^{(\tau_1)} \cdots S^{(\tau_k)}] = \frac{1}{pn^k} \mathrm{E}\,\mathrm{Tr}[X^{(\tau_1')\epsilon_1} X^{(\tau_2')\epsilon_2} \cdots X^{(\tau_{2k}')\epsilon_{2k}}],$$

where for $t = 1, \ldots, k$,

$$\tau_{2t-1}' = \tau_{2t}' = \tau_t \in \{1, 2\}, \ \epsilon_{2t-1} = 1 \ \text{ and } \ \epsilon_{2t} = *.$$

Now we use the moment-trace formula. As in the proof of Theorem 9.3.1, only the pair-matches survive and we have

$$\begin{aligned}
\frac{1}{pn^k} \mathrm{E}\,\mathrm{Tr}[X^{(\tau_1')\epsilon_1} \cdots X^{(\tau_{2k}')\epsilon_{2k}}] &= \frac{1}{pn^k} \sum_{I_{2k}'} \mathrm{E}[x_{i_1 i_2}^{(\tau_1')\epsilon_1} \cdots x_{i_{2k} i_1}^{(\tau_{2k}')\epsilon_{2k}}] \\
&= \frac{1}{pn^k} \sum_{\pi \in \mathcal{P}_2(2k)} \sum_{I_{2k}'(\pi)} \prod_{(r,s) \in \pi} \mathrm{E}[x_{i_r i_{r+1}}^{(\tau_r')\epsilon_r} x_{i_s i_{s+1}}^{(\tau_s')\epsilon_s}] + o(1),
\end{aligned}$$

where

$$I_{2k}' = \{(i_1, \ldots, i_{2k}) \ : \ 1 \leq i_{2t} \leq n, 1 \leq i_{2t-1} \leq p, m = 1, \ldots, k\}$$

and $I_{2k}'(\pi) \subset I_{2k}'$ consists of elements that obey the partition π.

Arguing as in the proof of Theorem 9.3.1, only pair-matches where all constraints are (C2) survive, and we get

$$\lim_{p \to \infty} \varphi_n(S^{(\tau_1)} \cdots S^{(\tau_k)}) = \sum_{\pi \in NC_2(2k)} \lim_{p \to \infty} \frac{1}{pn^k} \sum_{I_{2k}'(\pi)} \prod_{(r,s) \in \pi} \delta_{i_r i_{s+1}} \delta_{i_s i_{r+1}} \delta_{\tau_r' \tau_s'}. \tag{9.20}$$

As in the proof of Parts (a) and (b), it is clear that the limit is exactly the same as what it would be if the entries were independent within each matrix (that is, if they all had satisfied Assumption I).

Now we proceed to calculate the limit along the lines of the last part of the proof of Theorem 8.2.3 and conclude asymptotic freeness.

Suppose $S^{(1)}$ and $S^{(2)}$ appear in the monomial k_1 and k_2 times respectively. Then $k_1 + k_2 = k$. Let

$$M_i = \{t : 1 \leq t \leq 2k, \tau'_t = i\},\ i = 1, 2.$$

Note that

$$\# M_i = 2k_i,\ \ i = 1, 2.$$

Consider any $\pi \in NC_2(2k)$ in the above sum. Then since the pair-matches must occur only within the same index, and in a non-crossing way, only those π contribute which are of the form

$$\pi = \pi_1 \cup \pi_2 \text{ where } \pi_i \in NC_2(M_i),\ i = 1, 2.$$

For an index of type i, $i = 1, 2$, let

$$t_i + 1 = \text{ the number of odd first elements of } \pi_i, i = 1, 2.$$

Note that $0 \leq t_i \leq k_i - 1$, $i = 1, 2$. Then $t_1 + t_2 + 2 = t$ (say) is the total number of odd first elements of π.

Recall the definitions of $NC_{2,t}(2k) \subset NC_2(2k)$ and $A_{t,k} \subset NC(k)$ given respectively in (8.10) and (8.8) and the bijection between them given in Lemma 8.2.2.

We now simply "append indices" to this bijection, and then, clearly, the sets $NC_{2,t_1,t_2}(2k) \subset NC_{2,t}(2k) \subset NC_2(2k)$ and $A_{t_1,t_2,k} \subset A_{t,k} \subset NC(k)$ defined respectively as follows will be in bijection:

$$\{\pi = \pi_1 \cup \pi_2 \in NC_2(2k) : \pi_i \in NC_2(M_i) \text{ with } t_i{+}1 \text{ odd first elements}, i = 1, 2\},$$

$$\{\pi = \pi_1 \cup \pi_2 \in NC(k) : \pi_i \in NC(k_i) \text{ has } t_i + 1 \text{ blocks}, i = 1, 2\}.$$

Then arguing as in the proof of Theorem 8.2.3, for all $\pi \in NC_{2,t_1,t_2}(2k)$,

$$\begin{aligned}
&\lim_{p\to\infty} \frac{1}{pn^k} \sum_{I'_{2k}(\pi)} \prod_{(r,s)\in\pi} \delta_{i_r i_{s+1}} \delta_{i_s i_{r+1}} \delta_{\tau'_r \tau'_s} \\
&= \lim \frac{p^{k-(t_1+1+t_2+1-1)} n^{k-(k-t_1-t_2-2)}}{pn^k} \\
&= y^{k_1-t_1-1} y^{k_2-t_2-1}.
\end{aligned}$$

Hence joint convergence holds.

Let the limit variables be denoted by a_1 and a_2 respectively on some NCP $(\mathcal{A}, \varphi)$. We already know that both of these are MP_y variables.

Now, from the above calculations, the bijections discussed above, and the derivation in Remark 8.2.1 (c), it follows that

$$\begin{aligned}
\varphi(a_{\tau_1} a_{\tau_2} \cdots a_{\tau_k}) &= \sum_{t_1=0}^{k_1-1} \sum_{t_1=0}^{k_2-1} \sum_{\pi \in NC_{2,t_1,t_2}(2k)} y^{k_1-t_1-1} y^{k_2-t_2-1} \\
&= \sum_{t_1=0}^{k_1-1} \sum_{t_1=0}^{k_2-1} (\# A_{t_1,t_2,k}) y^{k_1-t_1-1} y^{k_2-t_2-1}.
\end{aligned}$$

But this is the moment formula given in Remark 9.4.1 for free MP_y variables. Hence $S^{(1)}$ and $S^{(2)}$ are asymptotically free and the proof is complete. ■

Remark 9.4.3. (a) In Section 11.4, we shall provide another proof of the asymptotic freeness of independent S-matrices based on the asymptotic freeness of the Wigner matrices with deterministic matrices and the technique of embedding.

(b) It is clear from the proof that if the entries of X are complex random variables, then the above argument would remain valid and hence the theorem would continue to hold.

(c) It is also clear that if $S^{(i)} = X^{(i)}_{p\times n_i} X^{(i)*}_{n_i \times p}$, where $X^{(i)}$ are elliptic with different ρ_i, $p \to \infty$ and $p/n_i \to y_i$, $1 \leq i \leq m$, even then, joint convergence and asymptotic freeness will hold. The proof of this is left as an exercise. ●

9.5 Symmetric Circulants: asymptotic independence

Definition 9.5.1. (Independence) Suppose $(\mathcal{A}, \varphi)$ is an NCP and $\{\mathcal{A}_i\}_{i\in J} \subset \mathcal{A}$ are unital sub-algebras. Then they are said to be *independent* with respect to the state φ, if they commute and $\varphi(a_1 \cdots a_n) = \varphi(a_1) \cdots \varphi(a_n)$ for all $a_i \in \mathcal{A}_{k(i)}$, where $i \neq j \implies k(i) \neq k(j)$. ◇

In particular, note that $\varphi(a_1 \cdots a_n)$ does not depend on the order in which the a_i's appear in the product.

It can be easily checked that Symmetric Circulant matrices of the same order commute. Thus, in view of Theorem 8.7.1, it is natural to anticipate the following result.

Theorem 9.5.1. Let $\{SC_n^{(i)}\}_{1\leq i\leq m}$ be m independent sequences of $n \times n$ Symmetric Circulant matrices which satisfy Assumption II. Then as elements of $(\mathcal{M}_n(\mathbb{C}), n^{-1} \operatorname{E} \operatorname{Tr})$, $\{n^{-1/2} SC_n^{(i)}\}_{1\leq i\leq m}$ converge to $\{a_i\}_{1\leq i\leq m}$ which are independent and each is distributed as standard Gaussian. ◆

Proof. First recall that the total number of non-indexed pair-matched words of length $2k$ equals $\frac{(2k)!}{k!2^k} = N_k$, say. Further, for *any* pair-matched word $w \in \mathcal{P}_2(2k)$,

$$p(w) = \lim_n \frac{1}{n^{1+k}} \#\Pi(w) = 1.$$

Now consider an order $2k$ monomial

$$q = n^{-1/2} SC_n^{(t_1)} n^{-1/2} SC_n^{(t_2)} \cdots n^{-1/2} SC_n^{(t_k)}$$

where the value of t_i is repeated $2n_i$ times in $t_1, \ldots, t_k$. From Theorem 9.1.1,

$$\lim_{n\to\infty} \varphi_n(q) = \sum_{w\in\mathcal{P}_2(2k,q)} p_q(w) = \#\mathcal{P}_2(2k,q).$$

Let l be the number of distinct indices (distinct matrices) in the monomial. Then the set of all indexed pair-matched words of length $2k$ is obtained by forming pair-matched sub-words of index (t_i) of lengths $2n_i$, $1 \le i \le l$. Hence

$$\varphi(a_{i_1}\cdots a_{i_{2k}}) = \prod_{i=1}^{l} N_{n_i}. \tag{9.21}$$

Now if $\{a_1, \ldots, a_m\}$ are i.i.d. standard normal random variables, then the above is the mixed moment $\mathrm{E}[\prod_{i=1}^{l} a_i^{2n_i}]$. This proves the theorem. ■

9.6 Reverse Circulants: asymptotic half-independence

It is easily checked that if A, B, C are arbitrary Reverse Circulant matrices of the same order, then

$$ABC = CBA.$$

This also implies that

$$A^2B^2 = B^2A^2.$$

This leads to the following definition.

Definition 9.6.1. (Half-commutativity) Let $\{a_i\}_{i\in J} \subset \mathcal{A}$. We say that they *half-commute* if

$$a_ia_ja_k = a_ka_ja_i \text{ for all } i, j, k \in J.$$

◇

Suppose $\{a_i\}_{i\in J} \subset \mathcal{A}$. For any $k \ge 1$ and any $\{i_j\} \subset J$, let

$$\begin{aligned} a &= a_{i_1}a_{i_2}\cdots a_{i_k} \in \mathcal{A}, \\ E_i(a) &= \text{number of times } a_i \text{ appears in the even positions of } a, \\ O_i(a) &= \text{number of times } a_i \text{ appears in the even positions of } a. \end{aligned}$$

The monomial a is said to be *symmetric* with respect to $\{a_i\}_{i\in J}$ if $E_i(a) = O_i(a)$ for all $i \in J$; otherwise it is called *non-symmetric*.

Definition 9.6.2. (Half-independence) Let $\{a_i\}_{i\in J}$ in $(\mathcal{A}, \varphi)$ be half-commuting. They are said to be *half-independent* if

(i) $\{a_i^2\}_{i\in J}$ are independent, and

(ii) for all a non-symmetric with respect to $\{a_i\}_{i\in J}$, we have $\varphi(a) = 0$. ◇

See Banica, Curran, and Speicher (2012)[9] for more details on half-independence. As pointed out in Speicher (1997)[93], the concept of half-independence does not extend to sub-algebras.

Example 9.6.1. (Example 2.4 of Banica, Curran, and Speicher (2012)[9]). Let $\{\eta_i\}$ be a family of independent complex Gaussian random variables. Define

$$a_i = \begin{bmatrix} 0 & \eta_i \\ \bar{\eta}_i & 0 \end{bmatrix}.$$

Then $\{a_i\}$ are half-independent as elements of $(\mathcal{M}_2(\mathbb{C}), 2^{-1}\,\mathrm{E}\,\mathrm{Tr})$. Moreover, the probability law of each a_i is the symmetrized Rayleigh whose density is given in (8.36). ▲

Theorem 9.6.1. Suppose $\{RC_n^{(i)}\}_{1\le i\le m}$ are independent $n \times n$ Reverse Circulant matrices whose entries satisfy Assumption II. Then as elements of $(\mathcal{M}_n(\mathbb{C}), n^{-1}\,\mathrm{E}\,\mathrm{Tr})$, $\{n^{-1/2}RC_n^{(i)}\}_{1\le i\le m}$ converge to $\{a_i\}_{1\le i\le m}$ which are half-independent and each a_i has the symmetrized Rayleigh distribution. ◆

Proof of Theorem 9.6.1. We already know the joint convergence and let φ denote the limit state. Fix $k \ge 2$. We will call a word $w \in \mathcal{P}_2(2k, q)$ *indexed symmetric* if each letter occurs once each in an odd and an even position *within the same index.*

Consider a monomial q of length $2k$ where each index appears an even number of times. From the single matrix case, it follows that $p(w) = 0$ if w is not a symmetric word.

Observe that if w is not an indexed symmetric word, then $\psi_q(w)$ is not a symmetric word. Hence for such w,

$$p_q(w) = p(\psi_q(w)) = 0.$$

Thus we may restrict our attention to only indexed symmetric words. In that case we have, by Theorem 9.1.1 (a), that

$$\lim_{n\to\infty} \varphi_n(q) = \#(S_2(2k, q)),$$

where $S_2(2k, q)$ is the collection of all indexed symmetric words of length $2k$.

The number of symmetric words of length $2k$ is $k!$. Let l be the number of distinct indices in the monomial, and let $2k_i$ be the number of matrices of the i-th index in the monomial q. All symmetric words are obtained by arranging the $2k_i$ letters of the i-th index in a symmetric way for $i = 1, 2, \ldots l$.

It is then easy to see that

$$\#(S_2(2k, q)) = k_1! \times k_2! \times \cdots \times k_l!. \tag{9.22}$$

Now we need to exhibit half-independent variables in some NCP whose moments coincide with the above. But this is provided by Example 9.6.1.

There each a_i has the symmetrized Rayleigh distribution and they are half-independent. To calculate the moment of any monomial in $\{a_i\}$, observe that if the monomial $a_{i_1}a_{i_2}\cdots a_{i_k}$ is non-symmetric, then

$$\mathrm{E}(\mathrm{Tr}[a_{i_1}a_{i_2}\cdots a_{i_k}]) = 0.$$

If, instead, $q(\{a_i\}) = a_{i_1}a_{i_2}\cdots a_{i_{2k}}$ is symmetric, then using half-independence,

$$\begin{aligned} \mathrm{E}[\mathrm{Tr}(a_{i_1}\cdots a_{i_{2k}}) &= k_1! \times k_2! \times \cdots \times k_l! \\ &= \lim_{n\to\infty} \varphi_n(q) \text{ by (9.22)}. \end{aligned}$$

Hence the moments match and this completes the proof. ■

9.7 Exercises

1. Verify the claim made in Remark 9.1.1.
2. Using Isserlis' formula, give a slightly simpler alternate proof of Theorems 9.2.1, 9.3.1 and 9.4.1 in the special case of Gaussian entries.
3. Show that Symmetric Circulant matrices of the same order commute.
4. Show that if A, B, C are arbitrary Reverse Circulant matrices of the same order, then $ABC = CBA$, and hence $A^2B^2 = B^2A^2$.
5. Verify the half-independence claim in Example 9.6.1.
6. Show that the Toeplitz limit is not free, independent or half-independent.
7. Show that for independent Hankel matrices, the indexed non-symmetric words *do not* contribute to the limit. Show that the Hankel limit is not free, independent or half-commuting.
8. Verify in detail that Theorems 9.2.1 and 9.3.1 remain valid respectively for Wigner and elliptic matrices with complex random variables as entries.

9. This problem extends Theorem 9.4.1. Suppose we have independent S-matrices $S^{(\tau)}, 1 \leq \tau \leq m$. Each $S^{(\tau)}$ is based on an $X_{p \times n_\tau}$ matrix with entries $\{x_{i,j,n}^{(\tau)}\}$ that satisfy Assumption Ie with possibly different correlation parameters ρ_τ. Suppose $\frac{p}{n_\tau} \to y_\tau > 0$ as $p \to \infty$. Then show that $S^{(1)}, \ldots, S^{(m)}$ jointly converge as elements of $(\mathcal{M}_p(\mathbb{C}), p^{-1} \mathrm{E}\,\mathrm{Tr})$ to $(MP_{y_1}, MP_{y_2} \ldots, MP_{y_m})$, which are asymptotically free Marčenko-Pastur variables.

10

Joint convergence II: multiple patterns

In Chapter 9, we demonstrated the joint convergence of independent copies of a *single* patterned matrix and showed that several matrices are asymptotically free. In this chapter, our goal is to show joint convergence when the sequences involved are not necessarily of the same pattern.

We begin with a general sufficient condition for joint convergence to hold and then show that it holds for the five matrices: Wigner, Toeplitz, Hankel, Reverse Circulant and Symmetric Circulant. As a consequence, joint convergence holds for copies of any two of these matrices. Moreover, the EESD of any symmetric polynomial involving independent copies of any two of these matrices converges.

10.1 Multiple patterns: colors and indices

Consider h different types of patterned matrices where type j has p_j independent copies. Within each type j, the p_j patterns are identical but the distribution of the entries can be different. The h different link functions shall be referred to as *colors* and p_j independent copies of color j shall be referred to as *indices*. There will be no loss of generality to assume that there are equal number of indices, say p, of each of the h colors.

So, let $\{X_{i,n}^{(j)}\}_{1\leq i\leq p}$ be $n \times n$ symmetric independent patterned matrices with link functions L_j, $j = 1, \ldots, h$. We suppress the dependence on n to simplify notation. Recall the notation $\Delta(L)$ introduced in (7.4). We shall work under the following obvious extension of Assumption I and Assumption II.

Assumption III The link functions $\{L_j\}$ satisfy $\Delta = \max_{1\leq j\leq h} \Delta(L_j) < \infty$. Further, the elements of all the input sequences are independent, have mean 0, variance 1, and satisfy the "all moments are uniformly bounded" assumption (Assumption I of Chapter 7). ⊡

Consider $\{\frac{1}{\sqrt{n}}X_{i,n}^{(j)}\}_{1\leq i\leq p, 1\leq j\leq h}$ as elements of $(\mathcal{M}_n(\mathbb{C}), \varphi_n = n^{-1}\operatorname{E}\operatorname{Tr})$. Then they jointly converge if and only if

$$\lim_{n\to\infty} \varphi_n\big(q(\{\frac{1}{\sqrt{n}}X_{i,n}^{(j)}\}_{1\leq i\leq p, 1\leq j\leq h})\big) \text{ exists for every monomial } q.$$

DOI: 10.1201/9781003144496-10

The case of a single patterned matrix, $h = 1$ and $p = 1$, was discussed in Chapter 8. The case of independent copies of a single patterned matrix, $h = 1$ and $p > 1$, was discussed in Chapter 9. In particular, we know that joint convergence holds for independent copies of any one of Wigner, elliptic, S, Topelitz, Hankel, Symmetric Circulant and Reverse Circulant matrices.

To deal with the case $h > 1$ and $p > 1$, the concepts of circuits, matches and indexed words will now be extended by adding color.

Since our primary aim is to show convergence for every monomial, we shall, from now on, *fix an arbitrary monomial q of length k*. We generally denote the colors and indices present in q by $(c_1, \ldots, c_k)$ and $(t_1, \ldots, t_k)$ respectively. Note that $1 \leq c_i \leq h$ and $1 \leq t_i \leq p$ for all $1 \leq i \leq k$. Then we may write

$$q\Big(\frac{1}{\sqrt{n}}\{X_{i,n}^{(j)}\}_{1\leq i\leq p, 1\leq j\leq h}\Big) = \frac{1}{n^{k/2}} Z_{c_1,t_1} \cdots Z_{c_k,t_k}, \tag{10.1}$$

where $Z_{c_m,t_m} = X_{t_m,n}^{(c_m)}$ for $1 \leq m \leq k$ and we have dropped the subscript n to simplify notation. Let $X_i^{(j)}(L_j(p,q))$ denote the (p,q)-th entry of $X_{i,n}^{(j)}$.

From (10.1) we get,

$$\begin{aligned}
\varphi_n(q) &:= \frac{1}{n}\mathrm{E}\Big[\mathrm{Tr}\Big[\frac{1}{n^{k/2}} Z_{c_1,t_1} Z_{c_2,t_2} \cdots Z_{c_k,t_k}\Big]\Big] \\
&= \frac{1}{n^{1+k/2}} \sum_{j_1,\cdots,j_k} \mathrm{E}\Big[\prod_{i=1}^{k} Z_{c_i,t_i}(L_{c_i}(j_i, j_{i+1}))\Big] \quad (\text{with } \ j_{k+1} = j_1) \\
&= \frac{1}{n^{1+k/2}} \sum_{\pi:\pi \text{ is a circuit}} \mathrm{E}\Big[\prod_{i=1}^{k} Z_{c_i,t_i}(L_{c_i}(\pi(i-1), \pi(i)))\Big] \\
&= \frac{1}{n^{1+k/2}} \sum_{\pi:\pi \text{ is a circuit}} \mathrm{E}\left[\mathbf{Z}_\pi\right]
\end{aligned} \tag{10.2}$$

where

$$\mathbf{Z}_\pi = \prod_{i=1}^{k} Z_{c_i,t_i}(L_{c_i}(j_i, j_{i+1})).$$

Due to the zero mean and independence imposition in Assumption III,

$$\mathrm{E}[\mathbf{Z}_\pi] = 0 \tag{10.3}$$

if at least one $(c_i, t_i, L_{c_i}(\pi(i-1), \pi))$ is not repeated in the product. If this is *not* the case, then we say the circuit is *matched* and only these circuits are relevant.

A circuit is said to be *color-matched* if all the L-values are repeated within the same color. A circuit is said to be *color-* and *index-matched* if, in addition, all the L-values are also repeated within the same index.

We can define an equivalence relation on the set of color- and index-matched circuits, extending the ideas of Chapters 8 and 9. We say $\pi_1 \sim \pi_2$

if and only if their matches take place at the same colors and at the same indices. In other words, $\pi_1 \sim \pi_2$ if and only if for all i, j,

$$c_i = c_j, t_i = t_j \text{ and } L_{c_i}(\pi_1(i-1), \pi_1(i)) = L_{c_j}(\pi_1(j-1), \pi_1(j))$$
$$\Updownarrow$$
$$c_i = c_j, t_i = t_j \text{ and } L_{c_i}(\pi_2(i-1), \pi_2(i))) = L_{c_j}(\pi_2(j-1), \pi_2(j)).$$

An equivalence class can be expressed as a colored and indexed word w: it is a string of letters in alphabetic order of their first occurrence, with a *subscript for index and a superscript for color.* The i-th position of w is denoted by $w[i]$.

For example, if

$$q = X_1^{(1)} X_2^{(1)} X_1^{(2)} X_1^{(2)} X_2^{(2)} X_2^{(2)} X_2^{(1)} X_1^{(1)},$$

then $a_1^1 b_2^1 c_1^2 c_1^2 d_2^2 d_2^2 b_2^1 a_1^1$ is *one of the* colored and indexed words corresponding to q. Conversely, given any colored and indexed word, by looking at the superscripts and the subscripts, one can reconstruct the unique monomial to which it corresponds.

A colored and indexed word is *pair-matched* if all its letters appear exactly twice. We shall see later that under Assumption III, only such circuits and words survive in the limits of (10.2).

Now we define the following useful subset of circuits. For a colored and indexed word w, let

$$\Pi_{CI}(w) = \{\pi : w[i] = w[j] \Leftrightarrow (c_i, t_i, L_{c_i}(\pi(i-1), \pi(i))) = (c_j, t_j, L_{c_j}(\pi(j-1), \pi(j))\}. \quad (10.4)$$

Let $\psi(q)$ be the monomial obtained by dropping the indices from q. For example,

$$q = X_1^{(1)} X_1^{(2)} X_2^{(1)} X_2^{(2)} X_2^{(2)} X_1^{(2)} \Rightarrow \psi(q) = X^{(1)} X^{(2)} X^{(1)} X^{(2)} X^{(2)} X^{(2)}.$$

In other words it corresponds to the case where $p = 1$. At the same time, every colored and indexed word w has a corresponding non-indexed version w' which is obtained by dropping the indices from the letters (i.e. the subscripts). For example, $a_1^1 b_2^1 c_1^2 c_1^2 d_2^2 d_2^2 b_2^1 a_1^1$ yields $a^1 b^1 c^2 c^2 d^2 d^2 b^1 a^1$. For any monomial q, dropping the indices amounts to essentially replacing for every j the independent copies $X_i^{(j)}$ by a single $X^{(j)}$ with link function L_j. Without any scope for confusion, we may denote this mapping also by ψ so that $w' = \psi(w)$. By this mapping, the set $\Pi_{CI}(w)$ of colored and indexed circuits gets mapped to the following subset of non-indexed circuits corresponding to the word w':

$$\Pi_C(w') = \{\pi : w'[i] = w'[j] \Leftrightarrow c_i = c_j \text{ and } L_{c_i}(\pi(i-1), \pi(i)) = L_{c_j}(\pi(j-1), \pi(j))\}.$$

Since pair-matched words are going to be crucial, let us define, for a fixed monomial of order k (which is necessarily even for the following),

$$\mathcal{CIP}_2(k) = \{w : w \text{ is indexed, colored and pair-matched corresponding to } q\},$$

$$\mathcal{CP}_2(k) = \{\psi(w) : w \in \mathcal{CIP}_2(k)\}.$$

Note that $\mathcal{CP}_2(k)$ is the set of all non-indexed colored pair-matched words and $w \to w'$ is a one-one mapping from $\mathcal{CIP}_2(k)$ to $\mathcal{CP}_2(k)$ and we continue to denote this mapping by ψ.

For any $w' \in \mathcal{CIP}_2(k)$ and $w' \in \mathcal{CP}_2(k)$ define, *whenever the respective limits exist,*

$$\begin{aligned} p_C(w') &= \lim_{n\to\infty} \frac{1}{n^{1+k/2}} \#\Pi_C(w') \text{ and} \\ p_{CI}(w) &= \lim_{n\to\infty} \frac{1}{n^{1+k/2}} \#\Pi_{CI}(w). \end{aligned}$$

10.2 Joint convergence

We now give a criterion for joint convergence which essentially says that if one copy each of several patterned matrices satisfies this criterion, then if we take several copies, they will automatically converge jointly. The burden then shifts to verifying this criterion for single copies. This will be done for our familiar patterned matrices—Wigner, Hankel, Toeplitz, Reverse Circulant and Symmetric Circulant, taken two at a time, in Theorem 10.3.1.

Theorem 10.2.1. Let $\{\frac{1}{\sqrt{n}} X_{i,n}^{(j)}\}_{1\le i\le p, 1\le j\le h}$ be a sequence of real symmetric patterned random matrices which satisfy Assumptions III. Fix a monomial q of length k and assume that, for all $w \in \mathcal{CP}_2(k)$,

$$p_C(w) = \lim_{n\to\infty} \frac{1}{n^{1+k/2}} \#\Pi_C(w) \quad \text{exists.} \tag{10.5}$$

Then,

(a) for all $w \in \mathcal{CIP}_2(k)$, $p_{CI}(w)$ exists and $p_{CI}(w) = p_C(\psi(w))$;

(b) if (10.5) holds for every q, then $\{\frac{1}{\sqrt{n}} X_{i,n}^{(j)}\}_{1\le i\le p, 1\le j\le h}$ converge jointly as elements of $(\mathcal{M}_n(\mathbb{C}), \varphi_n = n^{-1} \operatorname{E}\operatorname{Tr})$ and

$$\begin{aligned} \lim \varphi_n(q) &= \sum_{w\in\mathcal{CIP}_2(k)} p_{CI}(w) \\ &= \alpha(q) \text{ (say);} \end{aligned}$$

(c) the quantity $\alpha(q)$ satisfies

$$|\alpha(q)| \le \begin{cases} \frac{k!\Delta^{k/2}}{(k/2)!2^{k/2}} & \text{if } k \text{ is even and each index appears} \\ & \text{an even number of times,} \\ 0 & \text{otherwise;} \end{cases}$$

and so, if the monomial q yields a symmetric matrix, then its EESD converges weakly to a symmetric probability law. ◆

Remark 10.2.1. Theorem 10.2.1 asserts that if the joint convergence holds for $p = 1, j = 1, \ldots, h$ (that is, if condition (10.5) holds), then it continues to hold for $p > 1$. There is no general way of checking (10.5). However, see Theorem 10.3.1 where we prove that this condition holds for any pair from the following links: Wigner, Hankel, Toeplitz, Reverse Cirulant and Symmetric Circulant. ●

Proof of Theorem 10.2.1. (a) We first show that

$$\Pi_C(\psi(w)) = \Pi_{CI}(w) \quad \text{for all} \quad w \in \mathcal{CIP}_2(k). \tag{10.6}$$

Let $\pi \in \Pi_{CI}(w)$. As q is fixed,

$$\begin{aligned} &\psi(w)[i] = \psi(w)[j] \Rightarrow w[i] = w[j] \\ \Rightarrow &(c_i, t_i, L_{c_i}(\pi(i-1), \pi(i))) = (c_j, t_j, L_{c_j}(\pi(j-1), \pi(j))) \ (\text{as } \pi \in \Pi_{CI}(w)). \end{aligned}$$

Hence $L_{c_i}(\pi(i-1), \pi(i)) = L_{c_j}(\pi(j-1), \pi(j))$ and $\pi \in \Pi_C(\psi(w))$.

Now, conversely, let $\pi \in \Pi_C(\psi(w))$. Then we have

$$\begin{aligned} w[i] &= w[j] \\ \Rightarrow \psi(w)[i] &= \psi(w)[j] \\ \Rightarrow L_{c_i}(\pi(i-1), \pi(i)) &= L_{c_j}(\pi(j-1), \pi(j)) \\ \Rightarrow Z_{c_i,t_i}(L_{c_i}(\pi(i-1), \pi(i))) &= Z_{c_j,t_j}(L_{c_j}(\pi(j-1), \pi(j))) \end{aligned}$$

as $w[i] = w[j] \Rightarrow c_i = c_j$ and $t_i = t_j$. Hence $\pi \in \Pi_{CI}(w)$. So (10.6) is established.

As a consequence,

$$p_{CI}(w) = \lim_{n\to\infty} \frac{1}{n^{1+k/2}} \#\Pi_{CI}(w) = \lim_{n\to\infty} \frac{1}{n^{1+k/2}} \Pi_C(\psi(w)) = p_C(\psi(w))$$

for all $w \in \mathcal{CIP}_2(k)$. This proves part (a).

(b) Recall that $\mathbf{Z}_\pi = \prod_{j=1}^{k} Z_{c_j,t_j}(L_{c_i}(\pi(j-1), \pi(j))$ and (10.3) holds. Hence

$$\varphi_n(q) = \frac{1}{n^{1+k/2}} \sum_{w:\ w \text{ matched}} \ \sum_{\pi \in \Pi_{CI}(w)} \mathrm{E}(\mathbf{Z}_\pi). \tag{10.7}$$

By using Assumption III,

$$\sup_{\pi \in \Pi_C(w)} \mathrm{E}\,|\mathbf{Z}_\pi| < K < \infty. \tag{10.8}$$

By using arguments which are by now familiar, for any colored and indexed word w, we note that circuits that have at least one match of order three or more will contribute nothing to the limit. Hence

$$\lim_{n\to\infty} \frac{1}{n^{1+k/2}} \Big| \sum_{\pi \in \Pi_{CI}(w)} |\, \mathrm{E}(\mathbf{Z}_\pi)| \le \frac{K}{n^{1+k/2}} \#\Pi_{CI}(w) \to 0, \tag{10.9}$$

whenever w is matched but not pair-matched. Only pair-matched circuits (indexed or otherwise) will contribute to the limit.

By using (10.9) and the fact that $\mathrm{E}(\mathbf{Z}_\pi) = 1$ for every colored indexed pair-matched word, calculating the limit in (10.7) reduces to calculating $\lim \frac{1}{n^{1+k/2}} \sum_{w:\, w \in \mathcal{CIP}_2(k)} \#\Pi_{CI}(w)$.

As a consequence, since there are finitely many words,

$$\begin{aligned}\lim_{n\to\infty} \varphi_n(q) &= \lim_{n\to\infty} \sum_{w \in \mathcal{CIP}_2(k)} \frac{\#\Pi_{CI}(w)}{n^{1+k/2}} \\ &= \sum_{w \in \mathcal{CIP}_2(k)} p_{CI}(w) \\ &= \alpha(q).\end{aligned}$$

(c) Note that if either k is odd or some index appears an odd number of times in q then for that q, $\mathcal{CIP}_2(k)$ is empty and hence $\alpha(q) = 0$.

Now suppose that k is even and every index appears an even number of times. Then

$$\begin{aligned}\#\mathcal{CIP}_2(k) &\leq \#\mathcal{CP}_2(k) \\ &\leq \frac{k!}{(k/2)!2^{k/2}}.\end{aligned} \tag{10.10}$$

The first inequality above follows from the fact mentioned earlier that ψ is an injective map from $\mathcal{CIP}_2(k)$ to $\mathcal{CP}_2(k)$. The second inequality follows once we observe that the total number of *colored* pair-matched words of length k is less than the total number of pair-matched words of length k and that equals $\frac{k!}{(k/2)!2^{k/2}}$. Let

$$\Delta = \max_{1 \leq j \leq h} \Delta(L_j).$$

It then easily follows that

$$p_{CI}(w) \leq \Delta^{k/2}. \tag{10.11}$$

Combining (10.10) and (10.11), we conclude that $|\alpha(q)| \leq \frac{k!\Delta^{k/2}}{(k/2)!2^{k/2}}$. The rest of the argument is easy and is omitted. This completes the proof of the theorem. ■

Remark 10.2.2. (Almost sure convergence) As in Chapter 9, the convergence can be upgraded to almost sure joint convergence. We briefly sketch the argument. See Bose (2018)[20] for details. Let

$$\widetilde{\varphi_n}(q) = \frac{1}{n^{1+k/2}} \sum_{\pi:\pi \text{ is a circuit}} \mathbf{Z}_\pi. \tag{10.12}$$

Then $\varphi_n(q) = \mathrm{E}\left[\widetilde{\varphi_n}(q)\right]$. It can be shown that

$$\mathrm{E}[(\widetilde{\varphi_n}(q) - \varphi_n(q))^4] = O(n^{-2}). \tag{10.13}$$

Hence, by an application of the Borel–Cantelli lemma, $\widetilde{\mu_n}(q) \to \alpha(q)$ whenever $\varphi_n(q) \to \alpha(q)$. As a consequence, for any symmetric polynomial, the almost sure weak limit of the ESD exists and is the same as the weak limit of the EESD under the assumptions of Theorem 10.2.1. ●

10.3 Two or more patterns at a time

To apply Theorem 10.2.1 we need to identify situations where (10.5) holds. We state a result which verifies this when we take two patterns at a time out of a total of five patterns.

Theorem 10.3.1. Suppose Assumption III holds. Then $p_C(w)$ exists for all monomials q and $w \in \mathcal{CP}_2(k)$ for any of the matrices, Wigner, Toeplitz, Hankel, Symmetric Circulant and Reverse Circulant, taken two at a time. As a consequence, independent copies of any two these matrices converge jointly as elements of $(\mathcal{M}_n(\mathbb{C}), \varphi_n = n^{-1} \operatorname{E} \operatorname{Tr})$. ◆

We shall have more to say about the joint convergence of the Wigner matrix with other matrices in the next chapter.

Remark 10.3.1. It is only to simplify the notational aspects that we have restricted ourselves to the case $h = 2$. From the proof it will be clear that the theorem continues to hold when we have more than two patterns. ●

Proof of Theorem 10.3.1. Condition (10.5), which needs to be verified (only for even degree monomials), crucially depends on the type of the link function and hence we need to deal with every pair differently. Since we are dealing with only two link functions, we simplify the notation.

Let X and Y be any pair of the matrices, Wigner (W_n), Toeplitz (T_n), Hankel (H_n), Reverse Circulant (RC_n) and Symmetric Circulant (SC_n). Denote their link functions as L_1 and L_2, respectively.

Let $q(X, Y)$ be any monomial in X and Y of degree $2k$. To avoid trivialities, suppose both X and Y occur an even number of times in q, say $2k_1$ and $2k_2$, respectively. Thus $k = k_1 + k_2$. Then it is enough to show that (10.5) holds for every pair-matched colored word w of length $2k$ corresponding to q.

By abuse of notation, we denote the two colors also as X and Y. Fix a colored pair-matched word w (recall that other words are inconsequential). Let

$$S_X = \{i \wedge j : c_i = c_j = X, w[i] = w[j]\}, \;\; S_Y = \{i \wedge j : c_i = c_j = Y, \; w[i] = w[j]\}$$

be the generating vertices with letters of colors X and Y, respectively. Here $i \wedge j$ denotes the minimum of i and j.Then the set S of all generating vertices of w equals

$$S = \{0\} \cup S_X \cup S_Y.$$

Since each word is color pair-matched, for every $i \in S - \{0\}$,

$$w[i] = w[j_i], \quad j_i \neq i \text{ is unique.}$$

Let $\pi \in \Pi_C(w)$.

We have to treat each pair of matrices separately. However, the arguments are similar for the various different pairs, so we provide the details for the most involved cases and only sketches for the rest of the cases.

Case (i) Hankel and Toeplitz: Let $X = T$ and $Y = H$ be the Toeplitz and Hankel matrices. Recall the two link functions:

$$L_T(i,j) = |i-j| \text{ and } L_H(i,j) = i+j.$$

Let

$$I = \{l = (l_1, ..., l_{k_1}) : l_i \in \{-1, 1\}^{k_1} \text{ for all } i\}.$$

Let $\Pi_{C,l}(w)$ be the subset of $\Pi_C(w)$ such that

$$\pi(i-1) - \pi(i) = l_i(\pi(j_i - 1) - \pi(j_i)) \qquad \text{for all } i \in S_T,$$

$$\pi(i-1) + \pi(i) = \pi(j_i - 1) + \pi(j_i) \qquad \text{for all } i \in S_H.$$

Note that some of these sets may be vacuous and/or overlapping. In any case,

$$\Pi_C(w) = \cup_l \Pi_{C,l}(w). \tag{10.14}$$

We shall show that the following two relations hold:

$$p_{C,l}(w) := \lim_{n\to\infty} \frac{\#\Pi_{C,l}(w)}{n^{1+k}} \quad \text{exists for all } l \in I. \tag{10.15}$$

$$n^{-(k+1)}[\#(\Pi_{C,l}(w) \cap \Pi_{C,l'}(w))] \to 0. \tag{10.16}$$

Then it will follow that

$$\begin{aligned} p_C(w) &= \lim_{n\to\infty} \frac{1}{n^{1+k}} \#\Pi_C(w) \\ &= \sum_{l\in I} \lim_{n\to\infty} \frac{1}{n^{1+k}} \#\Pi_{C,l}(w) \\ &= \sum_{l\in I} p_{C,l}(w). \end{aligned}$$

To establish (10.15) and (10.16), we follow the arguments given in the proofs of Theorems 8.4.1 and 8.5.1.Define

$$v_i = \frac{\pi(i)}{n}, \qquad \text{and} \qquad U_n = \{0, \frac{1}{n}, ..., \frac{n-1}{n}\}. \tag{10.17}$$

Then

$$\begin{aligned}\#\Pi_{C,l}(w) &= \#\{(v_0, ..., v_{2k}) : v_i \in U_n\ \forall i,\ v_{i-1} - v_i = l_i(v_{j_i-1} - v_{j_i})\ \forall i \in S_T, \\ &\qquad \text{and } v_{i-1} + v_i = v_{j_i-1} + v_{j_i}\ \ \forall i \in S_H,\ \ v_0 = v_{2k}\}.\end{aligned}$$

Let us denote

$$v_S := \{v_i : i \in S\}.$$

It can easily be seen from the above equations on the vertices $\{v_j\}$ that each of the $\{v_i : i \notin S\}$ can be written uniquely as an integer linear combination $C_i^{(l)}(v_S)$ in addition to the circuit condition $v_0 = v_{2k}$. Moreover, $C_i^{(l)}(v_S)$ only contains $\{v_j : \ j \in S,\ j < i\}$ with non-zero coefficients. Clearly,

$$\#\Pi_{C,l}(w) = \#\{(v_0, ..., v_{2k}) : v_i \in U_n\ \forall i, v_0 = v_{2k}, v_i = C_i^{(l)}(v_S)\ \ \forall i \notin S\}.$$

Any integer linear combination of elements of U_n is again in U_n if and only if it is between 0 and 1. Hence

$$\#\Pi_{C,l}(w) = \#\{v_S : v_i \in U_n\ \forall i \in S, v_0 = C_{2k}^{(l)}(v_S), 0 \le C_i^{(l)}(v_S) < 1\ \ \forall i \notin S\}. \tag{10.18}$$

From (10.18), it follows that $\frac{\#\Pi_{C,l}(w)}{n^{1+k}}$ is nothing but the Riemann sum for the function $I(0 \le C_i^{(l)}(v_S) < 1, i \notin S, v_0 = C_{2k}^{(l)}(v_S))$ over $[0,1]^{k+1}$ and converges to the corresponding integral and hence

$$\begin{aligned}p_{C,l}(w) &= \lim_{n\to\infty} \frac{1}{n^{1+k}} \#\Pi_{C,l}(w) \\ &= \int_{[0,1]^{k+1}} I\big(0 \le C_i^{(l)}(v_S) < 1, i \notin S, v_0 = C_{2k}^{(l)}(v_S)\big) dv_S.\end{aligned}$$

This establishes (10.15). Incidentally, for some of the l, this value could be 0.

For (10.16), let $l \neq l'$. Without loss of generality, let us assume that $l_{i_1} = -l'_{i_1}$. Let $\pi \in \Pi_{C,l}(w) \cap \Pi_{C,l'}(w)$. Then $\pi(i_1 - 1) = \pi(i_1)$ and hence $C_{i_1-1}^{(l)}(v_S) = v_{i_1}$. It now follows along the lines of the preceding arguments that

$$\lim_{n\to\infty} \frac{1}{n^{1+k}} \#(\Pi_{C,l}(w) \cap \Pi_{C,l'}(w)) \le \int \cdots \int_{[0,1]^{k+1}} I(v_i = C_{i_1-1}^{(l)}(v_S)) dv_S.$$

$C_{i_1-1}^{(l)}(v_S)$ contains $\{v_j : \ \ j \in S,\ j < i_1\}$ and hence $\{C_{i_1-1}^{(l)}(v_S) = v_i\}$ is a k-dimensional subspace of $[0,1]^{k+1}$ and hence has Lebesgue measure 0. This establishes (10.16), and hence the proof for the case $X = T$ and $Y = H$ is complete.

Observe that the proof involved establishing a decomposition (equation (10.14) and then establishing two relations, (10.15) and (10.16). This is the idea that will be implemented for all the other pairs.

Case (ii) Hankel and Reverse Circulant: Let X and Y be Hankel (H) and Reverse Circulant (RC) respectively. Recall the link functions

$$L_H(i,j) = i+j-2, \quad \text{and} \quad L_{RC}(i,j) = (i+j) \mod n.$$

Note that

$$(\pi(i-1)+\pi(i)) \mod n = (\pi(j_i-1)+\pi(j_i)) \mod n \text{ for all } i \in S_{RC} \quad (10.19)$$

and this implies that

$$(\pi(i-1)+\pi(i)) - (\pi(j_i-1)+\pi(j_i)) = a_i n,$$

where $a_i \in \{0, 1, -1\}$. Hence define

$$I = \{a = (a_1, ..., a_{k_2}) : a_i \in \{-1, 0, 1\} \text{ for all } i\}.$$

Let $\Pi_{C,a}(w)$ be the subset of $\Pi_C(w)$ such that

$$\pi(i-1)+\pi(i) = \pi(j_i-1)+\pi(j_i) \text{ for all } i \in S_H \text{ and}$$

$$(\pi(i-1)+\pi(i)) - (\pi(j_i-1)+\pi(j_i)) = a_i n \text{ for all } i \in S_{RC}.$$

Now, clearly,

$$\Pi_C(w) = \cup_a \Pi_{C,a}(w).$$

Incidentally, the above is a disjoint union and hence we need to verify only the analog of (10.15). Now

$$\begin{aligned}\#\Pi_{C,a}(w) &= \#\{(v_0, ..., v_{2k}) : v_i \in U_n, 0 \le i \le 2k; \\ &\qquad v_{i-1}+v_i = v_{j_i-1}+v_{j_i} + a_i \text{ for all } i \in S_{RC}; \\ &\qquad v_{i-1}+v_i = v_{j_i-1}+v_{j_i} \text{ for all } i \in S_H, v_0 = v_{2k}\}.\end{aligned}$$

Other than $v_0 = v_{2k}$, each $\{v_i : i \notin S\}$ can be written uniquely as an affine linear combination $C_i^{(a)}(v_S) + b_i^{(a)}$ for some integer $b_i^{(a)}$. Moreover, $C_i^{(a)}(v_S)$ only contains $\{v_j : j \in S, j < i\}$ with non-zero coefficients. Arguing as in the previous case of the Toeplitz and Hankel pair,

$$\begin{aligned}\#\Pi_{C,a}(w) &= \#\{v_S : v_i \in U_n \text{ for all } i \in S, \\ &\qquad v_0 = C_{2k}^{(a)}(v_S) + b_{2k}^{(a)}, 0 \le C_i^{(a)}(v_S) + b_i^{(a)} < 1 \text{ for all } i \notin S\}.\end{aligned} \quad (10.20)$$

As before,

$$\begin{aligned}p_{C,a}(w) &= \lim_{n\to\infty} \frac{1}{n^{1+k}} \#\Pi_{C,a}(w) \\ &= \int_{[0,1]^{k+1}} I(0 \le C_i^{(a)}(v_S) + b_i^{(a)} < 1, i \notin S, v_0 = C_{2k}^{(a)}(v_S) + b_{2k}^{(a)}) dv_S,\end{aligned}$$

and the proof for this case is complete.

Case (iii) Hankel and Symmetric Circulant: Let X and Y be Hankel (H) and Symmetric Circulant (SC) respectively. Recall the Symmetric Circulant link function. This implies that

$$n/2 - |n/2 - |\pi(i-1) - \pi(i)|| = n/2 - |n/2 - |\pi(j_i - 1) - \pi(j_i)|| \;\; \forall i \in S_S.$$

Hence either $|\pi(i-1) - \pi(i)| = |\pi(j_i - 1) - \pi(j_i)|$, or $|\pi(i-1) - \pi(i)| + |\pi(j_i - 1) - \pi(j_i)| = n$, leading to six possible cases. These are (using the earlier notation):

1. $v_{i-1} - v_i - v_{j_i-1} + v_{j_i} = 0$,
2. $v_{i-1} - v_i + v_{j_i-1} - v_{j_i} = 0$,
3. $v_{i-1} - v_i + v_{j_i-1} - v_{j_i} = 1$,
4. $v_{i-1} - v_i - v_{j_i-1} + v_{j_i} = 1$,
5. $v_i - v_{i-1} + v_{j_i-1} - v_{j_i} = 1$, and
6. $v_i - v_{i-1} + v_{j_i} - v_{j_i-1} = 1$

We can write $\Pi_C(w)$ as the union (but not disjoint union) of 6^{k_2} possible $\Pi_{C,l}(w)$, where l denotes the combination of cases (1)-(6) above that is satisfied in the k_2 matches of the Symmetric Circulant. For each $\pi \in \Pi_{C,l}(w)$, each $\{v_i : i \notin S\}$ is a unique affine integer combination of v_S. Similar to the previous two pairs of matrices considered in Cases (i) and (ii), $\lim_{n\to\infty} \frac{1}{n^{1+k}} \#\Pi_{C,l}(w)$ exists as an integral.

Now (10.16) can be checked case by case for all pairs of the six cases. This is tedious but routine. As a typical case, suppose Case 1 and Case 3 hold. Then $\pi(i-1) - \pi(i) = n/2$ and $v_{j_i-1} - v_{j_i} = 1/2$. Since i is generating and v_{i-1} is a linear combination of $\{v_j : j \in S, \; j < i\}$, this implies a non-trivial linear relation between the generating vertices v_S. This, in turn, implies that the number of circuits π which satisfy the above conditions is $o(n^{1+k})$. This completes the proof for Case (iii).

Case (iv) Toeplitz and Symmetric Circulant: Let X and Y be Toeplitz (T) and Symmetric Circulant (SC), respectively. Again, note that

$$|\pi(i-1) - \pi(i)| = |\pi(j_i - 1) - \pi(j_i)| \text{ for all } i \in S_T, \text{ and}$$

$$\frac{n}{2} - |\frac{n}{2} - |\pi(i-1) - \pi(i)|| = \frac{n}{2} - |\frac{n}{2} - |\pi(j_i-1) - \pi(j_i)|| \text{ for all } i \in S_{SC}. \quad (10.21)$$

Now (10.21) implies either $|\pi(i-1) - \pi(i)| = |\pi(j_i - 1) - \pi(j_i)|$, or $|\pi(i-1) - \pi(i)| + |\pi(j_i - 1) - \pi(j_i)| = n$.

There are six cases for each Symmetric Circulant match as in Case (iii) above and two cases for each Toeplitz match.

As before we can write $\Pi_C(w)$ as the union (but not disjoint union) of $2^{k_1} \times 6^{k_2}$ possible $\Pi_{C,l}(w)$ where l denotes a combination of the cases (1)-(6)

for all SC matches (as in Case (iii)) and a combination of the two cases (as in Case (i)) for all T matches. As before, for each $\pi \in \Pi_{C,l}(w)$, each of the $\{v_i : i \notin S\}$ can be written uniquely as an affine integer combination of v_S. Thus, $\lim_{n\to\infty} \frac{1}{n^{1+k}} \#\Pi^*_{C,l}(w)$ exists as an integral which establishes (10.15).

Now suppose $l \neq l'$ and $\pi \in \Pi_{C,l}(w) \cap \Pi_{C,l'}(w)$. For $l \neq l'$, there must be one Toeplitz or Symmetric Circulant match such that two of the possible cases for the T-matches or six of the cases in (1)-(6) occur simultaneously. Then (10.16) must be checked case by case for each of these possibilities. This is tedious. We simply illustrate two cases:

Illustration 1. Suppose the two cases of Toeplitz match hold at the same locations. Then we have, for a generating vertex i, $\pi(i-1) - \pi(i) = (\pi(j-1) - \pi(j)$; $\pi(i-1) - \pi(i) = -(\pi(j-1) - \pi(j))$. This implies that $\pi(i-1) = \pi(i)$, and hence $v_{i-1} = v_i$. Since i is generating and v_{i-1} is a linear combination of $\{v_j : \ j \in S, \ j < i\}$, this implies that there exists a non-trivial relation between the generating vertices v_S. This, in turn, implies that the number of circuits π satisfying the above conditions is $o(n^{1+k})$.

Illustration 2. Suppose the Symmetric Circulant match happens for both Case (i) and Case (ii). Then again, we have $v_i = v_{i-1}$, and we can argue as before to conclude that (10.16) holds.

Case (v) Toeplitz and Reverse Circulant: Let X and Y be Toeplitz (T) and Reverse Circulant (RC), respectively. Let the number of Toeplitz and Reverse Circulant matches be k_1 and k_2, respectively, and let $S_T = \{i_1, i_2, ..., i_{k_1}\}$, $S_{RC} = \{i_{k_1+1}, i_{k_1+2}, ..., i_{k_1+k_2}\}$.
Let

$$I = \{l = (c, a) = (c_{i_1}, ..., c_{i_{k_1}}, a_{i_{k_1+1}}, ..., a_{i_{k_1+k_3}}) \in \{-1, 1\}^{k_1} \times \{-1, 0, 1\}^{k_2}\}.$$

Let $\Pi_{C,l}(w)$ be the subset of $\Pi_C(w)$ such that

$$\begin{aligned} \pi(i-1) - \pi(i) &= c_i(\pi(j_i - 1) - \pi(j_i)) \ \text{ for all } \ i \in S_T, \\ \pi(i-1) + \pi(i) &= \pi(j_i - 1) + \pi(j_i) + a_i n \ \text{ for all } \ i \in S_{RC}. \end{aligned}$$

Then, clearly,

$$\Pi_C(w) = \cup_{l \in I} \Pi_{C,l}(w),$$

and translating this into the language of the v_i's, we get

$$\begin{aligned} \#\Pi_{C,l}(w) = \ & \#\{(v_0, ..., v_{2k}) : \\ & v_i \in U_n \ \forall i, \ v_{i-1} + v_i = (v_{j_i - 1} + v_{j_i}) + a_i \ \text{ for all } \ i \in S_{RC}, \\ & \text{and } \ v_{i-1} - v_i = c_i(v_{j_i - 1} - v_{j_i}) \ \text{ for all } \ i \in S_T, \ \ v_0 = v_{2k}\}. \end{aligned}$$

As in the previous cases, $\lim_{n\to\infty} \frac{\#\Pi_{C,l}(w)}{n^{1+k}}$ exists which verifies (10.15).

It remains to verify (10.16). If $l = (c, a) \neq l' = (c', a')$, then either $c \neq c'$, or $a \neq a'$. If $c = c'$, then, clearly, $\Pi_{C,l}(w)$ and $\Pi_{C,l'}(w)$ are disjoint. Let $c \neq c'$. Without loss of generality, we assume that $c_{i_1} = -c_{i_1}$. Then, clearly, for every

$\pi \in \Pi_{C,l}(w) \cap \Pi_{C,l'}(w)$, we have $v_{i_1-1} = v_{i_1}$, which gives a non-trivial relation between $\{v_j : j \in S\}$. That, in turn, implies (10.16).

Case (vi) Reverse Circulant and Symmetric Circulant: There are now three cases for each Reverse Circulant match:

1. $v_{i-1} + v_i - v_{j_i-1} - v_{j_i} = 0$,
2. $v_{i-1} + v_i - v_{j_i-1} - v_{j_i} = 1$, and
3. $v_{i-1} + v_i - v_{j_i-1} - v_{j_i} = -1$.

Also, there are six cases for each Symmetric Circulant match as in Case (iii). As before, we can write $\Pi_C(w)$ as the union of $3^{k_1} \times 6^{k_2}$ possible $\Pi_{C,l}(w)$. Hence, arguing in a manner similar to the previous cases, $\lim_{n\to\infty} \frac{1}{n^{1+k}} \#\Pi_{C,l}(w)$ exists as an integral and that verifies (10.15).

Now, to check (10.16), we proceed case by case. Suppose $l \neq l'$ and $\pi \in \Pi_{C,l}(w) \cap \Pi_{C,l'}(w)$. Since $l \neq l'$, there must be one Reverse Circulant or Symmetric Circulant match such that two of the possible cases (1)-(3) given above, or (1)-(6) (which appear in Case (iii)) occur simultaneously. It is easily seen that such an occurrence is impossible for a Reverse Circulant match. So we assume that there is a Symmetric Circulant match. But this can be tackled by the arguments given in the previous cases.

Case (vii) Wigner and Hankel: Let X and Y be Wigner (W) and Hankel (H), respectively. Observe that for all $i \in S_W$,

$$(\pi(i-1), \pi(i)) = \begin{cases} (\pi(j_i-1), \pi(j_i)) & \text{(Constraint (C1)), or} \\ (\pi(j_i), \pi(j_i-1)) & \text{(Constraint (C2).} \end{cases} \tag{10.22}$$

Also, $\pi(i-1) + \pi(i) = \pi(j_i - 1) + \pi(j_i)$ for all $i \in S_H$. So, for each Wigner match, there are two possible constraints, and hence there are 2^{k_1} choices. Let λ be a typical choice of k_1 constraints and $\Pi_{C,\lambda}(w)$ be the corresponding subset of $\Pi_C(w)$. Hence

$$\Pi_C(w) = \cup_\lambda \Pi_{C,\lambda}(w)$$

which is *not* a disjoint union. Now, using equation (10.17), we have,

$$\begin{aligned} \#\Pi_{C,\lambda}(w) \quad = \quad & \#\{(v_0, \ldots, v_{2k}) : 0 \leq v_i \leq 1, v_0 = v_{2k}, \\ & v_{i-1} + v_i = v_{j_i-1} + v_{j_i}, \; i \in S_H, \\ & (v_{i-1}, v_i) \text{ equals } (v_{j_i-1}, v_{j_i}), \text{ or } (v_{j_i}, v_{j_i-1}), \; i \in S_W\}. \end{aligned}$$

It can be seen from the above equations that each v_j, $j \notin S$, can be written (may not be uniquely) as a linear combination C_j^λ of elements in v_S. Hence, as before,

$$\begin{aligned} \#\Pi_{C,\lambda}(w) \quad = \quad & \#\{v_S : v_i = C_i^\lambda(v_S), v_0 = v_{2k}, i \notin S, \\ & v_{i-1} + v_i = v_{j_i-1} + v_{j_i}, \; i \in S_H, \\ & (v_{i-1}, v_i) \text{ equals } (v_{j_i-1}, v_{j_i}), \text{ or } (v_{j_i}, v_{j_i-1}), \; i \in S_W\}. \end{aligned}$$

So the limit of $\#\Pi_{C,\lambda}(w)/n^{1+k}$ exists for every λ and can be expressed as an appropriate Riemann integral.

It remains to show (10.16) for any $\lambda_1 \neq \lambda_2$. Suppose $\pi \in \Pi_{C,\lambda_1}(w) \cap \Pi_{C,\lambda_2}(w)$. Then there is at least one $i \in S_W$, such that the Wigner matches given by λ_1 and λ_2 at i obey the (C1) and (C2) or vice versa. In any case, this implies that

$$(\pi(j_i), \pi(j_i - 1)) = (\pi(i-1), \pi(i)) = (\pi(j_i - 1), \pi(j_i)),$$

which implies that $\pi(i) = \pi(i-1)$. But since i is a generating vertex, this lowers the count order by a factor of n and hence the claim follows.

Case (viii) Wigner and other matrices: These cases follow by similar and repetitive arguments. ■

10.4 Sum of independent patterned matrices

The following result essentially follows from Theorem 10.2.1.

Corollary 10.4.1. Let A and B be two independent real symmetric patterned matrices which satisfy Assumption III.

(a) Suppose $p_C(w)$ exists for every q and every w. Then the EESD of $\frac{A+B}{\sqrt{n}}$ converges weakly. The limit probability law, say μ, is symmetric and does not depend on the underlying distribution of the entries. If either the LSD of $\frac{A}{\sqrt{n}}$ or that of $\frac{B}{\sqrt{n}}$ has unbounded support, then μ also has unbounded support.

(b) The conclusions of (a) hold for any two of the five matrices—Wigner, Toeplitz, Hankel, Reverse Circulant and Symmetric Circulant. ◆

Proof. (a) By Theorem 10.2.1, $\{\frac{A}{\sqrt{n}}, \frac{B}{\sqrt{n}}\}$ converge jointly. Let us fix k and let Q_k be the set of monomials such that $(A+B)^k = \sum_{q \in Q_k} q(A, B)$. Hence

$$\frac{1}{n} \mathrm{E}\left[\mathrm{Tr}\left(\frac{A+B}{\sqrt{n}}\right)^k\right] = \frac{1}{n^{1+k/2}} \mathrm{E}\left[\sum_{q \in Q_k} \mathrm{Tr}(q(A, B))\right] = \sum_{q \in Q_k} \varphi_n(q).$$

By Theorem 10.2.1(a), $\varphi_n(q) \to \alpha(q)$ and hence

$$m_k = \lim_{n \to \infty} \frac{1}{n} \mathrm{E}\left[\mathrm{Tr}\left(\frac{A+B}{\sqrt{n}}\right)^k\right] = \sum_{q \in Q_k} \alpha(q).$$

Using Theorem 10.2.1(b), we have

$$m_{2k} = \sum_{q \in Q_{2k}} \alpha(q) \leq \#(Q_{2k}) \frac{(2k)!}{k! 2^k} \Delta^k = 2^{2k} \frac{(2k)!}{k! 2^k} \Delta^k.$$

Thus $\{m_{2k}\}$ satisfies Riesz's condition, implying that the LSD exists.

To prove symmetry of the limit, let $q \in Q_{2k+1}$. Then from Theorem 10.2.1 (b), it follows that $\alpha(q) = 0$. Hence $m_{2k+1} = \sum_{q \in Q_{2k+1}} \alpha(q) = 0$ and the distribution is symmetric. We leave the proof of unboundedness as an exercise.

(b) From Theorem 10.3.1, $p_C(w)$ exists for these matrices, taken two at a time. ■

Example 10.4.1. The following figures provide the simulated distributions of the limits in the above corollary for some cases. ▲

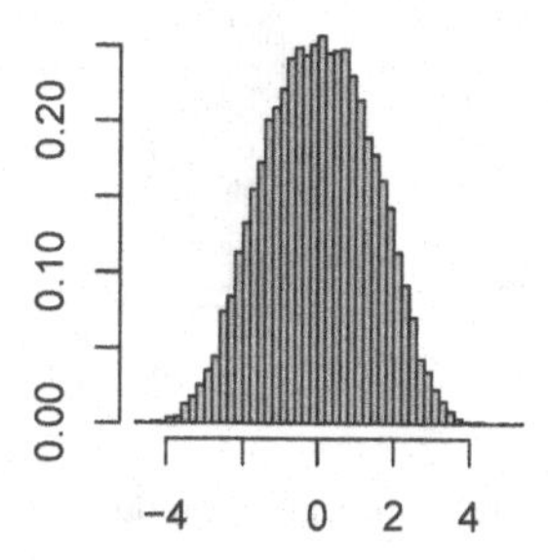

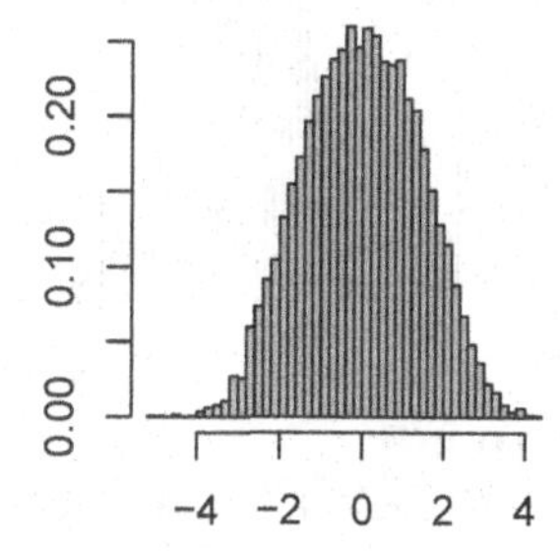

FIGURE 10.1
ESD of $n^{-1/2}(T_n + H_n)$, $n = 400$. Left: standard normal; right: standardized symmetric Bernoulli.

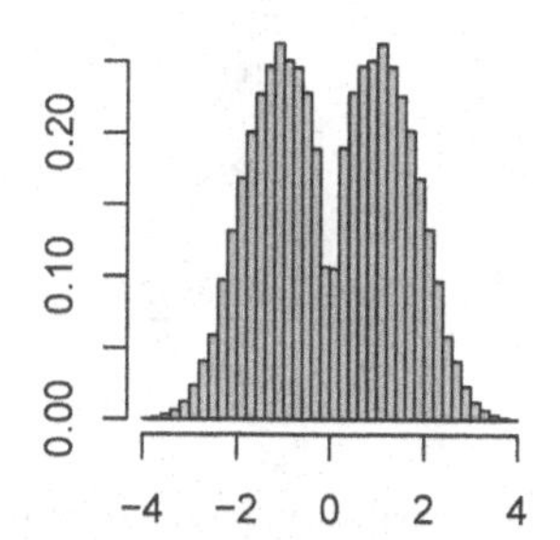

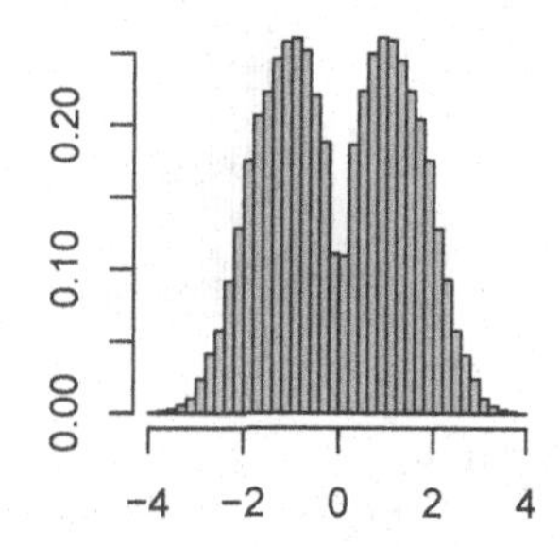

FIGURE 10.2
ESD of $n^{-1/2}(RC_n + H_n)$, $n = 1000$. Left: standard normal; right: standardized symmetric Bernoulli.

10.5 Discussion

Joint convergence of random matrices has many applications and we briefly mention a few. Traces of non-commutative polynomials of random matrices have applications in operator algebras and telecommunications. Joint convergence properties of high dimensional random matrices have been extremely useful in wireless communication theory and practice, especially in CDMA and MIMO systems. See Rashidi Far et al. (2008)[79], Couillet and Debbah (2011)[36], Couillet, Debbah, and Silverstein (2011)[37] and Tulino and Verdu (2004)[99]. See Bose and Bhattacharjee (2018)[21] for applications of joint convergence of sample autocovariance matrices in high dimensional time series analysis. We refer the reader to Male (2012)[67] for links between analysis of MIMO systems and joint convergence of certain patterned random matrices.

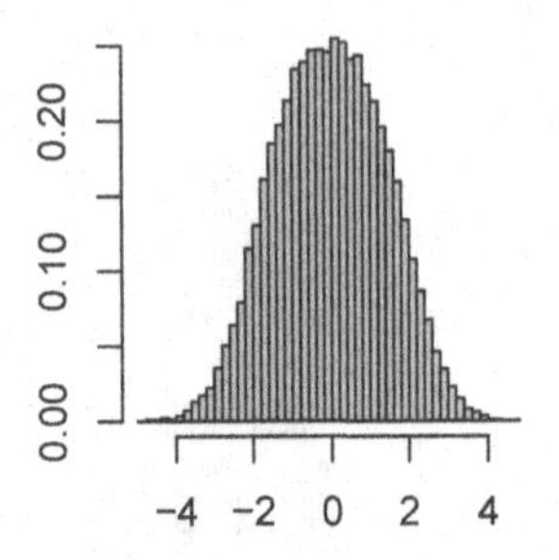

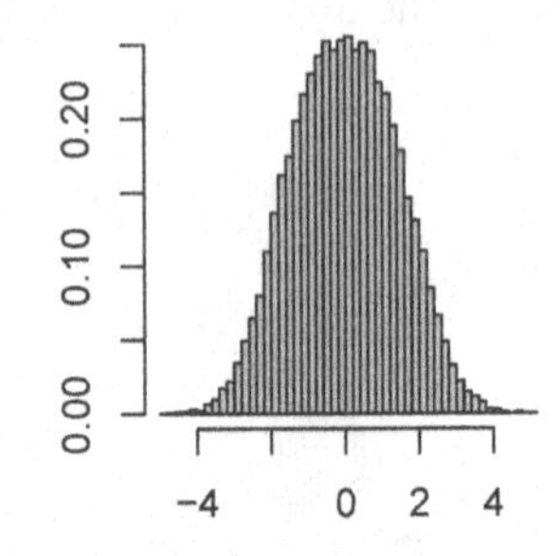

FIGURE 10.3
ESD of $n^{-1/2}(SC_n + H_n)$, $n = 1000$. Left: standard normal; right: standardized symmetric Bernoulli.

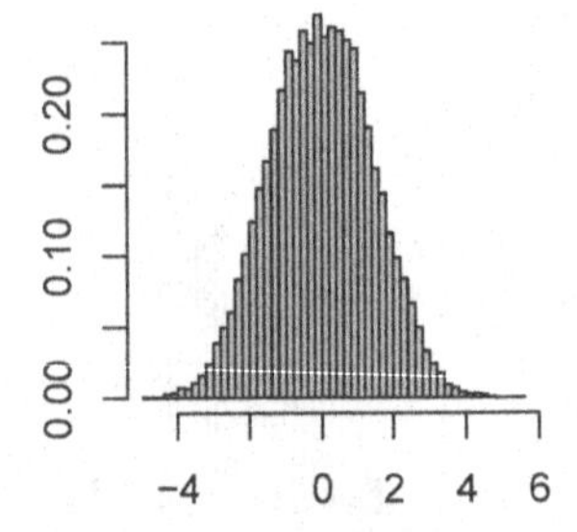

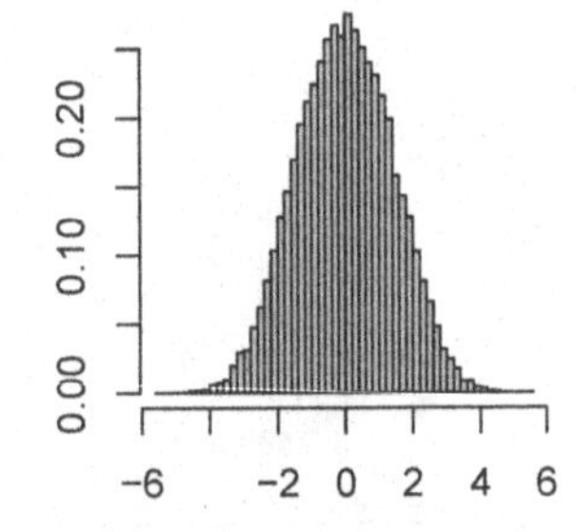

FIGURE 10.4
ESD of $n^{-1/2}(SC_n + RC_n)$, $n = 1000$. Left: standard normal; right: standardized symmetric Bernoulli.

10.6 Exercises

1. Tie up the loose ends in the proof of Theorem 10.2.1.
2. Establish the unboundedness claim in the proof of Corollary 10.4.1.
3. Work out the detailed proof of (10.13).
4. Simulate the ESD of sums of independent patterned random matrices considered in this chapter.

11

Asymptotic freeness of random matrices

We saw in Chapter 9 that independent elliptic matrices, which include IID and Wigner matrices, and also independent S-matrices are asymptotically free. We recall these results first.

Then we introduce deterministic matrices and show that they are free of independent elliptic matrices. When the elliptic matrices are Gaussian, no conditions on the deterministic matrices are needed, except for the necessary requirement that they should jointly converge. When Gaussianity is dropped, the entries of the deterministic matrices are additionally required to satisfy certain boundedness conditions. The joint asymptotic freeness of independent cross-covariance matrices where the component matrices are pair-correlated is also established.

In Chapter 9, we showed by direct moment calculations that independent S-matrices are asymptotically free. We now present an alternate proof of this result, by embedding these matrices into Wigner matrices of higher dimensions and using the asymptotic freeness of the Wigner and deterministic matrices.

We also show that the independent Wigner matrices and independent IID matrices are asymptotically free of other patterned random matrices.

11.1 Elliptic, IID, Wigner and S-matrices

We begin by recalling the two asymptotic freeness theorems that we have proved so far. The first result recollects Theorem 9.3.1 on the joint convergence of independent elliptic matrices. The IID and the Wigner matrices are special cases of the elliptic matrices, but we mention them individually in the statement for emphasis.

Theorem 11.1.1 (Wigner, elliptic and IID matrices). Suppose $\{W_n^{(i)}\}$, $\{IID_n^{(i)}\}$ and $\{E_n^{(i)}\}$ (with correlation parameter ρ_i), $1 \leq i \leq m$, are independent Wigner, IID and elliptic matrices respectively, with real or complex entries that satisfy, as the case may be, Assumption Ie or II of Chapter 9. Then as elements of $(\mathcal{M}_n(\mathbb{C}), n^{-1}\,\mathrm{E}\,\mathrm{Tr})$, these $3m$ matrices converge jointly and are asymptotically free. ◆

DOI: 10.1201/9781003144496-11

The second theorem recollects the asymptotic freeness of the S-matrices which is a consequence of Theorem 9.4.1 and Exercise 9 of Chapter 9.

Theorem 11.1.2 (Sample covariance matrices). Let $S^{(\tau)} := n_\tau^{-1} X_{p\times n_\tau}^{(\tau)} X_{n_\tau \times p}^{(\tau)*}$, $1 \leq \tau \leq m$, be independent S-matrices with entries $\{x_{i,j,n}^{(\tau)}\}$ of $X_{p\times n_\tau}^{(\tau)}$ that satisfy Assumption Ie of Chapter 9 with possibly different correlation parameters ρ_τ. If $\frac{p}{n_\tau} \to y_\tau (0 < y_\tau < \infty)$ as $p \to \infty$, then $S^{(1)}, \ldots, S^{(m)}$, as elements of $(\mathcal{M}_p(\mathbb{C}), p^{-1} \operatorname{E} \operatorname{Tr})$, are asymptotically free $MP_{y_1}, \ldots, MP_{y_m}$ variables. ◆

11.2 Gaussian elliptic, IID, Wigner and deterministic matrices

We now introduce deterministic matrices in the mix to investigate how far we can extend Theorem 11.1.1. We first tackle the case when the entries are Gaussian. Due to Isserlis' formula given in Section 1.5, it is relatively easier to deal with matrices that have Gaussian entries.

Recall the discussion from Section 3.8 on how any partition can be considered as a permutation. As a handy technical tool, we now extend our notion of multiplicative functions on partitions to multiplicative functions on permutations. Let S_n denote the group of permutations of $\{1, \ldots, n\}$.

Definition 11.2.1. (Multiplicative functions on permutations) Let $\mathcal{A}$ be an algebra, and for every k, let $\psi_k : \mathcal{A}^k \longrightarrow \mathbb{C}$, $k > 0$, be a multi-linear function which is also tracial in the sense that

$$\psi_k(A_1, \ldots, A_k) = \psi_k(A_k, A_1, \ldots, A_{k-1}) \text{ for all } A_1, \ldots, A_k \in \mathcal{A}^k.$$

Suppose $\alpha \in S_n$ and let $c_1, \ldots, c_r$ be the cycles of α. Then define

$$\psi_\alpha[A_1, \ldots, A_n] = \psi_{c_1}[A_1, \ldots, A_n] \cdots \psi_{c_r}[A_1, \ldots, A_n], \tag{11.1}$$

where

$$\psi_c[A_1, \ldots, A_n] = \psi_p\left(A_{i_1} \cdots A_{i_p}\right) \text{ if } c = (i_1, i_2, \ldots, i_p). \tag{11.2}$$

◇

For our purposes, $\mathcal{A}$ is the set of all $n \times n$ matrices and

$$\psi_k(A_1, \ldots, A_k) = \frac{1}{k}\text{Trace}(A_1 A_2 \cdots A_k) = \operatorname{tr}(A_1 A_2 \cdots A_k).$$

As an illustration, if $\sigma = (135)(2)(4)(6)$, then

$$\operatorname{tr}_\sigma[A_1, \ldots, A_6] = \operatorname{tr}(A_1 A_3 A_5)\operatorname{tr}(A_2)\operatorname{tr}(A_4)\operatorname{tr}(A_6). \tag{11.3}$$

We can now state our first result on the asymptotic freeness where deterministic matrices are also involved. This theorem includes the Gaussian Wigner matrices ($\rho = 1$) and the Gaussian IID matrices ($\rho = 0$) as special cases.

Theorem 11.2.1. (Gaussian elliptic and deterministic matrices) Let $\{E_n^{(i)}\}_{1 \le i \le m}$ be $n \times n$ independent Gaussian elliptic matrices. Let $\{D_n^{(i)}\}_{1 \le i \le \ell}$ be $n \times n$ deterministic matrices such that in $(\mathcal{M}_n(\mathbb{C}), \mathrm{tr})$,

$$(D_n^{(1)}, \ldots, D_n^{(\ell)}) \to (d_1, \ldots, d_\ell), \text{ as } n \to \infty, \tag{11.4}$$

for some $d_1, \ldots, d_\ell \in (\mathcal{A}, \varphi)$. Then as elements of $(\mathcal{M}_n(\mathbb{C}), \varphi_n := n^{-1} \mathrm{E}\,\mathrm{Tr})$,

$$(n^{-1/2} E_n^{(1)}, \ldots, n^{-1/2} E_n^{(m)}, D_n^{(1)}, \ldots, D_n^{(\ell)}) \to (e_1, \ldots, e_m, d_1, \ldots, d_\ell), \tag{11.5}$$

where $e_1, \ldots, e_m$ are free elliptic and are also free of $\{d_1, \ldots, d_\ell\}$. ◆

Proof. Note that by Theorem 11.1.1, $(n^{-1/2} E_n^{(1)}, \ldots, n^{-1/2} E_n^{(m)}$ converge jointly to $e_1, \ldots, e_m$ and are free elliptic. We need to show the joint convergence when the deterministic matrices are also involved. For notational simplicity, we give the proof only for $m = 1$ and $\ell = 1$. Later we shall comment upon how the general case can be tackled. To simplify notation further, we write E_n and D_n for $E_n^{(1)}$ and $D_n^{(1)}$ respectively. To show the joint convergence and the claimed freeness, we need to show the following two things:

(a) *The moment of any monomial converges.* Recalling traciality, we reduce this problem as follows. Fix $p \ge 1$. Let $\epsilon_1, \ldots, \epsilon_p \in \{1, *\}$. For $1 \le \ell \le p$, let

$$A_n^{(\ell)} = m_\ell(D_n, D_n^*) \tag{11.6}$$

be monomials of degrees q_ℓ, respectively. For $q_\ell = 0$, $A_n^{(\ell)}$ is the identity matrix. Then we need to show that for all choices of p and monomials $\{A_n^{(\ell)}\}$,

$$\lim \varphi_n(n^{-1/2} E_n^{\epsilon_1} A_n^{(1)} \cdots n^{-1/2} E_n^{\epsilon_p} A_n^{(p)}) \text{ exists.} \tag{11.7}$$

(b) *The limit moments obtained in (11.7) satisfy the necessary and sufficient criterion for freeness given in Lemma 3.9.1 (b).*

We label the main steps of the proof for clarity and for future reference.

Step 1. Trace formula with special indexing. Let $a_{ij}^{(\ell)}$ and e_{ij} denote the (i, j)-th element of $A^{(\ell)}$ and E_n respectively. We have suppressed the dependence of these quantities on n for notational convenience. While computing trace, we shall use a special type of indexing: let

$$\begin{aligned} I_p &:= \{i_1, \ldots, i_p : 1 \le i_t \le n, 1 \le t \le p\}, \quad i_{p+1} := i_1, \\ J_p &:= \{j_1, \ldots, j_p : 1 \le j_t \le n, 1 \le t \le p\}, \end{aligned}$$

and let $I_p(\pi)$ and $J_p(\pi)$ denote respectively the subsets of I_p and J_p which obey a given partition π.

As the elements of D_n are deterministic, we can write

$$\begin{aligned}\mathrm{E}[\mathrm{Trace}(E_n^{\epsilon_1}A_n^{(1)}\cdots E_n^{\epsilon_p}A_n^{(p)})] &= \sum_{I_p,J_p}\mathrm{E}[e_{i_1j_1}^{\epsilon_1}a_{j_1i_2}^{(1)}e_{i_2j_2}^{\epsilon_2}\cdots e_{i_pj_p}^{\epsilon_p}a_{j_pi_1}^{(p)}]\\ &= \sum_{I_p,J_p}\mathrm{E}[e_{i_1j_1}^{\epsilon_1}e_{i_2j_2}^{\epsilon_2}\cdots e_{i_pj_p}^{\epsilon_p}]\prod_{\ell=1}^{p}a_{j_\ell i_{\ell+1}}^{(\ell)}.\end{aligned}$$

Hence

$$\varphi_n(\bar{E}_n^{\epsilon_1}A_n^{(1)}\cdots\bar{E}_n^{\epsilon_p}A_n^{(p)}) = \frac{1}{n^{\frac{p}{2}+1}}\sum_{I_p,J_p}\mathrm{E}[e_{i_1j_1}^{\epsilon_1}e_{i_2j_2}^{\epsilon_2}\cdots e_{i_pj_p}^{\epsilon_p}]\prod_{\ell=1}^{p}a_{j_\ell i_{\ell+1}}^{(\ell)}, \tag{11.8}$$

where

$$\bar{E}_n := n^{-1/2}E_n.$$

Step 2. Monomials with odd number of factors do not contribute. Since we have assumed that the entries e_{ij} are Gaussian random variables with mean zero, by Isserlis' formula, if p is odd,

$$\varphi_n(\bar{E}_n^{\epsilon_1}A_n^{(1)}\cdots\bar{E}_n^{\epsilon_p}A_n^{(p)}) = 0. \tag{11.9}$$

Step 3. Only pair-partitions survive. Let $p = 2k$. Recalling the correlation structure of the entries of E_n, by Isserlis' formula,

$$\begin{aligned}\varphi_n(\bar{E}_n^{\epsilon_1}A_n^{(1)}\cdots\bar{E}_n^{\epsilon_{2k}}A_n^{(2k)}) &= \frac{1}{n^{k+1}}\sum_{\pi\in\mathcal{P}_2(2k)}\sum_{\substack{I_{2k}(\pi)\\ J_{2k}(\pi)}}\prod_{(r,s)\in\pi}\mathrm{E}[e_{i_rj_r}^{\epsilon_r}e_{i_sj_s}^{\epsilon_s}]\prod_{\ell=1}^{2k}a_{j_\ell i_{\ell+1}}^{(\ell)}\\ &= \frac{1}{n^{k+1}}\sum_{\pi\in\mathcal{P}_2(2k)}\sum_{\substack{I_{2k}(\pi)\\ J_{2k}(\pi)}}\prod_{(r,s)\in\pi}(u(r,s)+v(r,s))\prod_{\ell=1}^{2k}a_{j_\ell i_{\ell+1}}^{(\ell)},\end{aligned}$$

where $(r,s), r < s$ denotes a typical block of π, and

$$\begin{aligned}u(r,s) &:= (\rho+(1-\rho)\delta_{\epsilon_r\epsilon_s})\delta_{i_ri_s}\delta_{j_rj_s} \qquad (11.10)\\ &= \begin{cases}\delta_{i_ri_s}\delta_{j_rj_s} & \text{if } \epsilon_r = \epsilon_s,\\ \rho\delta_{i_ri_s}\delta_{j_rj_s} & \text{if } \epsilon_r \neq \epsilon_s;\end{cases}\\ v(r,s) &:= (1-(1-\rho)\delta_{\epsilon_r\epsilon_s})\delta_{i_rj_s}\delta_{j_ri_s}\\ &= \begin{cases}\rho\delta_{i_ri_s}\delta_{j_rj_s} & \text{if } \epsilon_r = \epsilon_s,\\ 1 & \text{if } \epsilon_r \neq \epsilon_s.\end{cases}\end{aligned}$$

Note that

$$|u(r,s)| \leq \delta_{i_ri_s}\delta_{j_rj_s} \text{ and } |v(r,s)| \leq \delta_{i_rj_s}\delta_{j_ri_s}. \tag{11.11}$$

The Gaussianity of the variables is used up to Step 3 and is not used in the rest of the proof. This observation will be important in the non-Gaussian case.

Step 4. In the above sum,

$$\textbf{Terms with at least one } u(r,s) \textbf{ factor vanish in the limit.} \quad (11.12)$$

This is established in Lemma 11.2.2. We continue with the theorem's proof assuming this for the moment.

Step 5. Verification of (a) (convergence). Since $D_n \to d$ (say), we have

$$(A_n^{(\ell)},\ 1 \le \ell \le 2k) \to (a^{(\ell)},\ 1 \le \ell \le 2k) \text{ (say), where } a^{(\ell)} = m_\ell(d, d^*).$$

Hence for any permutation σ of $\{1, 2, \ldots, 2k\}$, the following limit exists:

$$\varphi_\sigma[a^{(1)}, \ldots, a^{(2k)}] := \lim_{n\to\infty} \mathrm{tr}_\sigma[A_n^{(1)}, \ldots, A_n^{(2k)}]. \quad (11.13)$$

Let γ_{2k} be the cyclic permutation $1 \to 2 \to \cdots \to 2k-1 \to 2k \to 1$. Hence ($\pi$ viewed as a permutation satisfies $\pi(r) = s, \pi(s) = r$ and so $\{r, s\}$ is a cycle)

$$\begin{aligned}
&\lim_{n\to\infty} \varphi_n(\bar{E}_n^{\epsilon_1} A_n^{(1)} \cdots \bar{E}_n^{\epsilon_{2k}} A_n^{(2k)}) \\
&= \lim \frac{1}{n^{k+1}} \sum_{\pi\in\mathcal{P}_2(2k)} \sum_{I_{2k}(\pi),\ J_{2k}(\pi)} \prod_{(r,s)\in\pi} v(r,s) \prod_{\ell=1}^{2k} a^{(\ell)}_{j_\ell i_{\ell+1}} \\
&= \lim_{n\to\infty} \frac{1}{n^{k+1}} \sum_{\pi\in\mathcal{P}_2(2k)} \sum_{\substack{I_{2k}(\pi)\\ J_{2k}(\pi)}} \prod_{(r,s)\in\pi} (1-(1-\rho)\delta_{\epsilon_r \epsilon_s}) \prod_{t=1}^{2k} \delta_{i_t j_{\pi(t)}} \prod_{\ell=1}^{2k} a^{(\ell)}_{j_\ell i_{\gamma_{2k}(\ell)}}, \\
&= \sum_{\pi\in\mathcal{P}_2(2k)} \rho^{T(\pi)} \lim_{n\to\infty} \frac{1}{n^{k+1}} \sum_{J_{2k}(\pi)} \prod_{\ell=1}^{2k} a^{(\ell)}_{j_\ell j_{\pi\gamma_{2k}(\ell)}}, \quad (\text{since } i_t = j_{\pi(t)} \text{ for all } t) \\
&= \sum_{\pi\in\mathcal{P}_2(2k)} \rho^{T(\pi)} \lim_{n\to\infty} \frac{1}{n^{k+1}} \mathrm{tr}_{\pi\gamma_{2k}}[A_n^{(1)}, \ldots, A_n^{(2k)}] n^{|\pi\gamma_{2k}|} \quad (\text{by (11.1)– (11.3)}),
\end{aligned}$$

where $T(\pi)$ is given by

$$T(\pi) := \#\{(r,s) \in \pi : \delta_{\epsilon_r \epsilon_s} = 1\} \quad (\text{compare with (3.4)}).$$

We know from Lemma 3.8.1 (h) that, if $\pi \in \mathcal{P}_2(2k)$, then $|\pi\gamma_{2k}| \le k+1$. Moreover, equality holds if and only if $\pi \in NC_2(2k)$. Using (11.13), the above limit is 0 whenever $|\pi\gamma_{2k}| < k+1$. As a consequence, only the non-crossing pair-partitions survive in the above limit and we obtain,

$$\lim_{n\to\infty} \varphi_n(\bar{E}_n^{\epsilon_1} A_n^{(1)} \cdots \bar{E}_n^{\epsilon_{2k}} A_n^{(2k)}) = \sum_{\pi\in NC_2(2k)} \rho^{T(\pi)} \varphi_{\pi\gamma_{2k}}[a^{(1)}, \ldots, a^{(2k)}]. \quad (11.14)$$

Hence $(n^{-1/2}E_n, D_n) \to (e, d)$ (say), and the limit moments are given by (recall that $a^{(\ell)} = m^{(\ell)}(d, d^*)$ are polynomials in $\{d, d^*\}$)

$$\varphi(e^{\epsilon_1} a^{(1)} \cdots e^{\epsilon_{2k}} a^{(2k)}) = \sum_{\pi\in NC_2(2k)} \rho^{T(\pi)} \varphi_{\pi\gamma_{2k}}[a^{(1)}, \ldots, a^{(2k)}]. \quad (11.15)$$

Step 6. Verification of (b) (freeness in the limit). From the convergence of $n^{-1/2}E_n$, proved in Theorem 8.3.1, we know that for all $\pi \in NC_2(2k)$,

$$\kappa_\pi[e^{\epsilon_1}, \ldots, e^{\epsilon_{2k}}] = \rho^{T(\pi)}.$$

We also know from Lemma 3.8.1 (g) that if $\pi \in NC_2(2k)$ then its Kreweras complement $K(\pi) = \pi\gamma_{2k}$. Hence (11.15) can be rewritten as

$$\varphi(e^{\epsilon_1}a^{(1)} \cdots e^{\epsilon_{2k}}a^{(2k)}) = \sum_{\pi \in NC_2(2k)} \kappa_\pi[e^{\epsilon_1}, \ldots, e^{\epsilon_{2k}}]\, \varphi_{K(\pi)}[a^{(1)}, \ldots, a^{(2k)}].$$

Now we can invoke Lemma 3.9.1 (b) to complete the proof for the case $l = m = 1$, except that it remains to establish (11.12).

Step 7. The case of l or $m \geq 2$. A similar proof works but we need to then invoke Theorem 9.3.1 and prove an appropriate version of Lemma 11.2.2. The above proof and the proof of Lemma 11.2.2, both work, mutatis mutandis, with added notational complexity. We omit the tedious details. ■

It remains to verify claim (11.12) on the asymptotic negligibility of terms that involve at least one $u(r,s)$. Recalling the factor $\delta_{i_r i_s}\delta_{j_r j_s}$ in the definition of $u(r,s)$ from (11.10), any such term has at least one self-match within each of I_{2k} and J_{2k}. To state Lemma 11.2.2, we use the following notation. Let $\pi := \{(r_1, s_1), \ldots, (r_k, s_k), r_i < s_i, i = 1, \ldots k\}$ be a typical element of $NC_2(2k)$. Let $A_n^{(\ell)}$ be as defined in (11.6). Let

$$\begin{aligned} u_\ell &:= \delta_{i_{r_\ell} i_{s_\ell}}\delta_{j_{r_\ell} j_{s_\ell}},\ v_\ell := \delta_{i_{r_\ell} j_{s_\ell}}\delta_{j_{r_\ell} i_{s_\ell}} \\ S_{t,k} &:= \{\{\ell_1, \ldots, \ell_t\}, \{\ell_{t+1}, \ldots, \ell_k\}) : \{\ell_1, \ldots, \ell_t, \ell_{t+1}, \ldots, \ell_k\} = \{1, \ldots, k\}\}. \end{aligned} \tag{11.16}$$

Note that $S_{t,k}$ is the set of all two-block partitions of $\{1, \ldots, k\}$ with t and $k-t$ elements. Recall also the matrices $A_n^{(\ell)}$ defined in (11.6). Its (i,j)-the entry will be denoted by $a_{ij}^{(\ell)}$.

Lemma 11.2.2. Let $\{D_n\}$ be a sequence of deterministic matrices which converges in $*$-distribution to an element d. Then, for every $k \geq 1$, every $\pi \in \mathcal{P}_2(2k)$ and every $1 \leq t \leq k$,

$$\lim_{n\to\infty} \frac{1}{n^{k+1}} \sum_{I_{2k}(\pi), J_{2k}(\pi)} \Big(\sum_{S_{t,k}} u_{\ell_1} \cdots u_{\ell_t} v_{\ell_{t+1}} \cdots v_{\ell_k}\Big) \prod_{\ell=1}^{2k} a_{j_\ell i_{\ell+1}}^{(\ell)} = 0. \tag{11.17}$$

As a consequence,

$$\lim_{n\to\infty} \frac{1}{n^{k+1}} \sum_{\pi \in \mathcal{P}_2(2k)} \sum_{\substack{I_{2k}(\pi) \\ J_{2k}(\pi)}} \Big(\sum_{t=1}^{k} \sum_{S_{t,k}} u_{\ell_1} \cdots u_{\ell_t} v_{\ell_{t+1}} \cdots v_{\ell_k}\Big) \prod_{\ell=1}^{2k} a_{j_\ell i_{\ell+1}}^{(\ell)} = 0. \tag{11.18}$$

A similar result holds for deterministic matrices which converge jointly. ♦

Example 11.2.1. To provide insight, let us illustrate this in a particular case. Suppose $k = 2$ and $\pi = (12)(34)$ is a non-crossing partition.

(i) Consider the case $i_1 = i_2, i_3 = i_4$ and $j_1 = j_2, j_3 = j_4$ of self-matches within both I_4 and J_4. Then we have

$$\lim_{n\to\infty} \frac{1}{n^3} \sum_{I_4,J_4} \delta_{i_1i_2}\delta_{j_1j_2}\delta_{i_3i_4}\delta_{j_3j_4} \prod_1^4 a^{(\ell)}_{j_\ell i_{\ell+1}} = \lim_{n\to\infty} \frac{1}{n^3} \sum_{i_1,j_1,i_3,j_3} a^{(1)}_{j_1i_1} a^{(2)}_{j_1i_3} a^{(3)}_{j_3i_3} a^{(4)}_{j_3i_1}$$

$$= \lim_{n\to\infty} \frac{1}{n^3} \left\{ n \operatorname{tr}(A^{(1)*}_n A^{(2)}_n A^{(3)*}_n A^{(4)}_n) \right\} =0,$$

since D_n converge in $*$-distribution.

(ii) Consider the case of one self-match and one cross-match among the i's and j's: $i_1 = i_2, i_3 = j_4$ and $j_1 = j_2, j_3 = i_4$. Then we have

$$\lim_{n\to\infty} \frac{1}{n^3} \sum_{I_4,J_4} \delta_{i_1i_2}\delta_{j_1j_2}\delta_{i_3j_4}\delta_{j_3i_4} \prod_1^4 a^{(\ell)}_{j_\ell i_{\ell+1}} = \lim_{n\to\infty} \frac{1}{n^3} \sum_{i_1,j_1,j_3,j_4} a^{(1)}_{j_1i_1} a^{(2)}_{j_1j_4} a^{(3)}_{j_3j_3} a^{(1)}_{j_4i_1}$$

$$= \lim_{n\to\infty} \frac{1}{n^3} \left\{ n^2 \operatorname{tr}(A^{(1)*}_n A^{(2)}_n A^{(4)}_n) \operatorname{tr}(A^{(3)}_n) \right\} =0.$$

(iii) In the remaining case, there is no self-match among the i's and j's ($i_1 = j_2, i_3 = j_4$ and $j_1 = i_2, j_3 = i_4$). Then

$$\lim_{n\to\infty} \frac{1}{n^3} \sum_{I_4,J_4} \delta_{i_1j_2}\delta_{j_1i_2}\delta_{i_3j_4}\delta_{j_3i_4} \prod_1^4 a^{(\ell)}_{j_\ell i_{\ell+1}} = \lim_{n\to\infty} \frac{1}{n^3} \sum_{j_1,j_2,j_3,j_4} a^{(1)}_{j_1j_3} a^{(2)}_{j_2j_2} a^{(3)}_{j_3j_1} a^{(1)}_{j_4j_4}$$

$$= \lim_{n\to\infty} \frac{1}{n^3} \left\{ \operatorname{tr}(A^{(1)}_n A^{(3)}_n) \operatorname{tr}(A^{(2)}_n) \operatorname{tr}(A^{(4)}_n) \cdot n^3 \right\} = \varphi(d_1d_3)\varphi(d_2)\varphi(d_4).$$

Hence the limit is zero whenever there is at least one self-match among both the i's and j's. That this is also true when $k > 2$ is the contention of the lemma. ▲

Proof of Lemma 11.2.2. The proof given below for a single matrix can be used mutatis mutandis, when there are several matrices which converge jointly.

It is enough to prove (11.17) since (11.18) is obtained by adding (11.17) over all possible (independent of n) values of π and t.

We prove (11.17) by induction. So, first suppose $k = 1$. Then there is only one pair-partition π and the expression in the left side of (11.17) reduces to

$$\lim_{n\to\infty} \frac{1}{n^2} \sum_{I_2,J_2} \delta_{i_1i_2}\delta_{j_1j_2} a^{(1)}_{j_1i_2} a^{(2)}_{j_2i_1} = \lim_{n\to\infty} \frac{1}{n^2} \sum_{i_1,j_1} a^{(1)}_{j_1i_1} a^{(2)}_{j_1i_1}$$

$$= \lim_{n\to\infty} \frac{1}{n^2}\, n \operatorname{tr}(A^{(1)} A^{(2)*}) = 0,$$

since $\{A^{(1)}, A^{(2)}\} \to \{a^{(1)}, a^{(2)}\}$ implies that $\operatorname{tr}(A^{(1)}A^{(2)*}) \to \varphi(a^{(1)}a^{(2)*})$. Hence the result is true for $k = 1$.

So assume that (11.17) holds up to the integer k. We wish to show that it is true for $k+1$.

We write π as $\pi = (r_1, s_1)\cdots(r_{k+1}, s_{k+1})$, with the convention that $r_i \leq s_i$ for all i. Recall that we are working under the hypothesis that there is at least one u-factor in the left side of (11.17). Without loss of generailty, assume that u_1 appears as a factor. Recall from (11.16) that $u_1 = \delta_{i_{r_1} i_{s_1}} \delta_{j_{r_1} j_{s_1}}$. Hence $i_{r_1} = i_{s_1}$ and $j_{r_1} = j_{s_1}$. Then

$$\lim_{n\to\infty} \frac{1}{n^{k+2}} \sum_{\substack{I_{2k+2}(\pi)\\ J_{2k+2}(\pi)}} \sum_{S_{t,k+1}} u_{\ell_1}\cdots u_{\ell_t} v_{\ell_{t+1}}\cdots v_{\ell_{k+1}} \prod_{\ell=1}^{2k+2} a^{(\ell)}_{j_\ell i_{\ell+1}} \tag{11.19}$$

$$= \lim_{n\to\infty} \frac{1}{n^{k+2}} \sum_{\substack{I_{2k+2}(\pi)\\ J_{2k+2}(\pi)}} \sum_{S_{t,k+1}} \delta_{i_{r_1} i_{s_1}} \delta_{j_{r_1} j_{s_1}} u_{\ell_2}\cdots u_{\ell_t} v_{\ell_{t+1}}\cdots v_{\ell_{k+1}} \prod_{\ell=1}^{2k+2} a^{(\ell)}_{j_\ell i_{\ell+1}}.$$

Case I: Suppose $(r_1, s_1 = r_1+1) \in \pi$, $r_1 \neq 2k$. (A similar proof works if $(r_1 = 1, s_1 = 2k) \in \pi$). Then we have

$$\sum_{i_{r_1}, i_{s_1}, j_{r_1}, j_{s_1}} \delta_{i_{r_1} i_{s_1}} \delta_{j_{r_1} j_{s_1}} a^{(r_1-1)}_{j_{r_1-1} i_{r_1}} a^{(r_1)}_{j_{r_1} i_{r_1+1}} a^{(s_1)}_{j_{s_1} i_{s_1+1}}$$
$$= \sum_{i_{r_1}, j_{r_1}} a^{(r_1-1)}_{j_{r_1-1} i_{r_1}} a^{(r_1)*}_{i_{r_1} j_{r_1}} a^{(s_1)}_{j_{r_1} i_{s_1+1}} \quad \text{(note that } i_{r_1} = i_{s_1})$$
$$= (A^{(r_1-1)} A^{(r_1)*} A^{(s_1)})_{j_{r_1-1} i_{s_1+1}} = (B^{(r_1-1)})_{j_{r_1-1} i_{s_1+1}} \quad \text{(say)},$$

where

$$B^{(r_1-1)} := A^{(r_1-1)} A^{(r_1)*} A^{(s_1)}$$

and for any matrix M, M_{ij} denotes its (i,j)th entry.

Note that B is also a monomial in D_n and D_n^*. Now drop the pair (r_1, s_1) and after relabeling the indices, let π' denote the reduced pair-partition of $\{1, 2, \ldots 2k\}$. Also denote, for convenience, the other A matrices in (11.19) accordingly as B matrices. Note that three A-matrices have been reduced to one B-matrix. Further, the earlier set $S_{t,k+1}$ has now changed to $S_{t-1,k}$. In summary, (11.19) reduces to the form

$$\lim_{n\to\infty} \frac{1}{n^{k+2}} \sum_{I_{2k}(\pi'), J_{2k}(\pi')} \sum_{S_{t-1,k}} u_{\ell_1}\cdots u_{\ell_t} v_{\ell_{t+1}}\cdots v_{\ell_k} \prod_{\ell=1}^{2k} B^{(\ell)}_{j_\ell i_{\ell+1}}. \tag{11.20}$$

Suppose $t-1 \geq 1$. Then there exists at least one self-match within the i's and j's and the limit in (11.20) is zero by the induction hypothesis.

Now suppose $t-1=0$. Then there are no self-matches in the i's or j's. In this case, the limit (11.20) is also zero because the number of blocks among the $A^{(i)}$'s equals $|\pi'\gamma_{2k}| \leq k+1$.

Case II: Let $r_1 \neq s_1 - 1$ (but $r_1 = 1$ and $s_1 = 2k$ is not allowed. It has been dealt with in Case I).

Then we have

$$\sum_{i_{r_1}, i_{s_1}, j_{r_1}, j_{s_1}} \delta_{i_{r_1} i_{s_1}} \delta_{j_{r_1} j_{s_1}} a^{(r_1-1)}_{j_{r_1-1} i_{r_1}} a^{(s_1-1)}_{j_{s_1-1} i_{s_1}} a^{(r_1)}_{j_{r_1} i_{r_1+1}} a^{(s_1)}_{j_{s_1} i_{s_1+1}}$$

$$= (A^{(r_1-1)} A^{(s_1-1)*})_{j_{r_1-1} j_{s_1-1}} (A^{(r_1)*} A^{(s_1)})_{i_{r_1+1} i_{s_1+1}}.$$

Therefore now *four* A-matrices reduce to two B-matrices. We also have a reduction of four variable,s $i_{r_1}, i_{s_1}, j_{r_1}, j_{s_1}$. By renaming the variables, the expression in (11.19) is of the form of (11.20), which is zero as argued earlier. We omit the details. This completes the proof for one D_n matrix.

When we have more D_n matrices, a similar induction argument via reduction of the number of matrices works mutatis mutandis. We omit the details. ■

11.3 General elliptic, IID, Wigner and deterministic matrices

We shall now drop the Gaussian assumption. In the absence of Gaussianity, higher order expectations appear in the trace formula. Additional conditions on the deterministic matrices are now required to control the magnitude of the corresponding factors from the deterministic matrices, so that we can still restrict ourselves to the pair-matches. For any $n \times n$ matrix A_n, let

$$\mathrm{Tr}(|A_n|^k) := \sum_{1 \leq i_j \leq n, j=1,\ldots,k} |a_{i_1 i_2}| \cdots |a_{i_k i_1}|.$$

Theorem 11.3.1. Suppose that $\{E_n^{(i)}\}_{1 \leq i \leq m}$ are elliptic matrices whose entries satisfy Assumption Ie. Suppose $\{D_n^{(i)}\}_{1 \leq i \leq \ell}$ are deterministic matrices which satisfy condition (11.4), and

$$D := \sup_{k \in \mathbb{N}} \max_{1 \leq i \leq l} \sup_n \left[\frac{1}{n} \mathrm{Tr}(|D_n^{(i)}|^k)\right]^{\frac{1}{k}} < \infty. \tag{11.21}$$

Then also (11.5) holds. ◆

Proof. We shall make use of the notation and developments in the proof of Theorem 11.2.1. Assume first that $l = 1 = m$. As mentioned during the course of that proof, Gaussianity was needed to complete up to Step 3, that is, to reduce the number of terms in the trace formula to those involving only the pair-partitions. The rest of the proof there did not use Gaussianity. Hence, in the present case, if we can show that only pair-partitions contribute to the limit, the proof will be complete.

The trace formula (11.8) of Step 1 is still valid. However, (11.9) in Step 2 is not valid anymore. Instead, we can only say that for any $p \geq 2$,

$$\mathrm{E}[e_{i_1 j_1}^{\epsilon_1} e_{i_2 j_2}^{\epsilon_2} \cdots e_{i_p j_p}^{\epsilon_p}] = 0 \quad \text{if any variable appears exactly once.} \tag{11.22}$$

Hence such terms cannot contribute to the limit.

In Step 3 (note that $p \geq 2$ is not necessarily even anymore), the sum is now over all partitions each of whose block is of size at least 2. For any $p \geq 2$, let

$$\mathcal{P}_{2+}(p) := \{\pi \in \mathcal{P}(p) : \text{ every block of } \pi \text{ is of size at least } 2\}. \tag{11.23}$$

Suppose we could prove the following:

(i) For any $p \geq 3$ odd, the partitions $\pi \in \mathcal{P}_{2+}(p)$ do not contribute in the limit.

(ii) For $p = 2k, k \geq 2$ the partitions $\pi \in \mathcal{P}_{2+}(2k) \setminus \mathcal{P}_2(2k)$ do not contribute in the limit.

(i) would imply that for odd p, the limit of (11.8) is 0. Recalling Step 3, (ii) would imply that the even limit moments are as before. That would complete the proof of the theorem. Let

$$\mathcal{P}_{3+}(p) := \{\pi \in \mathcal{P}_{2+}(p) : \text{ at least one block of } \pi \text{ is of size at least } 3\}. \tag{11.24}$$

Note that (i) and (ii) would be established if we show that the contribution of any $\pi \in \mathcal{P}_{3+}(p)$ is 0.

Fix a typical $\pi \in \mathcal{P}_{3+}(p)$. Suppose it has s blocks of sizes $p_1, \ldots p_s$, respectively, $\sum_{\ell=1}^{s} p_\ell = p$. We write these blocks as

$$V_1 := \{v_{11}, \ldots, v_{1p_1}\}, \ldots, V_s := \{v_{s1}, \ldots, v_{sp_s}\},$$

where we have indexed the elements in increasing order within each block, so that for all l, $v_{lt} < v_{lq}$ if $1 \leq t < q \leq p_l$, and the blocks have been indexed in increasing order of their smallest element, so that $v_{t1} < v_{q1}$ if $1 \leq t < q \leq s$. Then the contribution of $\pi \in \mathcal{P}_{3+}(p)$ to (11.8) can be rewritten as

$$\frac{1}{n^{\frac{p}{2}+1}} \sum_{\pi \in \mathcal{P}_{3+}(p)} \sum_{I_p(\pi), J_p(\pi)} \Big(\prod_{t=1}^{s} \mathrm{E}_{V_t}\Big)\Big(\prod_{\ell=1}^{p} a_{j_\ell i_{\ell+1}}^{(\ell)}\Big)$$

where

$$\mathrm{E}_{V_t} := \mathrm{E}\Big(\prod_{k=1}^{p_l} e_{i_{v_{tk}}}^{\epsilon_{v_{tk}}}\Big). \tag{11.25}$$

It is enough to show that each term goes to zero as $n \to \infty$. We omit the details and only indicate how the previous proof of negligibility can be extended.

Recall that earlier, since we were working with $\mathcal{P}_{2k}$, we had only moments of order 2 and each such moment was one of two types: (i) $u(r,s)$ when the i-indices matched, and (ii) $v(r,s)$ when the i and j-indices were cross-matched. Then those partitions where there was at least one $u(r,s)$ contributed 0, and those that had only $v(r,s)$ survived in the limit.

Now instead, we are working with $\mathcal{P}_{3+}(p)$. In (11.25), $2 \leq p_l \leq p$ and there is at least one $p_l \geq 3$. We now have an obvious extension of u and v which involve higher moments. We need to estimate these higher moments. For a block V, let δ_{VI} and δ_{VJ} denote the (product of) indicators where the i-indices and the j-indices match, respectively. We now have a third category where some i-indices and j-indices cross-match. Denote this (product of) indicators by δ_{VIJ}. Using Assumption Ie,

$$|\,\mathrm{E}_{V_l}\,| \leq C_p \delta_{VI}\delta_{VJ}\delta_{VIJ} \text{ for some constant } C_p.$$

Hence for any fixed $\pi \in \mathcal{P}_{3+}(p)$,

$$\frac{1}{n^{\frac{p}{2}+1}}\Big|\sum_{\substack{I_p(\pi)\\ J_p(\pi)}}\Big(\prod_{l=1}^{s}\mathrm{E}_{V_l}\Big)\Big(\prod_{\ell=1}^{p} a^{(\ell)}_{j_\ell i_{\ell+1}}\Big)\Big| \leq \frac{C_p}{n^{\frac{p}{2}+1}}\sum_{I_p,J_p}\Big(\prod_{l=1}^{s}\delta_{V_{lI}}\delta_{V_{lJ}}\delta_{V_{lIJ}}\Big)\prod_{\ell=1}^{p}|a^{(\ell)}_{j_\ell i_{\ell+1}}|.$$

If there are no cross-matches between the i- and j-indices, then the argument is similar to the proof of Lemma 11.2.2 and we omit the details. If instead there is a block of size at least three and with at least one cross-match, then in this block, there must be at least one match between a pair of i-indices. Again, this is enough for the arguments of the proof of Lemma 11.2.2 to go through. We omit the tedious details. This completes the proof of the theorem. ■

11.4 S-matrices and embedding

Theorem 11.1.2 implies that independent S matrices are asymptotically free and the marginal limits are all Marčenko-Pastur laws. We now show how Theorem 11.3.1 can be used to prove this result via the method of embedding. This proof is based on Bhattacharjee and Bose (2021)[16].

For simplicity, we consider only a special case. Suppose $S^{(i)}$, $1 \leq i \leq m$, are independent S-matrices defined as

$$S^{(i)} := \frac{1}{n}X_i X_i^* \text{ for all } 1 \leq i \leq m,$$

where $X_1, X_2, \ldots, X_m$ are independent $p \times n_i$ matrices whose entries are independent and satisfy Assumption I. For simplicity, we assume that $n_i = n$ for all i. Let $\lim_{p\to\infty} n^{-1}p = y \in (0,\infty)$. Then the statement of Theorem 11.1.2 is equivalent to (see also Theorem 9.4.1 and Remark 9.4.1) the following equation: for all $k \geq 1$ and $\tau_1, \tau_2, \ldots, \tau_k \in \{1, 2, \ldots, m\}$,

$$\lim_{p\to\infty} \frac{1}{p} \operatorname{E}\operatorname{Tr}(S^{(\tau_1)} \cdots S^{(\tau_k)}) = \sum_{t_1=0}^{T_1-1} \cdots \sum_{t_m=0}^{T_m-1} \Big(\prod_{i=1}^{m} y_i^{T_i - t_i - 1}\Big) \# A_{t_1, t_2, \ldots, t_m, k}, \tag{11.26}$$

where

$$T_i := \# J_i, \ J_i := \{j \in \{1, 2, \ldots, k\} : \ \tau_j = i\}, \ 1 \leq i \leq m$$

and $A_{t_1, \ldots, t_m, k}$ equals

$$\{\pi \in NC(k) : \pi = \cup_{i=1}^{m} \pi_i, \pi_i \in NC(J_i) \ \text{ has } \ t_i + 1 \ \text{ blocks}, \ 1 \leq i \leq m\}.$$

We now show how the limit in (11.26) can be established by *embedding* X_i *matrices into larger Wigner matrices* and then invoking Theorem 11.3.1. For $1 \leq i \leq m$, let $\{\widetilde{W}_{2i-1}\}$ and $\{\widetilde{W}_{2i}\}$ be independent Wigner matrices of order p and n, respectively, which are also independent of $\{X_i\}$, and whose elements satisfy Assumption I. Using these, we construct the following Wigner matrices of order $(p+n)$:

$$W_j := \begin{bmatrix} \widetilde{W}_{2j-1} & X_j \\ X_j^* & \widetilde{W}_{2j} \end{bmatrix}, \ 1 \leq j \leq m.$$

Let I_p be the identity matrix of order p and $0_{p\times n}$ be the $p \times n$ matrix, all of whose entries are zeroes. Define

$$\bar{I} := \begin{bmatrix} I_p & 0_{p\times n} \\ 0_{n\times p} & 0_{n\times n} \end{bmatrix}, \ \ \underline{I} := \begin{bmatrix} 0_{p\times p} & 0_{p\times n} \\ 0_{n\times p} & I_n \end{bmatrix}.$$

It is easy to check that

$$\begin{bmatrix} S^{(i)} & 0_{p\times n} \\ 0_{n\times p} & 0_{n\times n} \end{bmatrix} = \frac{n+p}{n} \bar{I} \frac{W_i}{\sqrt{n+p}} \underline{I} \frac{W_i}{\sqrt{n+p}} \bar{I}, \ 1 \leq i \leq m.$$

Thus the X_i-matrices have been embedded in Wigner matrices of higher dimension. The right side is a monomial in independent Wigner matrices and deterministic matrices and the latter trivially satisfy the conditions of Theorem 11.3.1 and converge jointly. Hence $(\bar{I}, \underline{I}, (n+p)^{-1/2} W_i, 1 \leq i \leq m)$, as elements of $(\mathcal{M}_{n+p}(\mathbb{C}), (n+p)^{-1} \operatorname{E}\operatorname{Tr})$, converge in $*$-distribution to $(a_0, a_1, s_i, 1 \leq i \leq m)$, where $\{s_i, 1 \leq i \leq m\}$ are free semi-circular variables, and are free of $\{a_0, a_1\}$. Further, $a_0 \sim \text{Bernoulli}(y(1+y)^{-1})$ and $a_1 \sim \text{Bernoulli}((1+y)^{-1})$ (see Exercise 1). Their joint distribution will not be relevant to us. Hence

$$\begin{aligned} &\lim_{p\to\infty} \frac{1}{p} \operatorname{E}\operatorname{Tr}(S^{(\tau_1)} \cdots S^{(\tau_k)}) \\ &= \lim \frac{n+p}{p} \frac{1}{n+p} \operatorname{E}\operatorname{Tr}\Big(\prod_{j=1}^{k} \frac{n+p}{n} \bar{I} \frac{W_{\tau_j}}{\sqrt{n+p}} \underline{I} \frac{W_{\tau_j}}{\sqrt{n+p}} \bar{I}\Big) \end{aligned}$$

$$= (1+y)^{k+1}y^{-1}\lim \frac{1}{n+p}\,\mathrm{E}\,\mathrm{Tr}\Big(\prod_{j=1}^{k} \bar{I}\frac{W_{(\tau_j)}}{\sqrt{n+p}}\underline{I}\frac{W_{(\tau_j)}}{\sqrt{n+p}}\bar{I}\Big)$$

$$= (1+y)^{k+1}y^{-1}\varphi\Big(\prod_{j=1}^{k} a_0 s_{\tau_j} a_1 s_{\tau_j} a_0\Big). \tag{11.27}$$

For $A := \{2j_1-1, 2j_1, 2j_2-1, 2j_2, \ldots, 2j_r-1, 2j_r\}$ and $\pi \in NC_2(A)$, define

$$\begin{aligned}
S(\pi) &:= \text{ the number of odd first elements in } \pi, \qquad (11.28)\\
\tilde{J}_i &:= \cup_{j\in J_i}\{2j-1, 2j\},\\
NC_2(\tilde{J}_1,\ldots,\tilde{J}_m) &:= \{\pi \in NC_2(2k) : \pi = \cup_{i=1}^m \pi_i,\ \pi_i \in NC_2(\tilde{J}_i), 1\le i\le m\},\\
B_{t_1,\ldots,t_m,k} &:= \{\pi \in NC_2(\tilde{J}_1,\ldots,\tilde{J}_m) : S(\pi_i) = t_i+1, 1\le i\le m\}.
\end{aligned}$$

Using arguments similar to those used in the proof of Lemma 8.2.2 and in the last part of the proof of Theorem 9.4.1 (c), it follows that

$$\#B_{t_1,t_2,\ldots,t_m,k} := \#A_{t_1,t_2,\ldots,t_m,k} \ \text{ for all } \ 0\le t_i\le T_i-1,\ 1\le i\le m. \tag{11.29}$$

Also, it is easy to see that, for any $\pi \in NC_2(2k)$, every block of $K(\pi)$ contains either only even elements or only odd elements. Hence, for $\pi \in B_{t_1,t_2,\ldots,t_m,k}$,

$$\#(\text{blocks in } K(\pi) \text{ containing only even elements}) = \sum_{i=1}^{m}(t_i+1), \tag{11.30}$$

$$\#(\text{blocks in } K(\pi) \text{ containing only odd elements}) = (k+1) - \sum_{i=1}^{m}(t_i+1).$$

Therefore

$$\begin{aligned}
&\varphi\Big(\prod_{j=1}^{k} a_0 s_{\tau_j} a_1 s_{\tau_j} a_0\Big)\\
&= \sum_{\pi\in NC_2(\tilde{J}_1,\tilde{J}_2,\ldots,\tilde{J}_m)} \varphi_{K(\pi)}[a_0, a_1, a_0, a_1, \ldots, a_1]\\
&= \sum_{t_1=0}^{T_1-1}\cdots\sum_{t_m=0}^{T_m-1}\ \sum_{\pi\in B_{t_1,t_2,\ldots,t_m,k}} \Big(\frac{y}{1+y}\Big)^{k+1-\sum(t_i+1)}\Big(\frac{1}{1+y}\Big)^{\sum(t_i+1)} \text{(by (11.30))}\\
&= (1+y)^{-k-1}\sum_{t_1=0}^{T_1-1}\cdots\sum_{t_m=0}^{T_m-1}\ \sum_{\pi\in B_{t_1,t_2,\ldots,t_m,k}} y^{k+1-\sum_{i=1}^m(t_i+1)}\\
&= (1+y)^{-k-1}\sum_{t_1=0}^{T_1-1}\cdots\sum_{t_m=0}^{T_m-1} y^{k+1-\sum_{i=1}^m(t_i+1)}\#B_{t_1,t_2,\ldots,t_m,k}
\end{aligned}$$

$$= (1+y)^{-k-1} \sum_{t_1=0}^{T_1-1} \cdots \sum_{t_m=0}^{T_m-1} y^{k+1-\sum_{i=1}^{m}(t_i+1)} \# A_{t_1,t_2,\ldots,t_m,k} \text{ (by (11.29))}$$

$$= (1+y)^{-k-1} y \sum_{t_1=0}^{T_1-1} \cdots \sum_{t_m=0}^{T_m-1} \Big(\prod_{i=1}^{m} y_i^{T_i-t_i-1}\Big) \# A_{t_1,t_2,\ldots,t_m,k}. \tag{11.31}$$

Combining (11.27) and (11.31), the proof of (11.26) is complete.

Remark 11.4.1. The more general Theorem 11.1.2 on $S = XX^*$-matrices with elliptic X can also be proved by embedding X in elliptic matrices and then appealing to Theorem 11.3.1. This is left as an exercise. ●

11.5 Cross-covariance matrices

The *cross-covariance matrix* was introduced in Section 7.4. It is a $p \times p$ matrix $C := n^{-1}XY^*$ where X and Y are $p \times n$ random matrices. The pairs (x_{ij}, y_{ij}) are assumed to be independent across i, j and the variables within each pair have a common correlation ρ. The sample covariance matrix S is a special case of the cross-covariance matrix when $X = Y$ and so the correlation $\rho = 1$. We have not proved any convergence results for these matrices so far. Building on the ideas that we have developed in the study of the S-matrices and the elliptic matrices, we now prove results for independent cross-covariance matrices. The developments of this section are based on Bhattacharjee, Bose and Dey (2021)[17]. Suppose $\{(X_l, Y_l)\}_{l=1}^{t}$ are t pairs of $p \times n_l$ independent random matrices. Let $C_l = n^{-1}X_lY_l^*$, $1 \le l \le t$, be the corresponding cross-covariance matrices. We assume that the variables in X_l and Y_l are pair-correlated with common correlation ρ.

We shall need the following extension of Assumption Ie (and continue to call it by the same name). Let $x_{lij,n}$ and $y_{lij,n}$ denote the $(i,\ j)$th entry of X_l and Y_l respectively. We shall drop the suffix n for convenience in writing.

Assumption Ie. For every n, the variables $\{(x_{lij,n}, y_{lij,n}) : 1 \le i, j \le n\}$ form a collection of independent bivariate real random variables, and for $1 \le i, j \le n$,

$$\begin{aligned}
\mathrm{E}(x_{lij}) &= \mathrm{E}(y_{lij}) = 0, \\
\mathrm{Var}(x_{lij}) &= \mathrm{Var}(y_{lij}) = 1, \\
\mathrm{E}[x_{lij,n} y_{lij,n}] &= \rho, \\
\sup_{i,j,n} \mathrm{E}(|x_{lij,n}|^k + |y_{lij,n}|^k) &\le B_k < \infty \quad \text{for all } k \ge 1.
\end{aligned}$$

□

11.5.1 Pair-correlated cross-covariance; $p/n \to y \neq 0$

In this section we assume that for all l, $p \to \infty, p/n_l \to y_l \neq 0$. Figures 11.1–11.3 provide the simulated ESD of a few polynomials of two independent cross-covariance matrices. The following variable shall appear in the limit:

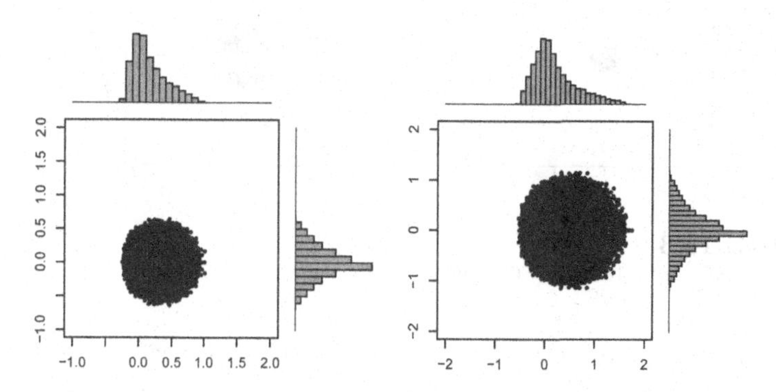

FIGURE 11.1
ESD of $C_1^2 + C_2^2$, $n = 1000$, $\rho = 0.25$. Left: $y = 0.25$; right: $y = 0.5$.

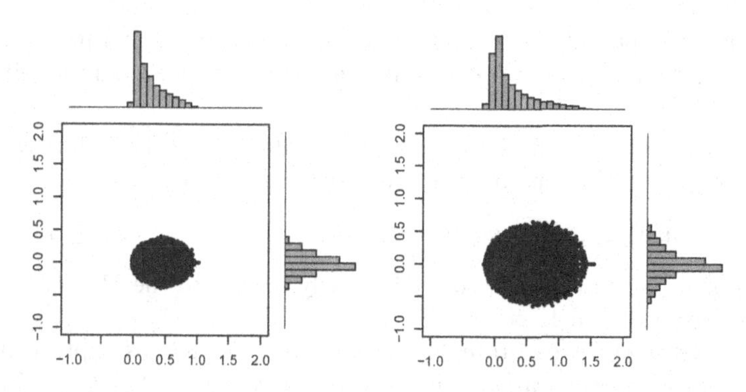

FIGURE 11.2
ESD of C_1C_2, $n = 1000$, $\rho = 0.5$. Left: $y = 0.25$; right: $y = 0.5$.

Definition 11.5.1. (Cross-covariance variable) A variable v will be called a *cross-covariance variable* with parameters ρ and y, $0 < y < \infty$, if its free cumulants are given by

$$\kappa_k(v^{\eta_1}, v^{\eta_2}, \ldots, v^{\eta_k}) = \begin{cases} y^{k-1}\rho^{S(\boldsymbol{\eta}_k)} & \text{if } \rho \neq 0, \\ y^{k-1}\delta_{S(\boldsymbol{\eta}_k)0} & \text{if } \rho = 0, \quad \text{where} \end{cases}$$

$$S(\boldsymbol{\eta}_k) := S(\eta_1, \ldots, \eta_k) := \sum_{1 \leq u \leq k,\ \eta_{k+1} = \eta_1} \delta_{\eta_u \eta_{u+1}}.$$

◇

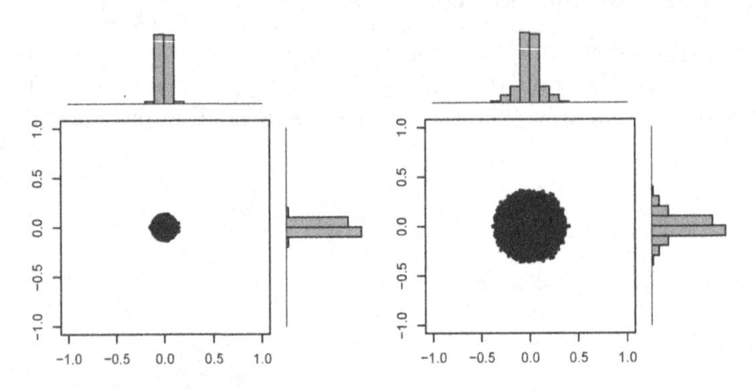

FIGURE 11.3
ESD of $C_1C_2C_1$, $n = 1000$, $\rho = 0$. Left: $y = 0.25$; right: $y = 0.5$.

It is interesting to note what happens for the two special cases $\rho = 1$ and $\rho = 0$.

(i) When $\rho = 1$, v is self-adjoint and its free cumulants are given by

$$\kappa_k(v) = y^{k-1}.$$

Thus v is a compound free Poisson variable with rate $1/y$ and jump distribution $\delta_{\{y\}}$. The moments of v determine the Marčenko-Pastur probability law with parameter y.

(ii) When $\rho = 0$, all odd order free cumulants of v vanish. Moreover, only alternating free cumulants of even order survive, and are given by

$$\kappa_{2k}(v, v^*, v, v^* \ldots, v^*) = \kappa_{2k}(v^*, v, v^*, v, \ldots, v) = y^{2k-1}, \ \forall\, k \geq 1. \quad (11.32)$$

It turns out that, in this case too, v is closely related to the Marcěnko-Pastur variable. See Exercise 4.

The following theorem states the joint convergence and asymptotic freeness of independent cross-covariance matrices when $y_l \neq 0$, $\forall\, 1 \leq l \leq t$. The case where $y_l = 0$ for all l, is treated in the next section.

Theorem 11.5.1. Suppose (X_l, Y_l), $1 \leq l \leq t$ are pairs of $p \times n_l$ random matrices with correlation parameters $\{\rho_l\}$ and whose entries satisfy Assumption Ie. Suppose $n_l, p \to \infty$ and $p/n_l \to y_l, 0 < y_l < \infty$ for all l. Then the following statements hold for the $p \times p$ cross-covariance matrices $\{C_l := n^{-1}X_lY_l^*\}$.

(a) As elements of the C^* probability space $(\mathcal{M}_p(\mathbb{C}), \varphi_p)$, $\{C_l\}$ converge in $*$-distribution to free variables $\{v_l\}$ where each v_l is a cross-covariance variable with parameters (ρ_l, y_l).

(b) Let $\Pi := \Pi(\{C_l : 1 \leq l \leq t\})$ be any finite degree real matrix polynomial in $\{C_l, C_l^* : 1 \leq l \leq t\}$ and which is symmetric. Then the EESD of Π converges weakly almost surely to the compactly supported probability law of the self-adjoint variable $\Pi(\{v_l : 1 \leq l \leq t\})$. ◆

Proof. (a) For simplicity, we will prove the result only for the special case where n_l, y_l and ρ_l do not depend on l. It will be clear from the arguments that the same proof works for the general case. Consider a typical monomial

$$(X_{\alpha_1}Y^*_{\alpha_1})^{\eta_1}\cdots(X_{\alpha_k}Y^*_{\alpha_k})^{\eta_k}.$$

This product has $2k$ factors. We shall write this monomial in a specific way to facilitate computation. Note that

$$(X_{\alpha_s}Y^*_{\alpha_s})^{\eta_s}=\begin{cases}X_{\alpha_s}Y^*_{\alpha_s} & \text{if } \eta_s=1,\\ Y_{\alpha_s}X^*_{\alpha_s} & \text{if } \eta_s=*.\end{cases}$$

For every index l, two types of matrices, namely X and Y are involved. To keep track of this, define

$$(\epsilon_{2s-1},\epsilon_{2s})=\begin{cases}(1,2) & \text{if } \eta_s=1,\\ (2,1) & \text{if } \eta_s=*.\end{cases}$$

Let $\lceil\cdot\rceil$ be the ceiling function. Note that

$$\epsilon_r\neq\epsilon_s\Leftrightarrow\begin{cases}(\epsilon_r,\epsilon_{r+1})=(\epsilon_{s-1},\epsilon_s)\Leftrightarrow\eta_{\lceil r/2\rceil}=\eta_{\lceil s/2\rceil} & \text{if } r \text{ is odd, } s \text{ is even,}\\ (\epsilon_{r-1},\epsilon_r)=(\epsilon_s,\epsilon_{s+1})\Leftrightarrow\eta_{\lceil r/2\rceil}=\eta_{\lceil s/2\rceil} & \text{if } r \text{ is even, } s \text{ is odd.}\end{cases} \tag{11.33}$$

Observe that $\delta_{\epsilon_{2s-1}\epsilon_{2s}}=0$ for all s. Define

$$A_l^{(\epsilon_{2s-1})}=\begin{cases}X_l & \text{if } \epsilon_{2s-1}=\eta_s=1,\\ Y_l & \text{if } \epsilon_{2s-1}=2 \text{ (or } \eta_s=*),\end{cases} \tag{11.34}$$

$$A_l^{(\epsilon_{2s})}=\begin{cases}X_l^* & \text{if } \epsilon_{2s}=1 \text{ (or } \eta_s=*),\\ Y_l^* & \text{if } \epsilon_{2s}=2 \text{ (or } \eta_s=1).\end{cases}$$

Extend the vector $(\alpha_1,\ldots,\alpha_k)$ of length k to the vector of length $2k$ as

$$(\beta_1,\ldots,\beta_{2k}):=(\alpha_1,\alpha_1,\ldots\alpha_k,\alpha_k).$$

We need to show that for all choices of $\alpha_s\in\{1,2,...,t\}$ and $\eta_s\in\{1,*\}$,

$$L_p:=\frac{1}{pn^k}\mathbb{E}\mathrm{Tr}(A_{\beta_1}^{(\epsilon_1)}A_{\beta_2}^{(\epsilon_2)}\cdots A_{\beta_{2k}}^{(\epsilon_{2k})})$$

converges to the appropriate limit. Upon expansion,

$$L_p=\frac{1}{pn^k}\sum_{I_{2k}}\mathbb{E}\prod_{\substack{1\le s\le 2k\\ i_{2k+1}=i_1}}A_{\beta_s}^{(\epsilon_s)}(i_s,i_{s+1})$$

where $A_\beta^{(\epsilon)}(i,j)$ denotes the (i,j)th element of $A_\beta^{(\epsilon)}$ for all choices of β, ϵ, i and j, and

$$I_{2k} = \{(i_1, i_2, \ldots, i_{2k}) : \ 1 \le i_{2s-1} \le p, \ 1 \le i_{2s} \le n, \ 1 \le s \le 2k\}.$$

Observe that the values of these i_j have different ranges p and n, depending on whether j is odd or even.

Note that the expectation of any summand is zero if there is at least one (i_s, i_{s+1}) whose value is not repeated elsewhere in the product. So, as usual, to split up the sum into indices that match, consider any connected bipartite graph between the *distinct* odd and even indices, $I = \{i_{2s-1} : 1 \le s \le k\}$ and $J = \{i_{2s} : 1 \le s \le k\}$. Then we need to consider only those cases where each edge appears at least twice. Hence there can be at most k distinct edges and since the graph is connected,

$$\#I + \#J \le \#E + 1 \le k + 1.$$

By Assumption Ie, there is a common bound for all expectations involved. Hence, the total expectation of the terms involved in this graph is of the order

$$O(\frac{p^{\#I} n^{\#J}}{pn^k}) = O(p^{\#I + \#J - (k+1)})$$

since $\frac{p}{n} \to y > 0$. As a consequence, only those terms can potentially contribute to the limit for which $\#I + \#J = k + 1$. This implies that $\#E = k$. So each edge is repeated exactly twice. Then each edges in E corresponds to some $\pi = \{(r,s) : r < s\} \in \mathcal{P}_2(2k)$. Let

$$a_{r,s} = \begin{cases} 1 & \text{if } \ r, s \text{ are both odd or both even,} \\ 0 & \text{otherwise.} \end{cases} \tag{11.35}$$

Then we have

$$\begin{aligned} \lim_{p\to\infty} L_p &= \lim_{n\to\infty} \frac{1}{pn^k} \sum_{I_{2k}} \mathbb{E}[\prod_{s=1}^{2k} A_{\beta_s}^{(\epsilon_s)}(i_s, i_{s+1})] \\ &= \sum_{\pi\in P_2(2k)} \lim_{n\to\infty} \frac{1}{pn^k} \sum_{I_{2k}} \prod_{(r,s)\in\pi} E(r,s) \ \ \text{(say)}, \end{aligned}$$

where, suppressing the dependence on other variables, $E(r,s)$ is given by

$$\begin{aligned} &\mathbb{E}\big[A_{\beta_r}^{(\epsilon_r)}(i_r, i_{r+1}) A_{\beta_s}^{(\epsilon_s)}(i_s, i_{s+1})\big] \\ &= \delta_{\beta_r\beta_s}\big(\rho(1-\delta_{\epsilon_r\epsilon_s}) + \delta_{\epsilon_r\epsilon_s}\big)\big(\delta_{i_r i_s}\delta_{i_{r+1} i_{s+1}} a(r,s) + \delta_{i_r i_{s+1}}\delta_{i_s i_{r+1}}(1 - a(r,s))\big) \\ &= \delta_{\beta_r\beta_s}\big(\rho^{1-\delta_{\epsilon_r\epsilon_s}}(1-\delta_{\rho 0}) + \delta_{\rho 0}\delta_{\epsilon_r\epsilon_s}\big)\big(\delta_{i_r i_s}\delta_{i_{r+1} i_{s+1}} a(r,s) \\ &\qquad + \delta_{i_r i_{s+1}}\delta_{i_s i_{r+1}}(1 - a(r,s))\big). \end{aligned}$$

Recall that $|\rho| \leq 1$. Hence each $E(r,s)$ is a sum of two factors—one of them is bounded by $\delta_{i_r i_s}\delta_{i_{r+1}i_{s+1}}$ and the other is bounded by $\delta_{i_r i_{s+1}}\delta_{i_s i_{r+1}}$. Hence when we expand $\prod_{(r,s)\in\pi} E(r,s)$, each term involves a product of these δ-values. Using arguments similar to those used in the proof of Theorem 3.2.6 in Bose (2018)[20], it is easy to see that the only term that will survive in $\prod_{(r,s)\in\pi} E(r,s)$ is

$$\prod_{(r,s)\in\pi} \delta_{\beta_r\beta_s}(\rho^{1-\delta_{\epsilon_r\epsilon_s}}(1-\delta_{\rho 0}) + \delta_{\rho 0}\delta_{\epsilon_r\epsilon_s})\delta_{i_r i_{s+1}}\delta_{i_s i_{r+1}}(1-a(r,s)).$$

Hence $\lim L_p$ is equal to the sum over all $\pi \in NC_2(2k)$ of the terms

$$\lim_{n\to\infty} \frac{1}{pn^k} \sum_{I_{2k}} \prod_{(r,s)\in\pi} \delta_{\beta_r\beta_s}(\rho^{1-\delta_{\epsilon_r\epsilon_s}}(1-\delta_{\rho 0}) + \delta_{\rho 0}\delta_{\epsilon_r\epsilon_s})(1-a(r,s)) \prod_{r=1}^{2k} \delta_{i_r i_{\gamma\pi(r)}}. \tag{11.36}$$

But for any $\pi \in NC_2(2k)$ if $(r,s) \in \pi$ then r and s have different parity and hence $a(r,s) = 0$. Let γ denote the cyclic permutation $1 \to 2 \to \cdots \to 2k \to 1$. Then (11.36) simplifies to

$$\prod_{(r,s)\in\pi} \delta_{\beta_r\beta_s}(\rho^{1-\delta_{\epsilon_r\epsilon_s}}(1-\delta_{\rho 0}) + \delta_{\rho 0}\delta_{\epsilon_r\epsilon_s}) \lim_{n\to\infty} \frac{\#\{I_{2k} : i_r = i_{\gamma\pi(r)} \text{ for all } r\}}{pn^k}. \tag{11.37}$$

Now note that as $\pi \in NC_2(2k)$, $\gamma\pi$ contains $k+1$ blocks. Moreover, each block of $\gamma\pi$ contains only odd or only even elements. Let

$$S(\gamma\pi) = \text{ the number of blocks in } \gamma\pi \text{ with only odd elements.}$$

Then the number of blocks of $\gamma\pi$ with only even elements is $k+1-S(\gamma\pi)$. Suppose $\pi \in NC_2(2k)$ such that $S(\gamma\pi) = m+1$. Then it is clear that

$$\#\{I_{2k} : i_r = i_{\gamma\pi(r)} \text{ for all } r\} = p^{m+1}n^{k+1-(m+1)}$$

and hence using (11.37), $\lim L_p$ equals

$$\begin{aligned}
&\sum_{\pi\in NC_2(2k)} \prod_{(r,s)\in\pi} \delta_{\beta_r\beta_s}(\rho^{1-\delta_{\epsilon_r\epsilon_s}}(1-\delta_{\rho 0}) + \delta_{\rho 0}\delta_{\epsilon_r\epsilon_s}) \lim_{n\to\infty} \frac{\#\{I_{2k} : i_r = i_{\gamma\pi(r)} \forall r\}}{pn^k} \\
&= \sum_{m=0}^{k-1} y^m \sum_{\substack{\pi\in NC_2(2k):\\ S(\gamma\pi)=m+1}} \prod_{(r,s)\in\pi} \delta_{\beta_r\beta_s}(\rho^{1-\delta_{\epsilon_r\epsilon_s}}(1-\delta_{\rho 0}) + \delta_{\rho 0}\delta_{\epsilon_r\epsilon_s}) y^m \\
&= \sum_{m=0}^{k-1} y^m \sum_{\substack{\pi\in NC_2(2k):\\ S(\gamma\pi)=m+1}} \prod_{(r,s)\in\pi} \delta_{\beta_r\beta_s}(\rho^{T(\pi)}(1-\delta_{\rho 0}) + \delta_{\rho 0}\delta_{T(\pi)0})
\end{aligned}$$

where

$$T(\pi) = \#\{(r,s) \in \pi : \delta_{\epsilon_r\epsilon_s} = 0\} \text{ for } \pi \in NC_2(2k).$$

Hence we have proved that $\{C_l : 1 \le l \le t\}$ converge jointly in $*$-distribution to say $\{v_l : 1 \le l \le t\}$ which are in the limit NCP $(\mathcal{A}, \varphi)$. We still have to identify the limit and prove the freeness. For this we need to go from $NC_2(2k)$ to $NC(k)$. Define

$$\begin{aligned} \tilde{J}_i &= \{j \in \{1,2,\ldots,2k\} : \beta_j = i\}, \ 1 \le i \le t, \\ \tilde{B}_k &= \{\pi \in NC_2(2k) : \pi = \cup_{i=1}^t \pi_i, \ \pi_i \in NC_2(\tilde{J}_i), \ 1 \le i \le t\}, \\ \tilde{B}_{m,k} &= \{\pi \in \tilde{B}_k : S(\gamma\pi) = m+1\}. \end{aligned}$$

Note that $\cup_{m=0}^{k-1} \tilde{B}_{m,k} = \tilde{B}_k$ and hence

$$\lim L_p = \sum_{m=0}^{k-1} y^m \sum_{\pi \in \tilde{B}_{m,k}} (\rho^{T(\pi)}(1-\delta_{\rho 0}) + \delta_{\rho 0}\delta_{T(\pi)0}). \tag{11.38}$$

Also define

$$\begin{aligned} J_i &= \{j \in \{1,2,\ldots,k\} : \alpha_j = i\}, \ 1 \le i \le t, \\ B_k &= \{\pi \in NC(k) : \pi = \cup_{i=1}^t \pi_i, \ \pi_i \in NC(J_i), \ 1 \le i \le t\}, \\ B_{m,k} &= \{\pi \in B_k : \pi \text{ has } m \text{ blocks}\}. \end{aligned}$$

Note that $\cup_{m=0}^{k-1} B_{m+1,k} = B_k$. For any finite subset $S = \{j_1, j_2, \ldots, j_r\}$ of positive integers, define

$$\tilde{T}(S) = \sum_{\substack{1 \le u \le r \\ j_{r+1} = j_1}} \delta_{\eta_{j_u}\eta_{j_{u+1}}}$$

and for $\pi = \{V_1, V_2, \ldots, V_m\} \in NC(k)$, define

$$\mathcal{T}(\pi) = \sum_{i=1}^m \tilde{T}(V_i).$$

Consider the bijection $f : NC_2(2k) \to NC(k)$ as follows. Take $\pi \in NC_2(2k)$. Suppose (r, s) is a block of π. Then $\lceil r/2 \rceil$ and $\lceil s/2 \rceil$ are put in the same block in $f(\pi) \in NC(k)$. Using arguments similar to those used in the proof of Lemma 8.2.2 (b) (or in the proof of Lemma 3.2 of Bhattacharjee, Bose and Dey (2021)[17]), it is easy to see that f is indeed a bijection and is also a bijection between $\tilde{B}_{m,k}$ and $B_{k-m,k}$. Moreover, using (11.33), it is immediate that $T(\pi) = \mathcal{T}(f(\pi)) \ \forall \ \pi \in \tilde{B}_{m,k}$ i.e. $T(f^{-1}(\pi)) = \mathcal{T}(\pi) \ \forall \ \pi \in B_{k-m,k}$.

As an example, let $\pi = \{(1,8),(2,5),(6,7),(3,4),(9,10)\}$. Then $\pi \in \tilde{B}_{2,5}$ and is mapped to $f(\pi) = \{(1,3,4),(3),(5)\} \in B_{3,5}$. Let $(\eta_1, \eta_2, \eta_3, \eta_4, \eta_5) = (1, *, 1, *, 1)$. Further, $(\epsilon_1, \epsilon_2, \ldots, \epsilon_{10}) = (1,2,2,1,1,2,2,1,1,2)$ and $T(\pi) = \mathcal{T}(f(\pi)) = 3$.

Now from (11.38), we have

$$\begin{aligned}\lim L_p &= \sum_{m=0}^{k-1} y^m \sum_{\pi \in B_{k-m,k}} \left(\rho^{T(f^{-1}(\pi))}(1-\delta_{\rho 0}) + \delta_{\rho 0}\delta_{T(f^{-1}(\pi))0}\right) \\ &= \sum_{m=0}^{k-1} y^m \sum_{\pi \in B_{k-m,k}} \left(\rho^{\mathcal{T}(\pi)}(1-\delta_{\rho 0}) + \delta_{\rho 0}\delta_{\mathcal{T}(\pi)0}\right) \\ &= \sum_{m=0}^{k-1} y^{k-m-1} \sum_{\pi \in B_{m+1,k}} \left(\rho^{\mathcal{T}(\pi)}(1-\delta_{\rho 0}) + \delta_{\rho 0}\delta_{\mathcal{T}(\pi)0}\right). \quad (11.39)\end{aligned}$$

Then, (11.39) implies

$$\begin{aligned}&\varphi(v_{\alpha_1}^{\eta_1} v_{\alpha_2}^{\eta_2} \cdots v_{\alpha_k}^{\eta_k}) \\ &= \sum_{m=0}^{k-1} \sum_{\substack{\pi \in B_{m+1,k} \\ \pi = \{V_1, V_2, \ldots, V_{m+1}\}}} \prod_{l=1}^{m+1} y^{\#V_l - 1}\left(\rho^{\tilde{T}(V_l)}(1-\delta_{\rho 0}) + \delta_{\rho 0}\delta_{\tilde{T}(V_l)0}\right) \\ &= \sum_{\pi \in B_k} \prod_{l=1}^{\#\pi} y^{\#V_l - 1}\left(\rho^{\tilde{T}(V_l)}(1-\delta_{\rho 0}) + \delta_{\rho 0}\delta_{\tilde{T}(V_l)0}\right).\end{aligned}$$

Hence, by moment-free cumulant relation, we have

$$\begin{aligned}\kappa_\pi[v_{\alpha_1}^{\eta_1}, v_{\alpha_2}^{\eta_2}, \ldots, v_{\alpha_k}^{\eta_k}] &= 0 \text{ for all } \pi \in NC(k) - B_k, \\ \kappa_\pi[v_{\alpha_1}^{\eta_1}, v_{\alpha_2}^{\eta_2}, \ldots, v_{\alpha_k}^{\eta_k}] &= \prod_{l=1}^{\#\pi} y^{\#V_l - 1}\left(\rho^{\tilde{T}(V_l)}(1-\delta_{\rho 0}) + \delta_{\rho 0}\delta_{\tilde{T}(V_l)0}\right) \text{ for all } \pi \in B_k.\end{aligned}$$

This implies that

$$\kappa_k(v_{\alpha_1}^{\eta_1}, v_{\alpha_2}^{\eta_2}, \ldots, v_{\alpha_k}^{\eta_k}) = \begin{cases} y^{k-1}\left(\rho^{\mathcal{T}(\mathbf{1}_k)}(1-\delta_{\rho 0}) + \delta_{\rho 0}\delta_{\mathcal{T}(\mathbf{1}_k)0}\right), \text{if } \alpha_1 = \cdots = \alpha_k, \\ 0 \ \text{ otherwise.} \end{cases}$$

Therefore $\{v_l : 1 \leq l \leq t\}$ are free across l, and the marginal free cumulant of order k is

$$\begin{aligned}\kappa_k(v_l^{\eta_1}, v_l^{\eta_2}, \ldots, v_l^{\eta_k}) &= y^{k-1}\left(\rho^{S(\boldsymbol{\eta}_k)}(1-\delta_{\rho 0}) + \delta_{\rho 0}\delta_{S(\boldsymbol{\eta}_k)0}\right) \\ \text{where } S(\boldsymbol{\eta}_k) &= \mathcal{T}(\mathbf{1}_k) = \sum_{\substack{1 \leq u \leq k \\ u_{k+1} = u_1}} \delta_{\eta_u \eta_{u+1}}.\end{aligned}$$

This completes the proof of (a) for the special case when the values of ρ_l and of y_l are same. It is easy to see that the above proof continues to hold for the general case, except for notational complexity. We omit the details.

(b) Now suppose that Π is symmetric. Then by the above argument, all moments of the EESD of Π converge and there is a $C > 0$, depending on Π such that the limiting k-th moment is bounded by C^k for all k. This implies that these moments define a unique probability law, say μ with support contained in $[-C, C]$, and hence the EESD of Π converges weakly to μ. Note that this argument works only if Π is a symmetric matrix. ■

11.5.2 Pair correlated cross-covariance; $p/n \to 0$

In this case, the matrices C_l need to be centered, and a different scaling compared to the previous case must be used. For the single S matrix, the corresponding result was stated in Exercise 2 of Chapter 8.

Theorem 11.5.2. Suppose $\{X_l, Y_l\}_{l=1}^q$ are q pairs of $p \times n$ random matrices which satisfy Assumption Ie and are independent across l. Suppose $p \to \infty$, $p/n \to 0$. Then $E_l = \sqrt{np^{-1}}(C_l - \rho I_p), 1 \leq l \leq q$, as elements of $(\mathcal{M}_p, \mathrm{E}\,\mathrm{tr})$, converge jointly to free elliptic variables $e_1, ..., e_q$, with common parameter ρ^2. ◆

Proof. Fix any $k \geq 1$ and $\epsilon_1, ..., \epsilon_k \in \{1, *\}$. We will consider the limit of the following as $p, n \to \infty$ with $p/n \to 0$,

$$\frac{1}{p}\mathbb{E}\mathrm{Tr}(E_{l_1}^{\epsilon_1} \cdots E_{l_k}^{\epsilon_k}) = \frac{1}{p(np)^{k/2}} \sum_{\substack{1 \leq i_1, \ldots, i_k \leq p \\ 1 \leq j_1, \ldots, j_k \leq n}} \mathbb{E}\Big[\prod_{t=1}^{k} \big(a_{l_t i_t j_t} a_{l_t i_{t+1} j_t} - \rho \delta_{i_t i_{t+1}}\big)\Big], \tag{11.40}$$

with the understanding that $i_{k+1} := i_1$, and as ordered pairs, for all $1 \leq r \leq k$,

$$(a_{l_r i_r j_r}, a_{l_r i_{r+1} j_r}) := \begin{cases} (x_{l_r i_r j_r}, y_{l_r i_{r+1} j_r}) & \text{if } \epsilon_r = 1, \\ (y_{l_r i_r j_r}, x_{l_r i_{r+1} j_r}) & \text{if } \epsilon_r = *. \end{cases}$$

Consider the following collection of all *ordered* pairs of indices that appear in the above formula:

$$P := \{(i_r, j_r), (i_{r+1}, j_r), 1 \leq r \leq k\}.$$

(i) Suppose there is a pair, say $(i_r, j_r) \in P$, that appears only once. Then, in particular, $(i_r, j_r) \neq (i_{r+1}, j_r)$, and hence $i_r \neq i_{r+1}$. As a consequence, the variable $a_{l_r i_r j_r}$ is independent of all the other variables and we get

$$\begin{aligned} &\mathbb{E}\Big[\prod_{t=1}^{k} \big(a_{l_t i_t j_t} a_{l_t i_{t+1} j_t} - \rho \delta_{i_t i_{t+1}}\big)\Big] \\ &= \mathbb{E}[a_{l_r i_r j_r}]\mathbb{E}\Big[a_{l_r i_{r+1} j_r} \times \prod_{t \neq r} \big(a_{l_t i_t j_t} a_{l_t i_{t+1} j_t} - \rho \delta_{i_t i_{t+1}}\big)\Big] \\ &= 0. \end{aligned} \tag{11.41}$$

The same conclusion holds if a pair (i_{r+1}, j_r) occurs only once in P. So we need to consider the subset P_1 of P where each pair is repeated at least twice.

(ii) Suppose in P_1, there is a j_r such that $j_r \neq j_s$ for all $s \neq r$. Then the pair $(a_{l_r i_r j_r}, a_{l_r i_{r+1} j_r})$ is independent of all other factors in the product. Hence

$$\begin{aligned}
&\mathbb{E}\Big[\prod_{t=1}^{k} \big(a_{l_t i_t j_t} a_{l_t i_{t+1} j_t} - \rho\delta_{i_t i_{t+1}}\big)\Big] \\
&\quad = \underbrace{\mathbb{E}[a_{l_r i_r j_r} a_{l_r i_{r+1} j_r} - \rho\delta_{i_r i_{r+1}}]}_{=0} \times \mathbb{E}\Big[\prod_{t\neq r} \big(a_{l_t i_t j_t} a_{l_t i_{t+1} j_t} - \rho\delta_{i_t i_{t+1}}\big)\Big] \\
&\quad = 0.
\end{aligned}$$

So, consider the subset P_2 of P_1 where each j_r occurs in at least four pairs, i.e. in $(i_r, j_r), (i_{r+1}, j_r)$ and also in $(i_s, j_s), (i_{s+1}, j_s)$ for some $s \neq r$. We call the corresponding pairs *edges*. If $j_r = j_s$, then they are said to be matched and likewise for the i-vertices. Let V_I and V_J be the sets of distinct indices from $\{i_1, ..., i_k\}$ and $\{j_1, ..., j_k\}$, respectively. Let E be the set of (distinct) edges between V_I and V_J. This defines a simple connected bi-partite graph. There are at most $2k$ edges in P_2 but each edge appears at least twice and hence $\#E \leq k$. Since every j-index was originally matched, $\#V_J \leq k/2$. We also know from the connectedness property that

$$\#V_I + \#V_J \leq \#E + 1 \leq k + 1. \tag{11.42}$$

Hence the contribution to (11.40) is bounded above by

$$O\Big(\frac{p^{\#V_I} n^{\#V_J}}{p^{k/2+1} n^{k/2}}\Big) = O\Big(\frac{p^{k+1-\#V_J} n^{\#V_J}}{p^{k/2+1} n^{k/2}}\Big) = O\Big(\big(\frac{p}{n}\big)^{k/2-\#V_J}\Big). \tag{11.43}$$

If $\#V_J < k/2$, then the above expression goes to 0. So for a non-zero contribution to the limit of (11.40), we assume that $\#V_J = k/2$. This immediately shows that when k is odd, the limit is 0, and hence

$$\frac{1}{p}\mathbb{E}\mathrm{Tr}(E_{l_1}^{\epsilon_1} \cdots E_{l_k}^{\epsilon_k}) \to 0.$$

So now consider the (contributing) case where k is even and $\#V_J = k/2$. Let $k =: 2m$ and $\#V_J = m$. Then

$$O\Big(\frac{p^{\#V_I} n^{\#V_J}}{p^{k/2+1} n^{k/2}}\Big) = O(p^{\#V_I - (k/2+1)}) = O(p^{\#V_I - (m+1)}).$$

On the other hand, from (11.42), when $k = 2m$ and $\#V_J = m$, we get that $\#V_I \leq m + 1$. So we can restrict our attention to the case $\#V_I = m + 1$. This implies that

$$\begin{aligned}
m + 1 + m &= \#V_I + \#V_J \\
&\leq \#E + 1 \leq 2m + 1,
\end{aligned}$$

and hence $\#E = 2m$. In other words, each edge must appear exactly twice.

Suppose $(i_r, j_r) = (i_{r+1}, j_r)$ for some r. Since each edge appears exactly twice, this pair will be independent of all others and therefore

$$\begin{aligned}
&\mathbb{E}\Big[\prod_{t=1}^{k}\big(a_{l_t i_t j_t}a_{l_t i_{t+1} j_t} - \rho\delta_{i_t i_{t+1}}\big)\Big] \\
&\quad = \underbrace{\mathbb{E}[a_{l_r i_r j_r}a_{l_r i_{r+1} j_r} - \rho\delta_{i_r i_{r+1}}]}_{=0} \times \mathbb{E}\Big[\prod_{t\neq r}\big(a_{l_t i_t j_t}a_{l_t i_{t+1} j_t} - \rho\delta_{i_t i_{t+1}}\big)\Big] \\
&\quad = 0.
\end{aligned}$$

Hence such a combination cannot contribute to (11.40). So we may assume from now on that for every r, $i_r \neq i_{r+1}$, and hence $\delta_{i_r i_{r+1}} = 0$ always. Let P_3 be this subset of P_2. As a consequence, (11.40) reduces to

$$\frac{1}{p}\mathbb{E}\mathrm{Tr}(E_{l_1}^{\epsilon_1}\cdots E_{l_k}^{\epsilon_k}) = \frac{1}{p^{m+1}n^m}\sum_{P_3}\mathbb{E}\Big(\prod_{r=1}^{2m} a_{l_r i_r j_r}a_{l_r i_{r+1} j_r}\Big).$$

Due to the preceding discussion, we have two situations. Suppose $(i_r, j_r) = (i_s, j_s)$ for some $s \neq r$. Note that j_r and j_s are also adjacent to i_{r+1} and i_{s+1} respectively. Due to the nature of the edge set P_3, this forces $i_{r+1} = i_{s+1}$. Similarly, if $(i_r, j_r) = (i_{s+1}, j_s)$, then it would force $i_{r+1} = i_s$. Recall that $P_2(2m)$ is the set of all possible pair-partitions of $\{1, ..., 2m\}$.

We will think of each block in the partition as representing the equal pairs of edges in the graph. Now recalling the definition (11.41), the moment structure and the above developments, it is easy to verify that the possibly contributing part of the above sum (and hence of (11.40)) can be re-expressed as

$$\begin{aligned}
\sum_{\substack{i_1,\ldots,i_{2m}\\ j_1,\ldots,j_{2m}}}\sum_{\pi\in P_2(2m)}\frac{1}{p^{m+1}n^m}\prod_{(r,s)\in\pi}\Big[&\delta_{l_r l_s}\delta_{\epsilon_r\epsilon_s}(\rho^2\delta_{i_r i_{s+1}}\delta_{j_r j_s}\delta_{i_{r+1} i_s} + \delta_{i_r i_s}\delta_{j_r j_s}\delta_{i_{r+1} i_{s+1}}) \\
&+\delta_{l_r l_s}(1-\delta_{\epsilon_r\epsilon_s})(\rho^2\delta_{i_r i_s}\delta_{j_r j_s}\delta_{i_{r+1} i_{s+1}} + \delta_{j_r j_s}\delta_{i_r i_{s+1}}\delta_{i_{r+1} i_s})\Big],
\end{aligned}$$

which equals

$$\begin{aligned}
\sum_{\substack{i_1,\ldots,i_{2m}\\ j_1,\ldots,j_{2m}}}\sum_{\pi\in P_2(2m)}\frac{1}{p^{m+1}n^m}\prod_{(r,s)\in\pi}\ &\delta_{l_r l_s}\delta_{j_r j_s}\Big[\delta_{i_r i_{s+1}}\delta_{i_s i_{r+1}}\big(\rho^2\delta_{\epsilon_r\epsilon_s} \\
&+(1-\delta_{\epsilon_r\epsilon_s})\big) + \text{other terms}\Big]. \qquad (11.44)
\end{aligned}$$

To further identify the negligible and the contributing terms, it helps to compare with the special case where $t = 1, \rho = 1$. Then $Y = X$ and $C = S$, an S-matrix. Due to symmetry, we can take ϵ_r to be the same for all r. In Exercise 2 of Chapter 8, we have noted that the LSD of $\sqrt{np^{-1}}(S - I_p)$ is the semi-circular law. This is proved by showing that the $2m$-th moment of

C converges to the the m-th Catalan number C_m. The details are available in the proof of Theorem 3.3.1 in Bose (2018)[20]. Thus, in this case, the limit of (11.40), and hence of (11.44), equals $\#NC_2(2m) = C_m$. In other words,

$$\sum_{\substack{i_1,\ldots,i_{2m}\\ j_1,\ldots,j_{2m}}} \sum_{\pi \in P_2(2m)} \frac{1}{p^{m+1}n^m} \prod_{(r,s)\in\pi} \delta_{j_r j_s}(\delta_{i_r i_{s+1}}\delta_{i_s i_{r+1}} + \delta_{i_r i_s}\delta_{i_{r+1} i_{s+1}}) \to \sum_{\pi\in NC_2(2m)} 1.$$

On the other hand, for $\pi \in NC_2(2m)$,

$$\sum_{\substack{i_1,\ldots,i_{2m},\\ j_1,\ldots,j_{2m}}} \frac{1}{p^{m+1}n^m} \prod_{(r,s)\in\pi} \delta_{j_r j_s}\delta_{i_r i_{s+1}}\delta_{i_s i_{r+1}} = 1.$$

This can be seen as follows. If $(r, s) \in \pi$, then $j_r = j_s$, and there are n^m ways of choosing the j−indices. Now let γ be the cyclic permutation. Then we note that

$$\prod_{(r,s)\in\pi} \delta_{i_r i_{s+1}}\delta_{i_s i_{r+1}} = \prod_{(r,s)\in\pi} \delta_{i_r i_{\gamma\pi(r)}}\delta_{i_s i_{\gamma\pi(s)}} = \prod_{r=1}^{2m} \delta_{i_r i_{\gamma\pi(r)}}.$$

So $i_r = i_{\gamma\pi(r)}$ for each r if the above product is 1. But this means that i is constant on each block of $\gamma\pi$. Since $\pi \in NC_2(2m)$, from Lemma 3.8.1 (h), $|\gamma\pi| = m + 1$, and thus there are p^{m+1} choices in total for the i's.

These arguments establish that

$$\lim_{n\to\infty} \sum_{\pi\in\mathcal{P}_2(2m)} \sum_{\substack{i_1,\ldots,i_{2m}\\ j_1,\ldots,j_{2m}}} \frac{1}{p^{m+1}n^m} \prod_{(r,s)\in\pi} \delta_{j_r j_s}(\delta_{i_r i_{s+1}}\delta_{i_s i_{r+1}} + \delta_{i_r i_s}\delta_{i_{r+1} i_{s+1}})$$
$$= \lim_{n\to\infty} \sum_{\pi\in NC_2(2m)} \frac{1}{p^{m+1}n^m} \prod_{(r,s)\in\pi} \delta_{j_r j_s}\delta_{i_r i_{s+1}}\delta_{i_s i_{r+1}}.$$

This implies that when we compare the terms on the two sides of the above display, the extra terms on the left side must go to 0.

Now we get back to the general case and use the above conclusion. Note that for each $\pi \in \mathcal{P}_2(2m)$ and for each $(r, s) \in \pi$,

$$|\delta_{l_r l_s}\delta_{j_r j_s}(\delta_{i_r i_{s+1}}\delta_{i_s i_{r+1}}(\rho^2\delta_{\epsilon_r\epsilon_s} + (1 - \delta_{\epsilon_r\epsilon_s})) + \text{other terms})| \leq 1.$$

As a consequence, the corresponding terms continue to be negligible in the general case. Hence it follows that the expression in (11.44) converges to

$$\sum_{\pi\in NC_2(2m)} \prod_{(r,s)\in\pi} \delta_{l_r l_s}(\rho^2\delta_{\epsilon_r\epsilon_s} + (1 - \delta_{\epsilon_r\epsilon_s})) \text{ as } p \to \infty.$$

Now note that $\rho^2\delta_{\epsilon_r\epsilon_s} + (1 - \delta_{\epsilon_r\epsilon_s}) = (\rho^2)^{\delta_{\epsilon_r\epsilon_s}}$. Let

$$T(\pi) := \#\{(r, s) \in \pi : \epsilon_r = \epsilon_s\}.$$

Then the limit can be re-expressed as

$$\sum_{\pi \in NC_2(2m)} \prod_{(r,s)\in\pi} (\rho^2)^{\delta_{\epsilon_r \epsilon_s}} \delta_{l_r l_s} = \sum_{\pi \in NC_2(2m)} (\rho^2)^{T(\pi)} \prod_{(r,s)\in\pi} \delta_{l_r l_s}.$$

If $e_1, \ldots, e_{2m}$ are freely independent elliptic elements each with parameter ρ^2 in some NCP $(\mathcal{A}, \varphi)$, then the above expression is nothing but $\varphi(e_{l_1}^{\epsilon_1} \ldots e_{l_{2m}}^{\epsilon_{2m}})$. This proves the $*$-convergence as well as asymptotic freeness. ■

11.6 Wigner and patterned random matrices

In Chapter 9 we have seen that independent Wigner matrices are asymptotically free. We have also seen in Theorem 10.3.1 that independent copies of patterned random matrices, at least when taken two patterns at a time, converge jointly. Moreover, independent Wigner matrices are asymptotically free of deterministic matrices, as seen from Theorem 11.3.1.

In view of this, it is natural to ask whether independent Wigner matrices are free of other independent patterned random matrices. We now show that if we take independent copies of a single pattern, such as Hankel, Toeplitz, Reverse Circulant and Symmetric Circulant, then they are asymptotically free of independent Wigner matrices.

Theorem 11.6.1. Suppose $\{W_{i,n}, 1 \leq i \leq p, A_{i,n}, 1 \leq i \leq p\}$ are independent matrices which satisfy Assumption II of Chapter 9, where $W_{i,n}$ are Wigner matrices and $A_{i,n}$ have any *one* of Toeplitz, Hankel, Symmetric Circulant or Reverse Circulant patterns. Then the two collections $\{\frac{W_{i,n}}{\sqrt{n}}, 1 \leq i \leq p\}$ and $\{\frac{A_{i,n}}{\sqrt{n}}, 1 \leq i \leq p\}$, as subsets of $(\mathcal{M}_n(\mathbb{C}), n^{-1}\,\mathrm{E}\,\mathrm{Tr})$, are asymptotically free. ◆

Remark 11.6.1. (a) It may be noted that even if we argue conditionally, Theorem 11.6.1 does not follow from Theorem 11.3.1 since the above four patterned matrices have unbounded limit spectrum.

(b) Theorem 11.6.1 is anticipated to hold if more than one pattern is allowed, but the notational complexity required in the proof would be daunting.

(c) Asymptotic freeness between the Gaussian unitary ensemble (GUE) and independent patterned matrices is much easier to establish. We provide a precise assertion in Theorem 11.6.5 along with a brief proof.

(d) The technique to prove Theorem 11.6.1 is close in spirit to the arguments that are to be found in Chapter 22 of Nica and Speicher (2006)[74]. It is quite plausible that the techniques in Collins (2003)[33] and in Capitaine and Casalis (2004)[30] may be extended to prove Theorem 11.6.1.

(e) Remark 10.2.2 implies that all the moments converge (with respect to the state tr) almost surely. Hence the asymptotic freeness in Theorem 11.6.1 also holds in the almost sure sense. ●

Before we begin the proof, we recall that $\mathcal{C}(2k)$ is in bijection with $NC_2(2k)$. Any $\pi \in NC_2(2k)$ will be written as $\{(r, \pi(r)),\ r = 1, \ldots, 2k\}$. We add an index to each letter of a word to keep track of which copy of the Wigner matrix appears at which positions. These indexed Catalan words are in bijection with $NC_2^{(l)}(2k) \subset NC_2(2k)$ which is defined as follows:

$$\begin{aligned} l &:= (l(1), \ldots, l(2k)), 1 \leq l(i) \leq p \text{ for all } 1 \leq i \leq 2k, \\ NC_2^{(l)}(2k) &:= \{\pi \in NC_2(2k) : l(\pi(r)) = l(r) \text{ for all } r = 1, \ldots, 2k\}. \end{aligned}$$

That is, each block of π carries the same index.

Since semi-circular variables are involved, the following criterion for freeness will be helpful: Recall the cyclic permutation γ_m, $1 \to 2 \to \cdots \to m \to 1$. Suppose $\{s_j\}_{1 \leq j \leq p}$ are free standard semi-circular and $\{a_\alpha\}_{\alpha \in I}$ are some other variables, all from $(\mathcal{A}, \varphi)$. Then $\{s_j\}_{1 \leq j \leq p}$ are free of $\{a_\alpha\}_{\alpha \in I}$, if and only if, for every $m \geq 1$ and non-negative integers $q(1), \ldots, q(m)$, and every choice of $\{s_{l(j)}\}_{1 \leq j \leq m}$ and $\{a_i\}_{1 \leq i \leq m}$ from these collections,

$$\begin{aligned} \varphi(s_{l(1)} a_1^{q(1)} \cdots s_{l(m)} a_m^{q(m)}) &= \sum_{\pi \in NC(m)} \kappa_\pi[s_{l(1)}, \ldots, s_{l(m)}] \varphi_{\pi\gamma_m}[a_1^{q(1)}, \ldots, a_m^{q(m)}] \\ &= \sum_{\pi \in NC_2^{(l)}(m)} \varphi_{\pi\gamma_m}[a_1^{q(1)}, \ldots, a_m^{q(m)}]. \end{aligned} \tag{11.45}$$

Proof of Theorem 11.6.1. From Theorem 10.3.1, the matrices converge jointly, and further, only colored pair-matched words contribute to the limit moments. It thus remains to prove asymptotic freeness, that is (11.45), and only for the case where m and $q(1) + \cdots + q(m)$ are even.

We will provide a detailed proof only for $p = 1$ and then discuss how this proof can be extended to the case $p > 1$. So there is one Wigner matrix (say W) and one other patterned random matrix (say A). We shall use the notation of the earlier chapters, but for simplicity, drop the suffixes that were used earlier to denote colors and indices. For example,

$$\begin{aligned} \Pi(w) = \{\pi : w[i] = w[j] \Leftrightarrow \\ (c_i, t_i, L_{c_i}(\pi(i-1), \pi(i))) = (c_j, t_j, L_{c_j}(\pi(j-1), \pi(j))\}. \end{aligned}$$

For a colored pair-matched word w of any given length k suppose $w[i] = w[j]$. This match is said to be a W or an A match according as the i-th and j-th positions have W or A. Let $w_{(i,j)}$ be the sub-word of length $j-i+1$ consisting of the letters from the i-th to the j-th positions of w:

$$w_{(i,j)}[t] := w[i - 1 + t] \text{ for all } 1 \leq t \leq j - i + 1.$$

Let $w_{(i,j)^c}$ be the complementary sub-word of length $k+i-j-1$ obtained by removing $w_{(i,j)}$ from w, that is

$$w_{(i,j)^c}[t] := \begin{cases} w[t] & \text{if } t < i, \\ w[t+j-i+1] & \text{if } i \le t \le k+i-j-1. \end{cases}$$

These sub-words need not be matched. If (i,j) is a W match, we will call $w_{(i,j)}$ a *Wigner string* of length $(j-i+1)$. For instance, if $q = WAAAAWWW$, then $w = abbccadd$ is a valid word and $abbcca$ and dd are Wigner strings of lengths six and two, respectively. For any word w, recall the (C2) constraint (see, for example, (10.22)) and define the following classes:

$$\begin{aligned}
\Pi^*(w) &:= \{\pi : w[i] = w[j] \Rightarrow \\
&\qquad (c_i, t_i, L_{c_i}(\pi(i-1), \pi(i))) = (c_j, t_j, L_{c_j}(\pi(j-1), \pi(j))\}, \\
\Pi^*_{(C2)}(w) &:= \{\pi \in \Pi^*(w) : (i,j)\ W-\text{match} \Rightarrow (\pi(i-1), \pi(i)) = (\pi(j), \pi(j-1))\}, \\
\Pi^*_{(i,j)}(w) &:= \{\pi \in \Pi^*(w) : (\pi(i-1), \pi(i)) = (\pi(j), \pi(j-1))\}.
\end{aligned}$$

Clearly,

$$\Pi^*_{(C2)}(w) = \cap_{(i,j)\ W-\text{match}} \Pi^*_{(i,j)}(w). \tag{11.46}$$

Now, for Wigner matrices, $p(w) \neq 0$ if and only if all constraints are (C2) (see (8.1) and the proof of Theorem 9.2.1). We now need an extension of this fact.

Lemma 11.6.2. Suppose w is a colored pair-matched word of length $2k$ and $p(w) \neq 0$.

(a) Then every Wigner string in w is a colored pair-matched word.

(b) For any (i,j) which is a W match we have

$$\lim_{n\to\infty} \frac{\#(\Pi^*(w) \setminus \Pi^*_{(i,j)}(w))}{n^{1+k}} = 0. \tag{11.47}$$

(c)

$$\lim_{n\to\infty} \frac{\#(\Pi^*(w) \setminus \Pi^*_{(C2)}(w))}{n^{1+k}} = 0. \tag{11.48}$$

By (11.46), (b) and (c) are equivalent. ♦

Lemma 11.6.3. Suppose X_n is any of the five patterned matrices of Theorem 11.6.1 and satisfies Assumption II of Chapter 9. Then for any $l \geq 1$ and integers $(k_1, \ldots, k_l)$, we have

$$\mathrm{E}\left[\prod_{i=1}^{l} \mathrm{tr}\left(\frac{X_n}{\sqrt{n}}\right)^{k_i}\right] - \prod_{i=1}^{l} \mathrm{E}\left[\mathrm{tr}\left(\frac{X_n}{\sqrt{n}}\right)^{k_i}\right] \to 0 \text{ as } n \to \infty. \tag{11.49}$$

♦

The proofs of Lemmas 11.6.2 and 11.6.3 shall be given later. Let

$$q := WA^{q(1)} \cdots WA^{q(m)}$$

be a typical monomial where the $q(i)$'s may equal 0. Further, m and $q(1) + \cdots + q(m)$ are even. Let $2k := m + q(1) + \cdots + q(m)$ be the length of q. From Theorem 10.3.1, $(W/\sqrt{n}, A/\sqrt{n}) \to (s, a)$ where s is a semi-circular variable and the state φ on the non-commutative polynomial algebra generated by $\{a, s\}$ is defined via

$$\varphi(sa^{q(1)} \cdots sa^{q(m)}) := \lim_{n\to\infty} \frac{1}{n^{k+1}} \operatorname{E}\operatorname{Tr}(q).$$

We must show that this φ satisfies (11.45).

Now we have the following string of equalities where the last equality follows from Lemma 11.6.2 (c) and Assumption II.

$$\lim_{n\to\infty} \frac{1}{n^{1+k}} \operatorname{E}[\operatorname{Tr}(WA^{q(1)} \cdots WA^{q(m)})] \tag{11.50}$$

$$= \lim_{n\to\infty} \frac{1}{n^{1+k}} \sum_{\substack{i(1),\ldots,i(m)\\ j(1),\ldots,j(m)=1}}^{n} \operatorname{E}[w_{i(1)j(1)} a^{q(1)}_{j(1)i(2)} w_{i(2)j(2)} a^{q(2)}_{j(2)i(3)} \cdots w_{i(m)j(m)} a^{q(m)}_{j(m)i(1)}]$$

$$= \lim_{n\to\infty} \frac{1}{n^{1+k}} \sum_{w\in\mathcal{CP}_2(k)} \sum_{\pi\in\Pi^*(w)} \operatorname{E}[\mathbb{X}_\pi] \text{ (say)}$$

$$= \lim_{n\to\infty} \frac{1}{n^{1+k}} \sum_{w\in\mathcal{CP}_2(k)} \sum_{\pi\in\Pi^*_{(C2)}(w)} \operatorname{E}[\mathbb{X}_\pi]. \tag{11.51}$$

Each $w \in \mathcal{CP}_2(2k)$ induces a pair-partition σ_w of $\{1, \ldots, m\}$, by only the W-matches (i.e. $(a, b) \in \sigma_w$ iff (a, b) is a W-match). Given any pair-partition σ of $\{1, \ldots, m\}$, let $[\sigma]_W$ be the class of all w which induce the partition σ. So the sum in (11.51) can be written as

$$\lim_{n\to\infty} \frac{1}{n^{1+k}} \sum_{\sigma\in\mathcal{P}_2(m)} \sum_{w\in[\sigma]_W} \sum_{\pi\in\Pi^*_{(C2)}(w)} E[\mathbb{X}_\pi]. \tag{11.52}$$

By the (C2) constraint imposed on $\Pi^*_{(C2)}(w)$, if (r, s) is a W match then

$$(i(r), j(r)) = (j(s), i(s)) \text{ or equivalently, } (\pi(r-1), \pi(r)) = (\pi(s), \pi(s-1)).$$

Therefore we have the string of equalities (11.53)–(11.56); (11.53) follows from (11.50) and (11.51) and the steps to arrive at (11.54)–(11.56) are by now familiar to us. The notation $\operatorname{tr}_{\sigma\gamma}$ was introduced in Definition 11.2.1.

$$\lim_{n\to\infty} \frac{1}{n^{k+1}} \mathrm{E}[\mathrm{Tr}(WA^{q(1)} \cdots WA^{q(m)})] \tag{11.53}$$

$$= \lim_{n\to\infty} \frac{1}{n^{k+1}} \sum_{\sigma\in\mathcal{P}_2(m)} \sum_{\substack{i(1),\ldots,i(m)\\ j(1),\ldots,j(m)=1}}^{n} \prod_{(r,s)\in\sigma} \delta_{i(r)j(s)}\delta_{i(s)j(r)} \mathrm{E}[a^{q(1)}_{j(1)i(2)} \cdots a^{q(m)}_{j(m)i(1)}]$$

$$= \lim_{n\to\infty} \frac{1}{n^{k+1}} \sum_{\sigma\in\mathcal{P}_2(m)} \sum_{\substack{i(1),\ldots,i(m)\\ j(1),\ldots,j(m)=1}}^{n} \prod_{(r,s)\in\sigma} \delta_{i(r)j(s)}\delta_{i(s)j(r)} \mathrm{E}[\prod_{k=1}^{m} a^{q(k)}_{j(k)i(\gamma(k))}] \tag{11.54}$$

$$= \lim_{n\to\infty} \frac{1}{n^{k+1}} \sum_{\sigma\in\mathcal{P}_2(m)} \sum_{\substack{i(1),\ldots,i(m)\\ j(1),\ldots,j(m)=1}}^{n} \prod_{r=1}^{m} \delta_{i(r)j(\sigma(r))} \mathrm{E}[a^{q(1)}_{j(1)i(\gamma(1))} \cdots a^{q(m)}_{j(m)i(\gamma(m))}] \tag{11.55}$$

$$= \lim_{n\to\infty} \frac{1}{n^{k+1}} \sum_{\sigma\in\mathcal{P}_2(m)} \sum_{j(1),\ldots,j(m)=1}^{n} \mathrm{E}[a^{q(1)}_{j(1)j(\sigma\gamma(1))} \cdots a^{q(m)}_{j(m)j(\sigma\gamma(m))}] \tag{11.56}$$

$$= \sum_{\sigma\in NC_2(m)} \lim_{n\to\infty} \mathrm{E}\left(\mathrm{tr}_{\sigma\gamma}[A^{(q_1)}, \ldots, A^{(q_m)})]\right) \text{ (using Lemma 3.8.1 (h))}$$

$$= \sum_{\sigma\in NC_2(m)} \lim_{n\to\infty} (\mathrm{E}\,\mathrm{tr})_{\sigma\gamma}[A^{(q_1)}, \ldots, A^{(q_m)})] \text{ (using Lemma 11.6.3)}$$

$$= \sum_{\sigma\in NC_2(m)} \varphi_{\sigma\gamma}[a^{(q_1)}, a^{(q_2)}, \ldots, a^{(q_m)})].$$

This establishes (11.45) and hence freeness in the limit. The above method can be easily extended to plug in more independent copies (indices) of W and A. The following details will be necessary.

(1) Extension of Lemmas 11.6.2 and 11.6.3. These can be easily obtained using the mapping ψ given in Section 10.2 and used in Theorem 10.2.1.

(2) When we have independent copies of the Wigner matrix, the product in (11.55) gets replaced by

$$\prod_{r=1}^{m} \delta_{i(r)j(\sigma(r))}\delta_{l(r)l(\sigma(r))}.$$

Here $l(1), \ldots, l(m)$ are the indices corresponding to the independent Wigner matrices. In addition, if we have independent copies of the other patterned random matrix, then $A^{q(1)}, \ldots, A^{q(m)}$ are replaced by products of several A-type matrices and then we can follow the steps given earlier. For example, the same type of argument was used by Nica and Speicher (2006)[74] to arrive at their Theorem 22.35. We skip the algebraic details. This completes the proof of Theorem 11.6.1 except that we still need to prove the two lemmas. ■

Lemma 11.6.4 is the essential ingredient in the proof of Lemma 11.6.2.

Lemma 11.6.4. Suppose w is any colored pair-matched word. Let $w_{(i,j)}$ be a Wigner string in w which is also a pair-matched word such that (11.47) is satisfied. Then

$$p(w) = p(w_{(i,j)})p(w_{(i,j)^c}). \tag{11.57}$$

Further, if $w_{(i+1,j-1)}$ and $w_{(i,j)^c}$ satisfy (11.48), then so does w. ♦

Proof. Given $\pi_1 \in \Pi^*(w_{(i+1,j-1)})$ and $\pi_2 \in \Pi^*(w_{(i,j)^c})$, construct $\pi \in \Pi^*_{(i,j)}(w)$ as:

$$(\pi_2(0), \ldots, \pi_2(i-1), \pi_1(0), \ldots, \pi_1(j-i-1), \pi_1(0), \pi_2(i-1), \ldots, \pi_2(2k-j+i-1)).$$

Conversely, from any $\pi \in \Pi^*_{(i,j)}(w)$, one can construct π_1 and π_2 by reversing the above construction.

So we have

$$\#\Pi^*_{(i,j)}(w) = \#\Pi^*(w_{(i+1,j-1)})\#\Pi^*(w_{(i,j)^c}). \tag{11.58}$$

Let

$$2l_1 := |w_{(i+1,j-1)}| \text{ and } 2l_2 := |w_{(i,j)^c}|$$

be the lengths of these words. Then

$$2l_1 + 2l_2 = (j - i - 1) + (2k + i - j - 1) = 2k - 2.$$

Now dividing equation (11.58) by n^{k+1} and using the assumption that $w_{(i,j)}$ satisfies (11.47), and then passing to the limit,

$$p(w) = p(w_{(i+1,j-1)})p(w^c_{(i,j)}). \tag{11.59}$$

We now claim that

$$\#\Pi^*(w_{(i,j)}) = n\#\Pi^*(w_{(i+1,j-1)}). \tag{11.60}$$

To prove this, note that given $\pi \in \Pi^*(w_{(i,j)})$, one can always get a $\pi' \in \Pi^*(w_{(i+1,j-1)})$, where $\pi(i-1)$ is arbitrary, and hence

$$\frac{\#\Pi^*(w_{(i,j)})}{n} \leq \#\Pi^*(w_{(i+1,j-1)}). \tag{11.61}$$

On the other hand, given a $\pi' \in \Pi^*(w_{(i+1,j-1)})$, one can choose $\pi(i-1)$ in n ways and also assign $\pi(j) = \pi(i-1)$ or $\pi(i)$, making j a dependent vertex. Hence we obtain

$$\#\Pi^*(w_{(i,j)}) \geq n\#\Pi^*(w_{(i+1,j-1)}). \tag{11.62}$$

Now (11.60) follows from (11.61) and (11.62). As a consequence, whenever $w_{(i,j)}$ is a Wigner string, we have

$$p(w_{(i,j)}) = p(w_{(i+1,j-1)}). \tag{11.63}$$

Now (11.57) follows from (11.59) and (11.63).

To prove the second part, note that from the first construction,

$$\#\Pi^*_{(C2)}(w) = \#\Pi^*_{(C2)}(w_{(i+1,j-1)})\#\Pi^*_{(C2)}(w_{(i,j)^c}).$$

Now suppose $w_{(i+1,j-1)}$ and $w_{(i,j)^c}$ satisfy (11.48). Then we have that

$$\begin{aligned}\#\Pi^*(w_{(i+1,j-1)}) &= \#\Pi^*_{(C2)}(w_{(i+1,j-1)}) + o(n^{l_1+1}) \text{ and}\\ \#\Pi^*(w_{(i,j)^c}) &= \#\Pi^*_{(C2)}(w_{(i,j)^c}) + o(n^{l_2+1}).\end{aligned}$$

Now we can multiply the above two equations and use the fact that (see (11.58))

$$\#\Pi^*(w) = \#\Pi^*(w_{(i+1,j-1)})\#\Pi^*(w_{(i,j)^c}) + o(n^{k+1})$$

to complete the proof. ■

Proof of Lemma 11.6.2. We use induction on the length l of the Wigner string. Let w be a colored pair-matched word of length $2k$ with $p(w) \neq 0$. First suppose $l = 2$. Without loss of generality, assume that this string starts at the beginning. So we have

$$(\pi(0), \pi(1)) = \begin{cases}(\pi(1), \pi(2)) & \text{or}\\ (\pi(2), \pi(1)).\end{cases}$$

In the first case, $\pi(0) = \pi(1) = \pi(2)$, and so $\pi(1)$ cannot be a generating vertex. But then this reduces the number of generating vertices, and as a consequence, $p(w)$ equals 0. Hence, the only possibility is $(\pi(0), \pi(1)) = (\pi(2), \pi(1))$, and the circuit is complete for the Wigner string, and so it is a pair-matched word, proving part (a). Further, this yields a (C2) constraint.

Now suppose the result holds for all Wigner strings of length strictly less than l. Consider a Wigner string of length l, say $w_{(1,l)}$, assumed without loss of generality to start from the first position. We then have two cases:

Case I: The Wigner string $w_{(1,l)}$ contains another Wigner string of length less than l at the position (p, q) with $1 \leq p < q \leq l$. Then by Lemma 11.6.4 we have,

$$p(w) = p(w_{(p,q)})p(w_{(p,q)^c}) \neq 0.$$

So, by the induction hypothesis and the fact that both $p(w_{(p,q)})$ and $p(w_{(p,q)^c})$ are not equal to zero we have that $w_{(p,q)}$ and $w_{(p,q)^c}$ are pair-matched words which satisfy (11.47). So $w_{(1,l)}$ is a pair-matched word as it is made up of $w_{(p,q)}$ and $w_{(p,q)^c}$ which are pair-matched. Also, from the second part of Lemma 11.6.4, we have that $w_{(1,l)}$ satisfies parts (b) and (c).

Case II: The Wigner string $w_{(1,l)}$ does not contain another Wigner string of smaller length inside it. Then there is a string of A-letters after a W-letter. We show that this string is pair-matched and the last Wigner letter before the

l-th position is at the first position. This would also imply that this A-string does not cross a W-letter.

Suppose that the last W-letter before the l-th position is at j_0. Since there is no Wigner string of smaller length, $\pi(j_0)$ is a generating vertex. Further, all letters from $j_0 + 1$ to $l - 1$ are A-letters.

From this point onward, the argument relies on the specific structure of the matrix A.

Case II (i): *Suppose A is a Toeplitz matrix.* For $i = 1, 2, \ldots, l - 1 - j_0$, let $s_i := \pi(j_0 + i) - \pi(j_0 + i - 1)$. Now consider the following equation:

$$s_1 + \cdots + s_{l-1-j_0} = \pi(l - 1) - \pi(j_0). \tag{11.64}$$

If for any j, $w[j]$ is the first appearance of that letter, then consider s_j to be an independent variable (that is it can be chosen freely). Note that if $w[k] = w[j]$, where $k > j$, then $s_k = \pm s_j$. Since $(1, l)$ is a W-match, $\pi(l-1)$ is either $\pi(0)$ or $\pi(1)$ and hence it is not a generating vertex. If $\pi(l - 1) \neq \pi(j_0)$, then (11.64) would be an additional constraint on the independent variables. This nontrivial constraint would lower the number of independent variables and hence the limit contribution would be zero, which is not possible as $p(w) \neq 0$. So we must have

$$\pi(l - 1) = \pi(j_0) \text{ and } j_0 = 1.$$

This also shows that $(\pi(l), \pi(l-1)) = (\pi(0), \pi(1))$ and hence $w_{(1,l)}$ is a colored word. As $s_1 + \cdots + s_{l-1-j_0} = 0$, all the independent variables occur twice with different signs in the left side, since otherwise it would again mean a non-trivial relation among them and thus would lower the order. Hence we conclude that the Toeplitz letters inside the first l positions are also pair-matched. Since the (C2) constraint is satisfied at the position $(1, l)$, part (b) also holds.

Case II (ii): *Suppose A is a Hankel matrix.* Let $t_i := \pi(j_0 + i) + \pi(j_0 + i - 1)$ and consider

$$-t_1 + t_2 - t_3 + \cdots (-1)^{l-j_0-1} t_{l-j_0-1} = (-1)^{l-j_0-1}(\pi(l-1) - \pi(j_0)). \tag{11.65}$$

Now, again as earlier, the t_i's are independent variables, and so this implies that to avoid a non-trivial constraint which would lower the order, both sides of (11.65) have to vanish, which automatically leads to the conclusion that $\pi(l - 1) = \pi(j_0) = \pi(1)$. So $j_0 = 1$, and again, the Wigner paired string of length l is pair-matched. Part (b) follows as the (C2) constraint holds.

Case II (iii): Suppose A is a Symmetric or Reverse Circulant matrix. Their link functions are similar to Toeplitz and Hankel, respectively, and the proofs are similar to the above two cases. Hence we skip them. ■

Proof of Lemma 11.6.3. We first show that

$$\mathrm{E}\left[\prod_{i=1}^{l}\left(\mathrm{tr}\frac{X_n^{k_i}}{n^{k_i/2}} - \mathrm{E}\left[\mathrm{tr}\frac{X_n^{k_i}}{n^{k_i/2}}\right]\right)\right] = O(n^{-1}) \text{ as } n \to \infty. \tag{11.66}$$

To prove (11.66), we observe that

$$\mathrm{E}\left[\prod_{i=1}^{l}\left(\operatorname{Tr}\left(X_n^{k_i}-\mathrm{E}\left[\operatorname{Tr}X_n^{k_i}\right]\right)\right)\right]=\sum_{\pi_1,\ldots,\pi_l}\mathrm{E}\left[\left(\prod_{j=1}^{l}\left(X_{\pi_i}-\mathrm{E}(X_{\pi_i})\right)\right)\right], \quad (11.67)$$

where for each i, π_i is a circuit of length k_i. There are two situations where the summand above vanishes:

(i) If there is a circuit, say π_i, which is self-matched, then clearly

$$\begin{aligned}\mathrm{E}[(\prod_{j=1}^{l}(X_{\pi_i}-\mathrm{E}(X_{\pi_i})))] &= \mathrm{E}[X_{\pi_i}-\mathrm{E}(X_{\pi_i})]\,\mathrm{E}[(\prod_{j\neq i}(X_{\pi_j}-\mathrm{E}(X_{\pi_j})))] \\ &= 0.\end{aligned}$$

(ii) If there is a circuit π_i which has an edge (an L-value) that appears exactly once across all circuits, then clearly, $\mathrm{E}\,X_{\pi_i}=0$, and hence

$$\begin{aligned}\mathrm{E}[(\prod_{j=1}^{l}(X_{\pi_i}-\mathrm{E}(X_{\pi_i})))] &= \mathrm{E}[X_{\pi_i}(\prod_{j\neq i}(X_{\pi_j}-\mathrm{E}(X_{\pi_j})))] \\ &= 0.\end{aligned}$$

So we need to consider only those circuits of which none are self-matched and each edge appears at least twice. Now the total number of such circuits $\{\pi_1,\ldots,\pi_l\}$, where there is at least one edge that appears at least thrice, is bounded above by $Cn^{\sum_{i=1}^{l}k_i/2+l-1}$ since $\Delta<\infty$. Hence using Assumption II such terms in (11.67), after dividing by $n^{\sum_{i=1}^{l}k_i/2+l}$, are of the order $O(n^{-1})$.

Now we are left with only those terms where $\{\pi_1,\ldots,\pi_l\}$ are jointly-matched but not self-matched. Moreover, all the edges appear exactly twice and hence $k_1+\cdots+k_l$ is even. In view of the uniformly bounded moment assumption (part of Assumption II), it is then enough to prove that the number of *generating vertices* is bounded by $\sum_{i=1}^{l}k_i/2+l-1$. To prove this, first note that the number of generating vertices is bounded any way by $\sum_{i=1}^{l}k_i/2+l$. Thus it is enough to show that there is at least one generating vertex that cannot be chosen freely.

Since π_1 is not self-matched, assume, without loss of generality, that the value of $(\pi_1(0),\pi_1(1))$ occurs exactly once in π_1. Construct π_1 as follows. Set $\pi_1(0)=\pi_1(k_1)$. Then choose the remaining vertices of $\pi)1$ in the order $\pi_1(k_1),\pi_1(k_1-1),\ldots,\pi_1(1)$, using the other circuits. Then the edge $(\pi_1(0),\pi_1(1))$ is determined and hence the generating vertex $\pi_1(1)$ cannot be chosen freely. So the number of generating vertices is reduced by one and the proof of (11.66) is complete.

Now we show that (11.49) follows from (11.66). For $l=2$, the two relations are the same. Suppose (11.49) is true for all $2\leq m<l$. We expand

$$\lim_{n\to\infty}\mathrm{E}\left[\prod_{i=1}^{l}\left(\operatorname{tr}\left(\frac{X_n}{\sqrt{n}}\right)^{k_i}\right)-\mathrm{E}\left(\operatorname{tr}\left(\left(\frac{X_n}{\sqrt{n}}\right)^{k_i}\right)\right)\right]=0$$

to get

$$\lim_{n\to\infty}\sum_{m=1}^{l}(-1)^m \sum_{i_1<\cdots<i_m} \mathrm{E}[\prod_{j=1}^{m}\mathrm{tr}((\frac{X_n}{\sqrt{n}})^{k_{i_j}})] \prod_{i\notin\{i_1,\ldots,i_m\}} \mathrm{E}[\mathrm{tr}((\frac{X_n}{\sqrt{n}})^{k_i})] = 0.$$

Now using the result for products of smaller orders successively,

$$\begin{aligned}\lim_{n\to\infty}(-1)^l\,\mathrm{E}[\prod_{j=1}^{l}\mathrm{tr}((\frac{X_n}{\sqrt{n}})^{k_j})] &= \lim_{n\to\infty}\sum_{m<l}(-1)^m \sum_{i_1<\cdots<i_m} \mathrm{E}[\prod_{j=1}^{m}\mathrm{tr}((\frac{X_n}{\sqrt{n}})^{k_{i_j}})] \\ &\quad\times \prod_{i\notin\{i_1,\ldots,i_m\}} \mathrm{E}[\mathrm{tr}((\frac{X_n}{\sqrt{n}})^{k_i})].\end{aligned}$$

Now, using the induction hypothesis that (11.49) holds for all $2 \le m < l$, it is easy to see that the right side is a telescopic sum and simplifies to $(-1)^l \lim_{n\to\infty}\prod_{i=1}^{l}\mathrm{E}[\mathrm{tr}((\frac{X_n}{\sqrt{n}})^{k_i})]$. Hence the lemma is proved. ■

We now elaborate on Remark 11.6.1 (c) as promised.

Theorem 11.6.5. Let $\{W_{i,n}, 1 \le i \le p\}$, be $n\times n$ Gaussian unitary matrices. Let $A_{i,n}, 1 \le i \le p$ be $n \times n$ patterned random matrices whose entries satisfy Assumption II, and all the matrices are independent of each other. Suppose $\{A_{i,n}\}$, as elements of $(\mathcal{M}_n(\mathbb{C}), \varphi_n = n^{-1}\,\mathrm{E}\,\mathrm{Tr})$, converge jointly. Then $\{W_{i,n}, A_{i,n}, 1 \le i \le p\}$ converge jointly to $\{s_i, a_i, 1 \le i \le p\}$, where $\{s_i, 1 \le i \le p\}$ are free standard semi-circular variables that are free of $\{a_i, 1 \le i \le p\}$. ◆

Proof. We provide a proof only for the case $p = 1$. This is similar to, but easier than, the proof of Theorem 11.6.1. We only point out the advantage gained with Gaussian entries. We follow the steps (11.52) onward in the proof of Theorem 11.6.1. Here the entries $a_{i,j}, 1 \le i, j \le n$ satisfy

$$\mathrm{E}(a_{i,j}a_{k,l}) = \delta_{i,l}\delta_{j,k}. \tag{11.68}$$

This automatically implies that only (C2) constraints survive. Hence

$$\begin{aligned}&\lim_{n\to\infty}\frac{1}{n^{k+1}}\mathrm{E}[\mathrm{Tr}(WA^{q(1)}WA^{q(2)}\cdots WA^{q(m)})] \\ &= \lim_{n\to\infty}\frac{1}{n^{k+1}}\sum_{\sigma\in\mathcal{P}_2(m)}\sum_{\substack{i(1),\ldots,i(m)\\ j(1),\ldots,j(m)=1}}^{n}\prod_{(r,s)\in\sigma}\delta_{i(r)j(s)}\delta_{i(s)j(r)}\,\mathrm{E}[a^{q(1)}_{j(1)i(2)}\cdots a^{q(m)}_{j(m)i(1)}] \\ &= \sum_{\sigma\in NC_2(m)}\lim_{n\to\infty}\mathrm{E}\left(\mathrm{tr}_{\sigma\gamma}[A^{q(1)},A^{q(2)},\ldots,A^{q(m)})]\right).\end{aligned}$$

The result now follows easily. ■

11.7 Discussion

(a) *GUE, GOE and GSE.* For information on freeness in the context of the Gaussian unitary ensemble (GUE), the Gaussian orthogonal ensemble (GOE) and the Gaussian symplectic ensemble (GSE), see Capitaine and Casalis (2004)[30], Capitaine and Martin (2007)[31], Collins, Guionnet and Segala (2009)[34], Schultz (2005)[85], Ryan (1998)[82] and Voiculescu (1998)[101].

Voiculescu (1991)[100] was the first to show that independent Hermitian Wigner matrices, which are distributed as GUE, are asymptotically free and are also free of non-random matrices matrices $\{D_{i,n}\}$ which converge. Later, Voiculescu (1998)[101] improved the result by allowing $\{D_{i,n}\}$ to be not necessarily diagonal and whose entries satisfy

$$\sup_n \|D_{i,n}\| < \infty \text{ for each } i, \tag{11.69}$$

where $\|\cdot\|$ denotes the operator norm. This inclusion of constant matrices had important implications in the factor theory of von Neumann algebras.

(b) *Perturbation of Wigner.* The limit spectrum of $\frac{W_n}{\sqrt{n}} + P_n$, where W_n is a Wigner matrix and P_n is a random matrix which is independent of W_n, has been of interest for a long time (see Fulton (1998)[43]). Theorem 11.6.1 is applicable in this case when appropriate assumptions are made on the matrices and the limit distribution is $\mu_s \boxplus \nu$ where μ is the semi-circular law and ν is the LSD of P_n. Results on the sum also follow from Pastur and Vasilchuk (2000)[77]. Incidentally, it follows from Biane (1997)[18] that $\mu_s \boxplus \nu$, for any ν, is continuous and has a density that can be expressed in terms of ν.

(c) *Wigner and other matrices.* Dykema (1993)[39] established that independent Wigner matrices are asymptotically free of block-diagonal constant matrices with bounded block size. The results were also shown to hold almost surely. See Hiai and Petz (2000)[57, 58] for details.

By the results of Collins (2003)[33] and Collins and Śniady (2006)[35] independent Wigner matrices are asymptotically free of deterministic matrices which converge jointly. It appears that the existing results in the literature on freeness of Wigner and other random matrices need some condition on the behavior of the trace of the matrices as pointed out in Remark 3.6 of Collins (2003)[33]. This condition (equation (3.4) therein) was studied in Capitaine and Casalis (2004)[30] and they showed that under some technical conditions on the other random matrices (see Condition C and C' there), there is asymptotic freeness between Wigner matrices and them. They also showed the asymptotic freeness of independent S-matrices when the entries are Gaussian. Condition (11.69) appears in other available criteria for freeness (see Anderson, Guionnet and Zeitouni (2010)[1] and Theorem 22.2.4 of Speicher (2011)[94]).

(d) *Haar unitary matrices.* Any unitarily invariant matrix (in particular matrices from the GUE) can be written as UDU^*, where D is a diagonal matrix and U is Haar distributed on the space of unitary matrices and independent of D. Voiculescu (1991)[100] showed that $\{U, U^*\}$ and D are asymptotically free. Hiai and Petz (2000)[57] showed that Haar unitaries and general deterministic matrices that satisfy (11.69) are almost surely asymptotically free.

Collins (2003)[33] showed that general deterministic matrices and Haar unitary matrices are asymptotically free almost surely, provided the deterministic matrices jointly converge.

For information on Haar unitary matrices, see Diaconis (2003)[38], Edelman and Rao (2005)[40] and Guionnet (2004)[52].

(e) *Operator-valued freeness.* The concept of freeness that we have discussed has been extended to *operator-valued freeness*, also called *freeness with amalgamation*, and in particular this is useful in describing the behavior of polynomials of random band matrices and rectangular random matrices. For more information, see Shlyakhtenko (1996)[87] and Benaych-Georges (2009)[12, 13]).

(f) *Traffic independence.* There is a notion of *traffic distribution* of random matrices, which is richer than the $*$-distribution that we have discussed. This leads to a notion of *traffic independence* At an algebraic level, traffic independence, in some sense, unifies the three canonical notions of independence, namely tensor (classical), free and Boolean. See Male (2020)[68] for details. A central limiting theorem is stated in this context, interpolating between the tensor, free and Boolean central limit theorems.

11.8 Exercises

1. Let I_p be the identity matrix of order p and $0_{p\times n}$ be the $p \times n$ matrix all whose entries are zero. Define

$$\bar{I} := \begin{bmatrix} I_p & 0_{p\times n} \\ 0_{n\times p} & 0_{n\times n} \end{bmatrix}, \quad \underline{I} := \begin{bmatrix} 0_{p\times p} & 0_{p\times n} \\ 0_{n\times p} & I_n \end{bmatrix}.$$

(a) Show that if $p \to \infty, p/n \to y \neq 0$, then $(\bar{I}, \underline{I})$ converge jointly to (a_0, a_1) where a_0 and a_1 are marginally Bernoulli with probability of success $y(1+y)^{-1}$ and $(1+y)^{-1}$, respectively, and all joint moments of a_1 and a_2 are 0.

(b) If s is a standard semi-circular variable, free of (a_0, a_1), then show that the self-adjoint variable $a_0 s a_1 s a_0$ has the Marcěnko-Pastur law with parameter y.

2. Prove (11.29) that was left unproved in the text.

3. Prove the extended version of Theorem 9.4.1 for $S = XX^*$ when X is elliptic. Use embedding and appeal to Theorem 11.3.1.

4. Let M_y be a Marčenko-Pastur variable. Let $\tilde{M}_y$ be a *symmetrized Marčenko-Pastur* variable with parameter y. That is,

$$\kappa_k(\tilde{M}_y) = \begin{cases} \kappa_k(M_y) = y^{k-1} & \text{if } k \text{ is even,} \\ 0 \text{ if } k \text{ is odd.} \end{cases}$$

   Suppose u is a Haar unitary varaible and is free of M_y and $\tilde{M}_y$. Show that the $*$-distributions of the cross-covariance variable v with parameters $\rho = 0$ and y, $u\tilde{M}_y$ and uM_y are all identical. [A variable u is said to be Haar unitary if it satisfies the conditions given in Exercise 19 of Chapter 2].

5. Show that by taking $\rho = 1$ in Theorem 11.5.1, we get back the statement of Theorem 9.4.1 on the joint convergence of independent S-matrices.

6. Suppose that C is a cross-covariance matrix with $\rho = 0$ where the elements of C satisfy Assumption Ie. Suppose that $p \to \infty$ such that $p/n \to y \neq 0$. Then show that the LSD of $C + C^*$ is the free additive convolution $\mu \boxplus \mu$ where μ is the symmetrized Marčenko-Pastur law with parameter y.

7. Extend Theorem 11.5.1 to the case where (X_l, Y_l), $1 \leq l \leq t$ are independent $p \times n_l$ random matrices whose entries satisfy Assumption Ie, and $p \to \infty$ such that $p/n_l \to y_l, 0 < y_l < \infty$ for all l.

12

Brown measure

This is a very brief chapter on the Brown measure, a concept particularly useful for non-Hermitian matrices. We have not established the LSD of any non-Hermitian matrix in this book, restricting ourselves to only moment convergence. It turns out that in many cases the LSD and the Brown measure are identical, though this is not true in general. We hope that some flavor will be obtained about the importance of the Brown measure and its relation to the LSD of matrices from the limited discussion in this chapter.

12.1 Brown measure

The Brown measure was introduced by Brown (1986)[28]. It is a probability measure that serves as a generalization of the ESD of matrices to the eigenvalue distribution of operators in a von Neumann algebra which is equipped with a trace. We take a brief look at this concept. The interested reader may consult Mingo and Speicher (2017)[72] for more information.

The starting point is the determinant defined by Fuglede and Kadison (1952)[42].

Definition 12.1.1. (Fuglede-Kadison determinant) Let $(\mathcal{A}, \varphi)$ be a C^* probability space where φ is faithful. If $a \in \mathcal{A}$ is invertible, then its *Fuglede-Kadison determinant* is defined by

$$\Delta(a) := \exp[\frac{1}{2}\varphi(\log(aa^*))] \in (0, \ \infty).$$

If a is not invertible, then it is defined as

$$\Delta(a) := \lim_{\epsilon \downarrow 0} \exp[\frac{1}{2}\varphi(\log(aa^* + \epsilon))] \in [0, \infty).$$

◇

Note that the above limit always exists. To simplify notation, let us write $(a - \lambda)$ for $(a - \lambda \mathbf{1}_{\mathcal{A}})$. Define $u_a : \mathbb{C} \to \mathbb{R}$ by

$$u_a(\lambda) := \log \Delta(a - \lambda), \ \lambda \in \mathbb{C}.$$

DOI: 10.1201/9781003144496-12

The following two statements are true:

(i) The function u_a is *upper-semicontinuous.* That is

$$\limsup_{y \to x} u_a(y) \leq u_a(x) \quad \text{for all } x \in \mathbb{C}.$$

(ii) The function u_a satisfies

$$u_a(w) \leq \frac{1}{2\pi} \int_0^{2\pi} u_a(w + re^{\iota t}) dt, \quad \text{for all } r > 0.$$

That is, u_a is a *subharmonic function.* Hence there is a unique measure, say $\mu_{B,a}$ known as the *Riesz measure* associated to u_a. The measure $\mu_{B,a}$ is a probability measure on $\mathbb{C}$ with support contained in $\mathbf{sp}(a)$. It is often written as

$$\mu_{B,a} = \frac{1}{2\pi} \left(\frac{\partial^2}{\partial x^2} + \frac{\partial^2}{\partial y^2} \right) \log \Delta(a - \lambda),$$

where x and y are the real and imaginary parts of λ.

Definition 12.1.2. (Brown measure) The measure $\mu_{B,a}$ is known as the *Brown measure* of a. It is the unique probability measure that satisfies

$$\int_{\mathbb{C}} \log |\lambda - z| d\mu_{B,a}(z) = \log \Delta(a - \lambda) \quad \text{for all } \lambda \in \mathbb{C}.$$

◇

Example 12.1.1. Suppose A_n is an $n \times n$ matrix, considered as an element of $(\mathcal{M}_n(\mathbb{C}), \mathrm{tr})$. Suppose its eigenvalues are $\lambda_1, \ldots, \lambda_n$. Then it can be shown that

$$\Delta(A_n) = \sqrt[n]{|\det A_n|} \quad \text{and} \quad \mu_{B,A_n} = \frac{1}{n} \sum_{i=1}^{n} \delta_{\lambda_i}.$$

From this it follows that the Brown measure of A_n is its ESD. ▲

The Brown measure of a variable a is often not easy to compute explicitly. However, there is one situation where a nice expression is available. We need the following definition of an R-diagonal element. For more details on this concept, see Nica and Speicher (2006)[74].

Definition 12.1.3. (R-diagonal element) An element $a \in \mathcal{A}$ is called *R-diagonal* if $\kappa_n(a_1, \ldots, a_n) = 0$ for all $n \in \mathbb{N}$ whenever the arguments $a_1, \ldots, a_n \in \{a, a^*\}$ are not alternating in a and a^*. ◇

Example 12.1.2. A circular variable is R-diagonal since the only non-zero free cumulants are $\kappa_2(c, c^*) = \kappa_2(c^*, c) = 1$. ▲

Recall the S-transform defined in Section 5.4. Let $S^{\langle -1 \rangle}$ denote the inverse of S with respect to composition. The Brown measure of any R-diagonal element is connected to $S^{\langle -1 \rangle}$ in a simple way as stated in the lemma below. We refer the reader to Haagerup (2000)[55] for a proof.

Lemma 12.1.1. Suppose a is any R-diagonal element in a C^*-probability space $(\mathcal{A}, \varphi)$. Then its Brown measure $\mu_{\mathrm{B},a}$ is rotationally invariant and hence can be described by the probabilities of the sets $B(t) = \{\lambda \in \mathbb{C} : |\lambda| \leq t\}$. These probabilities are given by

$$\mu_{\mathrm{B},a}(B(t)) = \begin{cases} 0 & \text{if} \quad t \leq \frac{1}{\sqrt{\varphi((aa^*)^{-1})}}, \\ 1 - S_{aa^*}^{\langle -1 \rangle}(t^{-2}) & \text{if} \quad \frac{1}{\sqrt{\varphi((aa^*)^{-1})}} \leq t \leq \sqrt{\varphi(aa^*)}, \\ 1 & \text{if} \quad t \geq \sqrt{\varphi(aa^*)}. \end{cases}$$

♦

Note that when $0 \in \mathbf{sp}(aa^*)$ then aa^* is not invertible. Hence $(aa^*)^{-1}$ and $\varphi((aa^*)^{-1})$ has to be given proper meaning. We shall not discuss this issue in detail. Suffice it to observe the following: since aa^* is a self-adjoint variable in a C^*- algebra, its probability law μ_{aa^*} exists and is supported on a subset of $[0, \infty)$. Let X be a random variable with the probability law μ_{aa^*}. Then

$$\varphi((aa^*)^{-1}) = \mathrm{E}\left(\frac{1}{X}\right) = \int_{[0,\ \infty)} x \mu_{aa^*}(dx).$$

This expectation can be equal to ∞ and we interpret $1/\infty$ as 0.

Example 12.1.3. Suppose c is a circular element. We claim that the Brown measure of c is the uniform distribution on the unit disc in $\mathbb{C}$. Since c is R-diagonal, its Brown measure is radially symmetric. Hence it suffices to show that

$$\mu_{\mathrm{B},c}(B(t)) = t^2, 0 \leq t \leq 1. \tag{12.1}$$

Recall Example 5.4.1 of Chapter 5. From that example it follows that

$$\varphi(cc^*) = 1, \tag{12.2}$$

and the S-transform of cc^* is given by

$$S_{cc^*}(z) = \frac{1}{1+z} \quad \text{(in the appropriate domain)}.$$

Using this,

$$S^{\langle -1 \rangle}(z) = \frac{z-1}{z} \quad \text{(in the appropriate domain)}. \tag{12.3}$$

Hence

$$1 - S^{\langle -1 \rangle}(t^{-2}) = 1 - \frac{t^{-2} - 1}{t^{-2}} \tag{12.4}$$

$$= t^2. \tag{12.5}$$

It can be easily verified that (Exercise 9 of Chapter 6) $\sqrt{cc^*}$ has the quarter-circular law with density given by

$$q(x) = \frac{1}{\pi}\sqrt{4-x^2},\ 0 \leq x \leq 2.$$

Using this, it easily follows that

$$\varphi\big((cc^*)^{-1}\big) = \infty. \tag{12.6}$$

Now we can use (12.2), (12.4), (12.6) and apply Lemma 12.1.1 to arrive at (12.1). ▲

It is interesting to note that even when the ESD of A_n converges to, say, μ, and A_n converges in $*$-distribution to, say, a, there is no guarantee that $\mu_{B,a} = \mu$. A very simple example is given in Śniady (2002)[89].

Though, in general, the LSD and the Brown measure of the $*$-limit variable are not equal, there are many situations in which they are indeed equal.

Example 12.1.4. If C_n is the i.i.d. matrix, then $n^{-1/2}C_n$ converges in $*$-distribution to a circular element c. Further, the LSD of $n^{-1/2}C_n$ is the uniform distribution on the unit disc though we have not proved this fact. A proof is available in Tao and Vu (2010)[98]. See also Bordenave and Chafai (2012)[19]. In Example 12.1.3 we have verified that this uniform distribution is indeed the Brown measure of c. ▲

Example 12.1.5. The LSD of the elliptic matrix $n^{-1/2}E_n$ is the uniform probability measure on the interior of an ellipse, again a fact we have not proved. A proof is available in Nguyen and O'Rourke (2015)[73]. We have shown in Theorem 8.3.1 (a) that $n^{-1/2}E_n$ converges to an elliptic element e. Unfortunately, e is not R-diagonal and so Lemma 12.1.1 cannot be applied. Nevertheless, it can be shown that the Brown measure of e is also the same uniform probability measure. See Belinchi, Śniady and Speicher (2015)[11] and Larsen (1999)[63]. ▲

Example 12.1.6. In Guionnet, Krishnapur and Zeitouni (2011)[54], it has been shown that the LSD of a bi-unitarily invariant random matrix is actually the Brown measure of the $*$-distributional limit. ▲

We give a flavor of the connection between the LSD and the Brown measure through a relatively easy example—that of the product of independent elliptic matrices.

Suppose $E_1, \ldots, E_k$ are $n \times n$ independent elliptic matrices whose entries satisfy Assumption Ie. From Theorem 9.3.1 we know that $(n^{-1/2}E_1, \ldots, n^{-1/2}E_k)$ converges jointly to free elliptic elements $e_1, \ldots, e_k$, say. As a consequence, the product $n^{-1/2}E_1 \cdots n^{-1/2}E_k$ converges in $*$-distribution to $e_1 \cdots e_k$.

On the other hand, the LSD of $n^{-1/2}E_1 \cdots n^{-1/2}E_k$ exists and the limit has been calculated in O'Rourke, Renfrew, Soshnikov and Vu (2015)[75]. This limit is rotationally invariant and is given by

$$\mu_k(\{z : |z| \le t\}) = \begin{cases} t^{\frac{2}{k}} & \text{if } 0 \le t \le 1, \\ 0 & \text{otherwise.} \end{cases} \tag{12.7}$$

We will now show that this LSD μ_k is the same as the Brown measure $\mu_{\mathrm{B},e_1\cdots e_k}$ of the $*$-distributional limit, $e_1 \cdots e_k$. We need the following lemma.

Lemma 12.1.2. Let $e_1, \ldots, e_k$ be k free elliptic elements with parameters $0 \le \rho_1, \ldots, \rho_k \le 1$, respectively, and $c_1, \ldots, c_k$ be k free circular elements. Then $e_1 \cdots e_k e_k^* \cdots e_1^*$ and $c_1 \cdots c_k c_k^* \cdots c_1^*$ have the same distribution. ♦

Proof. Since $e_1 \cdots e_k e_k^* \cdots e_1^*$ and $c_1 \cdots c_k c_k^* \cdots c_1^*$ are self-adjoint variables, it is enough to show that all their moments agree. From the proof of Lemma 3.6.1, we have

$$\kappa_2(e, e) = \kappa_2(e^*, e^*) = \rho, \quad \text{and} \quad \kappa_2(e, e^*) = \kappa_2(e^*, e) = 1.$$

We also know that

$$\kappa_2(c, c) = \kappa_2(c^*, c^*) = 0, \quad \text{and} \quad \kappa_2(c, c^*) = \kappa_2(c^*, c) = 1.$$

We write

$$(e_1 \cdots e_k e_k^* \cdots e_1^*)^n = e_{\tau_1}^{\epsilon_1} e_{\tau_2}^{\epsilon_2} \cdots e_{\tau_{2kn}}^{\epsilon_{2kn}},$$

where $\tau_{2mk+j} = \tau_{2(m+1)k-j+1} = j$, $\epsilon_{2mk+j} = 1$ and $\epsilon_{(2m+1)k+j} = *$ for $j = 1, \ldots, k$ and $m = 0, \ldots, n-1$.

Applying the moment-free cumulant formula, as the higher order free cumulants are zero, we have

$$\varphi((e_1 \cdots e_k e_k^* \cdots e_1^*)^n) = \sum_{\pi \in NC_2(2nk)} \prod_{(r,s) \in \pi} \kappa_2(e_{\tau_r}^{\epsilon_r}, e_{\tau_s}^{\epsilon_s}),$$

where $\tau_r, \tau_s \in \{1, \ldots, k\}$ and $\epsilon_r, \epsilon_s \in \{1, *\}$. Observe that if $(r, s) \in \pi$, then one of $\{r, s\}$ is odd and the other is even. Further, if e_j appears at an odd place, then e_j^* appears at an even place and vice-versa. Indeed, consider the variable $e_1 \cdots e_k e_k^* \cdots e_1^*$. Here e_r appears at the r-th place and e_r^* appears at the $(2k + 1 - r)$-th place. Therefore $\kappa_2(e_{\tau_r}^{\epsilon_r}, e_{\tau_s}^{\epsilon_s})$ will be of the form $\kappa_2(e, e^*)$ or $\kappa_2(e^*, e)$ and so $\kappa_2(e_{\tau_r}^{\epsilon_r}, e_{\tau_s}^{\epsilon_s}) = 1$. Thus the mixed free cumulants of $e_1 \cdots e_k e_k^* \cdots e_1^*$ are the same as that of $c_1 \cdots c_k c_k^* \cdots c_1^*$ and hence we have

$$\begin{aligned} \varphi((e_1 \cdots e_k e_k^* \cdots e_1^*)^n) &= \sum_{\pi \in NC_2(2nk)} \prod_{(r,s) \in \pi} \kappa_2(c_{\tau_r}^{\epsilon_r}, c_{\tau_s}^{\epsilon_s}) \\ &= \varphi((c_1 \cdots c_k c_k^* \cdots c_1^*)^n). \end{aligned}$$

This completes the proof. ■

Theorem 12.1.3. Let $k \geq 2$ and $e_1, \ldots, e_k$ be k free elliptic elements in $(\mathcal{A}, \varphi)$ with possibly different correlation parameters. Then the Brown measure $\mu_{\mathrm{B}, e_1 \cdots e_k}$ does not depend on the correlations and is rotationally invariant. It can be described by the probabilities

$$\mu_{B, e_1 \cdots e_k}(\{z : |z| \leq t\}) = \begin{cases} t^{\frac{2}{k}} & \text{if } 0 \leq t \leq 1, \\ 1 & \text{otherwise.} \end{cases}$$

This Brown measure coincides with the rotationally invariant LSD of $n^{-1/2}E_1 \cdots n^{-1/2}E_k$. ◆

Remark 12.1.1. Theorem 12.1.3 does not hold for $k = 1$. As we have noted in Example 12.1.5, the Brown measure of an elliptic variable is the uniform distribution on the interior of an ellipse which is obviously not rotationally invariant. So the condition $k \geq 2$ is crucial. ●

Proof of Theorem 12.1.3. It is known that the product of free elliptic elements is R-diagonal (see O'Rourke, Renfrew, Soshnikov and Vu (2015, page 9)[75]. Therefore $e_1 \cdots e_k$ is an R-diagonal element for $k \geq 2$, and hence, by Lemma 12.1.1, its Brown measure is determined by the distribution of $e_1 \cdots e_k e_k^* \cdots e_1^*$.

By Lemma 12.1.2, this distribution is the same as the distribution of $c_1 \cdots c_k c_k^* \cdots c_1^*$. Let S_k denote its S-transform. By Lemma 5.4.1 and freeness, we have

$$\begin{aligned} S_k(z) &= S_{c_1 c_1^*}(z) \cdots S_{c_k c_k^*}(z) \\ &= \left(\frac{1}{1+z}\right)^k \quad \text{in an appropriate domain.} \end{aligned}$$

Therefore, for $t > 0$, we get

$$S_k^{\langle -1 \rangle}(t) = t^{-\frac{1}{k}} - 1. \tag{12.8}$$

Since $\varphi(c_i c_i^*) = 1$ for all i and $\{c_i\}$ are free, it follows that

$$\varphi(c_1 \cdots c_k c_k^* \cdots c_1^*) = 1.$$

On the other hand, we have observed in (12.6) of Example 12.1.3 that $\varphi((c_i c_i^*)^{-1}) = \infty$. Using this and the freeness of $\{c_i\}$, it then follows that

$$\varphi((c_1 \cdots c_k c_k^* \cdots c_1^*)^{-1}) = \infty.$$

Therefore, using (12.8) in Lemma 12.1.1, we have

$$\mu_{\mathrm{B}, e_1 \cdots e_k}(\{z : |z| \leq t\}) = \begin{cases} t^{\frac{2}{k}} & \text{if } 0 \leq t \leq 1, \\ 1 & \text{otherwise.} \end{cases}$$

Thus $\mu_{\mathrm{B}, e_1 \cdots e_k}$ coincides with μ_k given in (12.7). ■

12.2 Exercises

1. Suppose $(\mathcal{A}, \varphi)$ is a C^* probability space where φ is faithful. Show that for any $a \in \mathcal{A}$,

$$\Delta(a) = \lim_{\epsilon \downarrow 0} \exp[\frac{1}{2}\varphi(\log(aa^* + \epsilon^2))]$$

exists.

2. Show that $\Delta(a)$ defined above satisfies

   (a) $\Delta(ab) = \Delta(a)\Delta(b)$;

   (b) $\Delta(a) = \Delta(a^*)$;

   (c) $\Delta(u) = 1$ if u is unitary;

   (d) $\Delta(\lambda a) = |\lambda|\Delta(a)$ for all $\lambda \in \mathbb{C}$.

3. Show that for any $n \times n$ matrix A, as an element of $(\mathcal{M}_n(\mathbb{C}), \text{tr})$,

$$\Delta(A) = |\det(A)|^{1/n}.$$

Use this to verify that the Brown measure of A is the ESD of A.

4. Suppose c is a circular element. Find the probability law μ_{cc^*} of cc^*. If X is a random variable with probability law μ_{cc^*}, then show that $\text{E}(\frac{1}{X}) = \infty$, thereby verifying (12.6).

13

Tying three loose ends

We finally tie the three loose ends that we have left so far. First we calculate the Möbius function on $NC(n)$. This is quite interesting in its own right. In addition, it can be used to simplify the proof of the free CLT for non-identically distributed variables, as indicated in an Exercise 5 of Chapter 4. Then we show the equivalence of the definition of free independence (that is, the vanishing of mixed free cumulants) which we have used, and the traditional definition that uses moments. Finally, we develop the free product construction of $*$-probability spaces that we have stated in Theorem 3.2.1.

13.1 Möbius function on $NC(n)$

This section shows how to calculate the Möbius function on $NC(n)$. For this, a few additional facts about the Möbius function and the inversion formula are needed. This section is built along the lines of Lecture 10 of Nica and Speicher (2006)[74].

Even though some of the developments below hold for POSETS, we shall restrict our attention to only finite lattices. Suppose $(P, \leq)$ is a finite lattice. Recall the following sets from Chapter 1:

$$\begin{aligned} P^{(2)} &= \{[\pi, \sigma] : \pi \leq \sigma\}, \text{ where} \\ [\pi, \sigma] &= \{\tau \in P : \pi \leq \tau \leq \sigma\}, \ \pi, \sigma \in P. \end{aligned}$$

Let μ be the Möbius function on $P^{(2)}$. Recall the two formulae (1.16) and (1.17) from Chapter 1 to move between moments and free cumulants. The following more general version was stated in Exercise 27 of Chapter 1.

Suppose $f, g : P \to \mathbb{C}$ are two functions. Then the following two relations are equivalent. Relation (13.2) is known as the *inversion formula.*

$$f(\pi) = \sum_{\sigma \in P, \sigma \leq \pi} g(\sigma) \text{ for all } \pi \in P, \tag{13.1}$$

$$g(\pi) = \sum_{\sigma \in P, \sigma \leq \pi} f(\sigma)\mu[\sigma, \pi] \text{ for all } \pi \in P. \tag{13.2}$$

We need the following more general *partial inversion formula.*

DOI: 10.1201/9781003144496-13

Lemma 13.1.1 (Partial inversion formula). Let P be a finite lattice and let μ be the Möbius function on $P^{(2)}$. Suppose $f, g : P \to \mathbb{C}$ are two functions such that

$$f(\pi) = \sum_{\sigma \in P, \sigma \leq \pi} g(\sigma). \tag{13.3}$$

Then for all $\eta, \tau \in P$, $\eta \leq \tau$,

$$\sum_{\sigma \in P,\ \eta \leq \sigma \leq \tau} f(\sigma)\mu[\sigma,\ \tau] = \sum_{\pi \in P,\ \pi \vee \eta = \tau} g(\pi). \tag{13.4}$$

♦

Proof. Note that $\pi \leq \tau$ and $\eta \leq \tau$ if and only if $\pi \vee \eta \leq \tau$. Hence

$$\begin{aligned}
\sum_{\substack{\sigma \in P \\ \eta \leq \sigma \leq \tau}} f(\sigma)\mu[\sigma,\ \tau] &= \sum_{\substack{\sigma \in P \\ \eta \leq \sigma \leq \tau}} \sum_{\substack{\pi \in P \\ \pi \leq \sigma}} g(\pi)\mu[\sigma,\ \tau] \\
&= \sum_{\substack{\pi \in P \\ \pi \leq \tau}} g(\pi)\Big[\sum_{\substack{\sigma \in P \\ \pi \vee \eta \leq \sigma \leq \tau;}} \mu[\sigma,\ \tau]\Big].
\end{aligned}$$

Consider the sum within [ ] on the right for the two possible cases of π:

Case (i) $\pi \vee \eta = \tau$. Then there is only one term in the sum which equals

$$\mu[\sigma,\ \tau] = \mu[\tau,\ \tau] = 1.$$

Case (ii) $\pi \vee \eta < \tau$. In this case, (1.19) of Chapter 1 tells us that,

$$\sum_{\substack{\sigma \in P \\ \pi \vee \eta \leq \sigma \leq \tau}} \mu[\sigma,\ \tau] = 0 \text{ since } \pi \vee \eta < \tau.$$

Putting the two cases together,

$$\begin{aligned}
\sum_{\substack{\sigma \in P \\ \eta \leq \sigma \leq \tau}} f(\sigma)\mu[\sigma,\ \tau] &= \sum_{\substack{\pi \in P \\ \pi \leq \tau, \pi \vee \eta = \tau}} g(\pi) \\
&= \sum_{\substack{\pi \in P \\ \pi \vee \eta = \tau}} g(\pi).
\end{aligned}$$

This completes the proof. ■

We shall need the following important corollary.

Corollary 13.1.2. Let P be a finite lattice and let μ be the Möbius function on $P^{(2)}$. Then, for every $\eta \neq \mathbf{0}_P$, we have that

$$\sum_{\substack{\pi \in P \\ \pi \vee \eta = \mathbf{1}_P}} \mu[\mathbf{0}_P,\ \pi] = 0. \tag{13.5}$$

♦

Proof. Define the function $g : P \to \mathbb{C}$ by

$$g(\pi) = \mu[\mathbf{0}_P,\ \pi], \pi \in P.$$

Consider the function $f = g * \zeta$ where ζ, as introduced in Definition 1.4.2, is the inverse of μ. Then

$$\begin{aligned} f(\sigma) &= \sum_{\substack{\pi \in P \\ \pi \leq \sigma}} g(\pi) \\ &= \sum_{\substack{\pi \in P \\ \mathbf{0}_P \leq \pi \leq \sigma}} \mu[\mathbf{0}_P,\ \pi]\zeta[\pi, \sigma] \\ &= \mu * \zeta[\mathbf{0}_P,\ \sigma] \\ &= \begin{cases} 1 & \text{if } \sigma = \mathbf{0}_P, \\ 0 & \text{otherwise.} \end{cases} \end{aligned}$$

Now we apply (13.4) to the above (f, g), choosing $\tau = \mathbf{1}_P$. Then we get

$$\sum_{\sigma \in P,\ \eta \leq \sigma \leq \mathbf{1}_P} f(\sigma)\mu[\sigma,\ \mathbf{1}_P] = \sum_{\pi \in P,\ \pi \vee \eta = \mathbf{1}_P} g(\pi). \tag{13.6}$$

But since $\eta \neq \mathbf{0}_P$, each term $f(\sigma)$ in the left side of (13.6) equals 0. Hence the right side vanishes, and that completes the proof. ■

Suppose $(P_i,\ \leq_i), 1 \leq i \leq n$, are finite lattices. Then the ordering

$$(\pi_1, \ldots, \pi_n) \leq (\sigma_1, \ldots, \sigma_n) \Longleftrightarrow \pi_i \leq_i \sigma_i, i = 1, 2, \ldots n$$

makes $(P = P_1 \times \cdots \times P_n,\ \leq)$ a finite lattice and we will call it the *direct product* of $\{P_i\}$.

Recall the operations $\wedge$ and $\vee$ on finite lattices defined in Section 1.4. On P we define them by the component-wise operations on P_i. That is

$$\begin{aligned} (\pi_1, \ldots, \pi_n) \wedge (\sigma_1, \ldots, \sigma_n) &= (\pi_1 \wedge \sigma_1, \ldots, \pi_n \wedge \sigma_n), \\ (\pi_1, \ldots, \pi_n) \vee (\sigma_1, \ldots, \sigma_n) &= (\pi_1 \vee \sigma_1, \ldots, \pi_n \vee \sigma_n). \end{aligned}$$

Before we can state the factorization property for intervals in $NC(n)$ we need a definition.

Definition 13.1.1. (Lattice isomorphism) Let $(P_1,\ \leq_1)$ and $(P_2,\ \leq_2)$ be two lattices. A one-to-one onto map $h : P_1 \to P_2$ is said to be a *lattice isomorphism* if and only if

$$h(a \vee b) = h(a) \vee h(b) \text{ and } h(a \wedge b) = h(a) \wedge h(b) \text{ for all } a, b \in P_1.$$

In that case we say P_1 and P_2 are isomorphic.

The one-to-one onto map h is called an *anti-isomorphism* if

$$h(a \vee b) = h(a) \wedge h(b) \text{ and } h(a \wedge b) = h(a) \vee h(b) \text{ for all } a, b \in P.$$

Then we say that P_1 and P_2 are anti-isomorphic.

When h is an isomorphism or an anti-isomorphism, we write $P_1 \cong P_2$. ◇

Note that if h is a lattice isomorphism between P_1 and P_2 then it preserves the ordering:

$$a \leq_1 b \text{ implies } h(a) \leq_2 h(b).$$

On the other hand, if h is a lattice anti-isomorphism, then

$$a \leq_1 b \text{ implies } h(b) \leq_2 h(a).$$

For example the Kreweras complement $K : NC(n) \to NC(n)$ is an anti-isomorphism.

Also note that if P is a lattice, then so is $([\pi, \sigma], \leq)$ for any $\pi, \sigma \in P$, $\pi \leq \sigma$. Now we are ready to state the *canonical factorization result.*

Theorem 13.1.3 (Factorization of intervals in $NC(n)$)**.** For any $\pi, \sigma \in NC(n), \pi \leq \sigma$, there exists a sequence $k_1, \ldots, k_n$ of non-negative integers such that we have the lattice-isomorphism

$$[\pi, \sigma] \cong NC(1)^{k_1} \times NC(2)^{k_2} \times \cdots \times NC(n)^{k_n}.$$

◆

It may be noted that the order in which the product is written does not matter as far as the isomorphism is concerned.

Proof. Recall the notation $\pi|_V$ for the partition restricted to any subset V. Clearly, the following isomorphism holds:

$$[\pi, \sigma] \cong \prod_{V \in \sigma} [\pi|_V, \sigma|_V].$$

Now consider the order-preserving bijection between V and $\{1, \ldots, |V|\}$. This bijection identifies the interval $[\pi|_V, \sigma|_V]$ with an interval $[\tau, \mathbf{1}_{|V|}]$ for some $\tau \in NC(|V|)$. Thus we need to provide factorization for intervals of the form $[\tau, \mathbf{1}_k], k \geq 1$.

So, consider an interval $[\tau, \mathbf{1}_k]$. Consider the Kreweras complementation map K on $NC(k)$. By the properties of this map, $[\tau, \mathbf{1}_k]$ is *anti-isomorphic* to $[K(\mathbf{1}_k), K(\tau)] = [\mathbf{0}_k, K(\tau)]$.

But then we know that

$$\begin{aligned}
[\mathbf{0}_k, K(\tau)] &\cong \prod_{W \in K(\tau)} [\mathbf{0}|_W, K(\tau)|_W] \\
&\cong \prod_{W \in K(\tau)} NC(W) \\
&\cong \prod_{W \in K(\tau)} NC(|W|).
\end{aligned}$$

By applying the product of complementation map K, the above product is anti-isomorphic to itself. That is, we have obtained a factorization of $[\tau,\ \mathbf{1}_k]$ and the result is proved. ■

The proof of the following fact is left as an exercise: Suppose $P_1, \ldots, P_k$ are finite lattices with Möbius function $\mu_i, i = 1, \ldots, k$, respectively. Then the Möbius function μ on the finite lattice $P_1 \times \cdots \times P_k$ is given by

$$\mu[(\pi_1, \ldots, \pi_k), (\sigma_1, \ldots, \sigma_k)] = \prod_{i=1}^{k} \mu_i[\pi_i,\ \sigma_i].$$

Now consider the lattice $NC(n)$ and denote its Möbius function by μ_n. Let

$$s_n = \mu_n[\mathbf{0}_n,\ \mathbf{1}_n],\ n \geq 1. \tag{13.7}$$

It can be checked easily that

$$s_1 = 1, s_2 = -1, s_3 = 2.$$

We now state the following theorem.

Theorem 13.1.4 (Computation of Möbius function of $NC(n)$). For every $NC(n)$, $n \geq 1$,

$$s_n = \mu_n[\mathbf{0}_n,\ \mathbf{1}_n] = (-1)^{n-1} C_{n-1},$$

where $\{C_n\}$ are the Catalan numbers. As a consequence, for any $\pi, \sigma \in NC(n), \pi \leq \sigma$,

$$\mu_n[\pi,\ \sigma] = s_1^{k_1} \times s_2^{k_2} \times \cdots \times s_n^{k_n}$$

where $\{k_i\}$ are the exponents in the canonical factorization of $[\pi, \sigma]$. ◆

Proof. As noted above, a direct verification establishes the result for $n = 1, 2, 3$. Suppose $n \geq 4$. We shall use Corollary 13.1.2 with the following choice of η. Let

$$\eta = \{\{1\}, \{2\}, \ldots, \{n-2\}, \{n-1, n\}\}.$$

We first consider the set of partitions

$$\mathcal{S} = \{\pi \in NC(n) : \pi \vee \eta = \mathbf{1}_n\}.$$

Suppose $\pi \in \mathcal{S}$. Then it is easy to see that π cannot contain a block which is completely contained in $\{1, 2, \ldots, n-2\}$. So every block of π contains either $n-1$ or n. If $n-1$ and n belong to the same block, then it follows that $\pi = \mathbf{1}_n$.

If $n-1$ and n belong to different blocks, then π has exactly two blocks, say V_1 and V_2 where $n-1 \in V_1$ and $n \in V_2$. Let i be the smallest element of V_1. Since the blocks must be non-crossing, we must have

$$V_1 = \{i, \ldots, n-1\},\quad V_2 = \{1, \ldots i-1, n\}.$$

This implies that

$$S = \{\mathbf{1}_n, \pi_1, \ldots, \pi_{n-1}\},$$

where

$$\begin{aligned} \pi_1 &= \{\{1, \ldots, n-1\}, \{n\}\}, \\ \pi_i &= \{\{i, \ldots, n-1\}, \{1, \ldots, i-1, n\}, \ 2 \leq i \leq n-1. \end{aligned}$$

Now let us apply Corollary 13.1.2. Then

$$\mu_n[\mathbf{0}_n, \mathbf{1}_n] + \sum_{i=1}^{n} \mu_n[\mathbf{0}_n, \pi_i] = 0. \tag{13.8}$$

On the other hand, the canonical factorization of the intervals $[\mathbf{0}_n, \ \pi_i]$ are obviously given by

$$[\mathbf{0}_n, \ \pi_i] \cong NC(i) \times NC(n-i). \tag{13.9}$$

Hence, using (13.8) and (13.9),

$$s_n + \sum_{i=1}^{n-1} s_i s_{n-i} = 0, \ n \geq 4.$$

By direct computation, it is easily checked that this equation holds for $n = 2, 3$ also. To link these equations with those for the Catalan numbers, let

$$c_n := (-1)^n s_{n+1}.$$

Then the equations can be rewritten as

$$c_n - \sum_{i=1}^{n-1} c_{i-1} c_{n-i-1} = 0, \ n \geq 2.$$

But these are the same recursive equations that we have seen for the Catalan numbers in (2.5) of Chapter 2 with the boundary condition $c_1 = -s_2 = 1$. Hence $c_n = C_n$ and $s_n = (-1)^{n-1} C_{n-1}$. This completes the proof. ■

An easy but important consequence of the canonical factorization and the formula for s_n is the following.

Corollary 13.1.5. For any $\pi, \sigma \in NC(n), \ \pi \leq \sigma$,

$$|\mu_n[\pi, \ \sigma]| \leq 4^n$$

◆

Proof. Note that

$$\begin{aligned} |\mu_n[\pi, \ \sigma]| &= |s_1^{k_1} \cdots s_n^{k_n}| \\ &= C_2^{k_3} \cdots C_{n-1}^{k_n} \\ &\leq 4^n \end{aligned}$$

where the last inequality follows Exercise 1 of Chapter 2) and the fact that $k_2 + \cdots + k_n \leq n$. We omit the details. ■

13.2 Equivalence of two freeness definitions

In Chapter 3, we defined free independence in terms of the vanishing of the mixed free cumulants. Since there is a one-to-one correspondence between moments and free cumulants, there must be an equivalent definition that utilizes only the moments. The traditional way of defining free independence is indeed by using the moments.

We now give this definition and then show the equivalence of the two definitions.

Definition 13.2.1. (Free independence by moments) Let $(\mathcal{A}, \varphi)$ be an NCP and let I be a fixed index set. For each $i \in I$, let $\mathcal{A}_i$ be a unital sub-algebra of $\mathcal{A}$. Then $(\mathcal{A}_i, i \in I)$ are called *freely independent*, if

$$\varphi(a_1 \cdots a_k) = 0$$

for every $k \geq 1$ and for every $1 \leq j \leq k$, whenever (a) $a_j \in \mathcal{A}_{i(j)}$ $(i(j) \in I)$, (b) $\varphi(a_j) = 0$ for every $1 \leq j \leq k$ and (c) neighboring elements are from different sub-algebras, that is $i(1) \neq i(2) \neq \cdots i(k-1) \neq i(k)$. ◇

Note that NCP here refers to both types of non-commutative probability spaces—on algebras and $*$-algebras. As usual, variables in an NCP are said to be freely independent if the sub-algebras or, the $*$-algebras, as the case may be, generated by them are freely independent.

We now proceed to verify the equivalence of the concepts of freeness espoused in the above definition via moments and the one given earlier via free cumulants.

Theorem 13.2.1. Suppose $(\mathcal{A}, \varphi)$ is an NCP and $\mathcal{A}_i, i \in I$ are sub-algebras of $\mathcal{A}$. Then the following are equivalent:

(a) The sub-algebras are free according to Definition 3.1.1

(b) The sub-algebras are free according to Definition 13.2.1. ◆

Proof. (a) $\Rightarrow$ (b). Fix n and choose $a_j \in \mathcal{A}_{i(j)}$, $1 \leq j \leq n$ with $i(1) \neq i(2) \neq \cdots \neq i(n)$ and $\varphi(a_j) = 0$ for all $1 \leq j \leq n$. Recall the moment-free cumulant relation (2.17):

$$\begin{aligned}
\varphi(a_1 \cdots a_n) &= \sum_{\pi \in NC(n)} \kappa_\pi[a_1, \ldots, a_{n-1}, a_n] \\
&= \sum_{\pi \in NC(n)} \prod_{V \text{ is a block of } \pi} \kappa_{|V|}(a_{v_1}, \ldots, a_{v_k})
\end{aligned}$$

where $a_{v_1}, \ldots, a_{v_k}$ denote the elements of a typical block V of π.

Note that

$$\kappa_1(a_j) = \varphi(a_j) = 0, \quad \text{for all } 1 \leq j \leq n.$$

Hence,

$$\kappa_\pi[a_1, \ldots, a_{n-1}, a_n] = 0,$$

whenever π has at least one singleton block.

On the other hand, if all blocks of π have at least two elements, then there has to be at least one block, say V, with at least two consecutive integers. But since consecutive elements come from different sub-algebras, the mixed cumulant $\kappa_{|V|}(\cdot)$ vanishes. As a result, in this case too,

$$\kappa_\pi[a_1, \ldots, a_{n-1}, a_n] = 0.$$

Now by moment-free cumulant relation, $\varphi(a_1 \cdots a_n) = 0$.

(b) $\Rightarrow$ (a). We begin by noting that constants are trivially free of *any* sub-algebras, according to Definition 13.2.1, that is when (b) is assumed to hold. The proof of vanishing of mixed free cumulants stipulated in (a) is achieved in three steps.

Claim (i) *For all $n \geq 2$ and any $\{a_i, 1 \leq i \leq n\} \in \mathcal{A}$, $\kappa_n(a_1, \ldots, a_n) = 0$ whenever at least one of the $a_i = \mathbf{1}_\mathcal{A}$.*

To prove the above, for simplicity of writing, we assume that $\mathbf{1}_\mathcal{A}$ appears at the n-th position, that is, $a_n = \mathbf{1}_\mathcal{A}$.

We will use induction on n. For $n = 2$,

$$\kappa_2(a_1, \mathbf{1}_\mathcal{A}) = \varphi(a_1) - \varphi(a_1)\varphi(\mathbf{1}_\mathcal{A}) = 0.$$

Suppose the claim is true for all $k < n$. By the moment-free cumulant relation (2.17), we have

$$\begin{aligned}
\varphi(a_1 \cdots a_{n-1}\mathbf{1}_\mathcal{A}) &= \sum_{\pi \in NC(n)} \kappa_\pi[a_1, \ldots, a_{n-1}, \mathbf{1}_\mathcal{A}] \\
&= \kappa_n(a_1, \ldots, a_{n-1}, \mathbf{1}_\mathcal{A}) + \sum_{\substack{\pi \in NC(n) \\ \pi \neq \mathbf{1}_n}} \kappa_\pi[a_1, \ldots, a_{n-1}, \mathbf{1}_\mathcal{A}].
\end{aligned}$$

By induction hypothesis, a partition $\pi \neq \mathbf{1}_n$ cannot contribute unless $\{n\}$ is a singleton block. Hence continuing the above string of equalities, by the moment-free cumulant relation (2.15) and the fact that $\kappa_1(\mathbf{1}_\mathcal{A}) = 1$,

$$\begin{aligned}
\varphi(a_1 \cdots a_{n-1}\mathbf{1}_\mathcal{A}) &= \kappa_n(a_1, \ldots, a_{n-1}\mathbf{1}_\mathcal{A}) + \sum_{\sigma \in NC(n-1)} \kappa_\sigma[a_1, \ldots, a_{n-1}, \mathbf{1}_\mathcal{A}]\kappa_1(\mathbf{1}_\mathcal{A}) \\
&= \kappa_n(a_1, \ldots, a_{n-1}, \mathbf{1}_\mathcal{A}) + \varphi(a_1 \cdots a_{n-1}).
\end{aligned}$$

Hence $\kappa_n(a_1, \ldots, a_{n-1}, \mathbf{1}_\mathcal{A}) = 0$. This establishes Claim (i).

Claim (ii) *Suppose the sub-algebras are free according to Definition 13.2.1. Suppose $a_j \in \mathcal{A}_{i(j)}, 1 \leq j \leq n$ where $i(1) \neq i(2) \neq \cdots \neq i(n)$. Then $\kappa_n(a_1, \ldots, a_n) = 0$.*

By what we have shown above in Claim (i), we may assume without loss of generality that $\{a_j\}$ are centered. By the moment–free cumulant formula,

$$\kappa_n(a_1, \ldots, a_n) = \sum_{\pi \in NC(n)} \varphi_\pi[a_1, \ldots, a_n]\mu[\pi, \mathbf{1}_n], \tag{13.10}$$

where μ is the Möbius function on $NC(n)$.

If $\pi = \mathbf{1}_n$ then due to the alternating condition assumed, $\varphi_\pi[a_1, \ldots, a_n] = \varphi(a_1 \cdots a_n) = 0$. If $\pi \neq \mathbf{1}_n$, then there is at least one block, say V of π, of size at least two, and which consists of only consecutive integers. But then by condition (b), $\varphi_V(\cdot) = 0$, and hence $\varphi_\pi[\,\cdot\,] = 0$.

Thus all terms on the right side of the equality (13.10) vanish and hence $\kappa_n(a_1, \ldots, a_n) = 0$. This establishes Claim (ii).

Claim (iii) *Suppose the sub-algebras are free according to Definition 13.2.1. Suppose $\{a_i, 1 \leq i \leq n\}$, $n \geq 2$, are such that at least two of them come from different sub-algebras. Then $\kappa_n(a_1, \ldots, a_n) = 0$.*

By Claim (i), without loss of any generality, we may assume that $\{a_i\}$ are centered, so that $\varphi(a_i) = 0$ for all i.

Suppose $n = 2$ and a_1, a_2 are free. Then we know that $\varphi(a_1 a_2) = \varphi(a_1)\varphi(a_2)$. Hence

$$\kappa_2(a_1, a_2) = \varphi(a_1 a_2) - \varphi(a_1)\varphi(a_2) = 0.$$

So, (a) holds for $n = 2$. Now we shall use an induction argument, for which we shall need the following lemma.

This lemma is essentially Theorem 11.12 of Nica and Speicher (2006)[74]. Historically speaking, a special case of this result first appeared in Speicher (1994)[91]. It shall be also useful in the proof of Theorem 3.2.1 given in the next section.

Lemma 13.2.2. Fix $n \geq 2$ and $a_1, \ldots, a_n \in \mathcal{A}$. Fix $1 \leq i \leq n-1$ and let $\sigma \in NC(n)$ be the partition where $\{i, i+1\}$ is a block of σ and all other blocks are singletons. Then for all $\eta \in NC(n-1)$,

$$\kappa_\eta[a_1, \ldots, a_i a_{i+1}, \ldots, a_n] = \sum_{\substack{\pi \in NC(n) \\ \pi \vee \sigma = \eta}} \kappa_\pi[a_1, \ldots a_i, a_{i+1}, \ldots a_n]. \tag{13.11}$$

♦

Proof. To prove (13.11), it is enough to prove its cumulative version—for all $\zeta \in NC(n-1)$ we shall show that,

$$\sum_{\eta \leq \zeta} \kappa_\eta[a_1, \ldots, a_i a_{i+1}, \ldots, a_n] = \sum_{\eta \leq \zeta} \sum_{\substack{\pi \in NC(n) \\ \pi \vee \sigma = \eta}} \kappa_\pi[a_1, \ldots a_i, a_{i+1}, \ldots a_n]. \tag{13.12}$$

But then, by the moment-free cumulant relation, the left side of (13.12) yields

$$\sum_{\eta \leq \zeta} \kappa_\eta[a_1, \ldots, a_i a_{i+1}, \ldots, a_n] = \varphi_\zeta[a_1, \ldots, a_i a_{i+1}, \ldots, a_n]. \tag{13.13}$$

Let us now explore the right side of (13.12). Let

$$f : \{1, \ldots, n\} \to \{1, \ldots, n-1\}$$

be the function defined by $f(i) = f(i+1) = i$, $f(j) = j$ for all $j \leq i$ and $f(j) = j-1$ for all $j \geq i+2$. Note that f is monotone and onto. Let $\zeta' \in NC(n)$ be the unique partition whose blocks are of the form $f^{-1}(V)$ with V a block of ζ. Now the sum on the right side of (13.12) simplifies to

$$\begin{aligned} \sum_{\eta \leq \zeta} \sum_{\substack{\pi \in NC(n) \\ \pi \vee \sigma = \eta}} \kappa_\pi[a_1, \ldots a_i, a_{i+1}, \ldots a_n] &= \sum_{\substack{\pi \in NC(n), \\ \pi \leq \zeta'}} \kappa_\pi[a_1, \ldots a_i, a_{i+1}, \ldots a_n] \\ &= \varphi_{\zeta'}(a_1, \ldots, a_n). \end{aligned} \tag{13.14}$$

However, by construction,

$$\varphi_\zeta[a_1, \ldots, a_i a_{i+1} \cdots a_n] = \varphi_{\zeta'}[a_1, \ldots, a_i, a_{i+1} \cdots a_n].$$

Hence the quantities in (13.13) and (13.14) are equal and, as a consequence, (13.12) holds and the lemma is proved. ∎

Now we continue with the proof of Claim (iii). We assume that the claim has been proved up to $n-1$.

In view of Claim (ii), it remains to consider only the case where two sub-algebras are consecutive—that is, there exists an $i \in \{1, \ldots, n-1\}$ such that $j(i) = j(i+1)$. Let $\sigma \in NC(n)$ be as in Lemma 13.2.2— it has one block as $\{i, i+1\}$ and all other blocks are singletons. We use (13.11) with choice of $\eta = \mathbf{1}_{n-1}$. After rearranging terms, it yields,

$$\begin{aligned} \kappa_n(a_1, \ldots, a_i, a_{i+1}, \ldots, a_n) &= \kappa_{n-1}(a_1, \ldots, a_i a_{i+1}, \ldots, a_n) \\ &\quad - \sum_{\substack{\mathbf{1}_n \neq \pi \in NC(n) \\ \pi \vee \sigma = \mathbf{1}_n}} \kappa_\pi[a_1, \ldots a_i, a_{i+1}, \ldots a_n]. \end{aligned}$$

By induction, the first term on the right side vanishes.

Now it is not hard to argue, as we do below, that each product term of the sum in the second term also vanishes. Since $\pi \neq \mathbf{1}_n$, it has at least two blocks.

If i and $i+1$ are in different blocks of π, then by multiplicativity of κ_π and induction hypothesis, at least one of the sub-products equals 0, and hence this π does not contribute to the sum.

If i and $i+1$ are in the same block of π, call this block V. If there is another index in this block, then again by induction this block contributes 0 by induction hypothesis. If $\{i, i+1\}$ is a doubleton block of π, then $\pi \vee \sigma$ cannot equal $\mathbf{1}_n$. Hence this case does not arise.

Thus the entire right side of the above equation vanishes and then we have estabsihed Claim (iii). This completes the proof of the theorem. ∎

13.3 Free product construction

The following theorem was stated in Section 3.2. We now outline a proof which follows the arguments of Bose and Mukherjee (2020)[25] and uses the free cumulant-based Definition 3.1.1 of free independence as the starting point. See Nica and Speicher (2006)[74] for a detailed proof which uses the moment-based based Definition 13.2.1.

Theorem 13.3.1. (Theorem 3.2.1) Let $(\mathcal{A}_i, \varphi^{(i)})_{i\in I}$ be a family of NCP ($*$ or otherwise). Then accordingly, there exists an NCP $(\mathcal{A}, \varphi)$, called the $*$-*free* (or free) product of $(\mathcal{A}_i, \varphi^{(i)})_{i\in I}$, such that there is a copy of $\mathcal{A}_i$ in $\mathcal{A}, i \in I$ which are freely independent in $(\mathcal{A}, \varphi)$ and $\varphi|\mathcal{A}_i = \varphi^{(i)}$ for al $i \in I$. ◆

Proof. We demarcate the proof into a few steps:

Step 1. Construction of the free product algebra $\mathcal{A}$. This method is quite standard. Let the set of centered elements of $\mathcal{A}_i$ be defined as

$$\mathcal{A}_i^\circ = \{a^\circ : a^\circ = a - \varphi^{(i)}(a)\mathbf{1}_{\mathcal{A}_i}, a \in \mathcal{A}_i\},\ i \in I.$$

Identify all the identities $\mathbf{1}_{\mathcal{A}_i}$ as a single identity $\mathbf{1}_{\mathcal{A}} = \mathbf{1}$. Define the algebra $\mathcal{A}$ which is the *free product* of $\{\mathcal{A}_i\}$ as

$$\mathcal{A} = \mathbb{C}\mathbf{1} \bigoplus_{k\geq 1} \bigoplus_{\substack{i_1\neq i_2\neq\cdots\neq i_k \\ i_j\in I, 1\leq j\leq k}} \mathcal{A}_{i_1}^\circ \otimes \mathcal{A}_{i_2}^\circ \otimes \cdots \otimes \mathcal{A}_{i_k}^\circ.$$

As an example, if $a_1 \in \mathcal{A}_1$ and $a_2 \in \mathcal{A}_2$, then a_1a_2 will be defined to be the element

$$\varphi^{(1)}(a_1)\varphi^{(2)}(a_2) \oplus \varphi^{(1)}(a_1)a_2^\circ \oplus \varphi^{(2)}a_1^\circ \oplus a_1^\circ \otimes a_2^\circ$$

of $\mathcal{A}$. Note that *the algebras $\mathcal{A}_i$ are identified as sub-algebras of $\mathcal{A}$ in the obvious way and shall be referred to as such from now on. Moreover, they now share the same identity element.*

Multiplication on $\mathcal{A}$. This is defined by concatenation and then reduction as follows:

(a) multiplication by scalars is defined in the obvious way.

(b) If $a_{i_1}^\circ \otimes \cdots \otimes a_{i_k}^\circ \in \mathcal{A}_{i_1}^\circ \otimes \cdots \otimes \mathcal{A}_{i_k}^\circ$, and $a_{j_1}^\circ \otimes \cdots \otimes a_{j_\ell}^\circ \in \mathcal{A}_{j_1}^\circ \otimes \cdots \otimes \mathcal{A}_{j_\ell}^\circ$, then define

$$(a_{i_1}^\circ \otimes \cdots \otimes a_{i_k}^\circ)(a_{j_1}^\circ \otimes \cdots \otimes a_{j_\ell}^\circ) =$$
$$\begin{cases} a_{i_1}^\circ \otimes \cdots \otimes a_{i_k}^\circ \otimes a_{j_1}^\circ \otimes \cdots \otimes a_{j_\ell}^\circ \in \mathcal{A}_{i_1}^\circ \otimes \cdots \otimes \mathcal{A}_{i_k}^\circ \otimes \mathcal{A}_{j_1}^\circ \otimes \cdots \otimes \mathcal{A}_{j_\ell}^\circ \\ \qquad \text{if } i_k \neq j_1, \\ a_{i_1}^\circ \otimes \cdots \otimes (a_{i_k}a_{j_1})^\circ \otimes \cdots \otimes a_{j_\ell}^\circ + \varphi_{i_k}(a_{i_k}a_{j_1})(a_{i_1}^\circ \otimes \cdots a_{i_{k-1}}^\circ)(a_{j_2}^\circ \cdots \otimes a_{j_\ell}) \\ \qquad \text{if } i_k = j_1, \end{cases}$$

where the smaller order term $(a^\circ_{i_1} \otimes \cdots \otimes a^\circ_{i_{k-1}})(a^\circ_{j_2} \otimes \cdots \otimes a^\circ_{j_\ell})$ is defined via the same rule.

$*$-*operation.* If the $\mathcal{A}_i$ are $*$-algebras, then the $*$-operation on $\mathcal{A}$ is defined in the natural way. Also, note that we always work with the obvious *minimal representation* of a product of variables.

Step 2. Construction of the free cumulant functions κ_n. Let $\kappa_n^{(l)}$ denote the free cumulant functions corresponding to $(\mathcal{A}_l, \varphi^{(l)})$, $l \in I$. We shall now define free cumulant functions $\kappa_n : \mathcal{A}^n \to \mathbb{C}$, $n \geq 1$ making sure that κ_n agrees with $\kappa_n^{(l)}$ on each $\mathcal{A}_l^n$, $l \in I$. Moreover, all mixed free cumulants must vanish. Thus, we are compelled to begin our definition of free cumulants as follows: for every n,

$$\kappa_n(a_1, \ldots a_n) := \begin{cases} \kappa_n^{(l)}(a_1, \ldots, a_n) & \text{if all } \; a_i \in \mathcal{A}_l \; \text{ for some } \; l, \\ 0 \text{ if at least two } a_i\text{'s are from different sub-algebras.} \end{cases} \tag{13.15}$$

Note that these κ_n satisfy the obvious restricted multi-linearity.

Call $\{a_i\}$, $1 \leq i \leq n$ "pure", if all of them belong to one of the sub-algebras $\mathcal{A}_j$. For pure $\{a_i\}$ define κ_π by the obvious multiplicative extension of κ_n defined above, as

$$\kappa_\pi[a_1, \ldots, a_n] = \prod_{V=\{i_1,\ldots,i_s\}\in\pi} \kappa_{|V|}(a_{i_1}, \ldots, a_{i_s}), \tag{13.16}$$

where the right side is already defined due to (13.15).

We shall now use (13.15) and (13.16) to define all free cumulants. Let $b_1, \ldots, b_m$ be monomials in pure elements. For simplicity, write

$$b_j = a_{s_{j-1}+1} \cdots a_{s_j},$$

i.e. b_j is the product of $s_j - s_{j-1}$ pure elements, with $i_u \neq i_{u+1}$ for all $s_{j-1} \leq u \leq s_j - 1$. Now if we recall equation (13.11) from Lemma 13.2.2, the following definition is again compelled upon us. Define

$$\kappa_m(b_1, \ldots, b_m) := \sum_{\substack{\pi \in NC(s_m) \\ \pi \vee \sigma = \mathbf{1}_{s_m}}} \kappa_\pi[a_{i_{s_1}}, \ldots, a_{i_{s_1}}, \ldots, a_{i_{s_{m-1}+1}}, \ldots, a_{i_{s_m}}], \tag{13.17}$$

where $\sigma := \{\{1, \ldots, s_1\}, \ldots, \{s_{m-1}+1, \ldots, s_m\}\} \in NC(s_m)$ and $\mathbf{1}_{s_m}$ is the single block partition of $NC(s_m)$. Note that each κ_π on the right side of (13.17) has already been defined by (13.16) so this is a valid definition.

Now for arbitrary $\pi \in NC(m)$ we define

$$\kappa_\pi[b_1, \ldots, b_m] := \prod_{V=\{i_1,\ldots,i_s\}\in\pi} \kappa_{|V|}(b_{i_1}, \ldots, b_{i_s}). \tag{13.18}$$

Again, each term on the right side of (13.18) has already been defined in (13.17).

Now, by multi-linearity, we extend this definition to $\kappa_\pi[b_1, \ldots, b_m]$ where b_i are polynomials instead of monomials.

In particular, the functions $\kappa_n : \mathcal{A}^n \to \mathbb{C}$ are now completely defined for all n. Moreover, by construction, they are multi-linear and agree with $\kappa_n^{(l)}$ on $\mathcal{A}_l^n$ for all $l \in I$. It is also clear that

$$k_1(a^*) = \overline{k_1(a)} \text{ for all } a \in \mathcal{A}. \tag{13.19}$$

Step 3. Construction of the state φ on $\mathcal{A}$: Now, in view of equation (2.18), we are compelled to define the moment functional (state) φ on $\mathcal{A}$ by

$$\varphi_\pi[a_1, a_2, \ldots, a_n] = \sum_{\sigma \in NC(n),\ \sigma \leq \pi} \kappa_\sigma[a_1, a_2, \ldots, a_n],\ \pi \in NC(n),\ n \geq 1. \tag{13.20}$$

In particular $\varphi(a) = \kappa_1(a)$. Clearly φ is a state since it is linear and $\varphi(\mathbf{1}) = \kappa_1(\mathbf{1}) = 1$. Hence $(\mathcal{A}, \varphi)$ is an NCP. Moreover, by construction

$$\varphi(a) = \kappa_1(a) = \kappa_1^{(l)}(a) = \varphi^{(l)}(a) \text{ for all } a \in \mathcal{A}_l,\ l \in I.$$

Also note that κ_n are indeed the free cumulant functions corresponding to φ due to the relation (13.20).

Now since all mixed free cumulants vanish (see (13.15)), the algebras $\mathcal{A}_i$, as sub-algebras of $\mathcal{A}$, are free with respect to the state φ. This proves the theorem except that we need to check positivity when $*$-algebras are involved.

Step 4. Positivity for $*$-probability space. For $*$-algebras, we have to show that $\varphi(a^*a) \geq 0$ for all $a \in \mathcal{A}$.

By the moment-free cumulant relation we have, for all $a \in \mathcal{A}$,

$$\varphi(a^*a) = \kappa_2(a^*, a) + \kappa_1(a^*)\kappa_1(a).$$

As noted in (13.19), $\kappa_1(a^*) = \overline{\kappa_1(a)}$. Hence the second term on the right side above is non-negative. Hence we only need to check that the "variance" $\kappa_2(a^*, a) \geq 0$. We now prove two lemmas which will help us accomplish this.

Lemma 13.3.2. For any $a = a_{i_1}^\circ \otimes \cdots \otimes a_{i_k}^\circ \in \mathcal{A}_{i_1}^\circ \otimes \cdots \otimes \mathcal{A}_{i_k}^\circ$, we have $\kappa_2(\mathbf{1}, a) = \kappa_2(a, \mathbf{1}) = 0$. Here $\mathbf{1}$ denotes the identity of $\mathcal{A}$. ♦

Proof. By (13.17), we have

$$\begin{aligned}\kappa_2(\mathbf{1}, a) &= \kappa_2(\mathbf{1}, a_{i_1}^\circ \otimes \cdots \otimes a_{i_k}^\circ) \\ &= \sum_{\substack{\pi \in NC(k+1) \\ \pi \vee \sigma = \mathbf{1}_{k+1}}} \kappa_\pi[\mathbf{1}, a_{i_1}^\circ, \ldots, a_{i_k}^\circ],\end{aligned}$$

where $\sigma = \{\{1\}, \{2, \ldots, k+1\}\}$.

Note that each π in the above sum must have $\mathbf{1}$ in a block, say $V_{\pi,1}$, of length ≥ 2. Suppose

$$V_{\pi,1} = \{1, j_1, \ldots, j_t\}.$$

If $i_{j_1} = \cdots = i_{j_t}$, then

$$\kappa_{|V_{\pi,\mathbf{1}}|}(\mathbf{1}, a^{\circ}_{i_{j_1}}, \ldots, a^{\circ}_{i_{j_t}}) = \kappa^{(i_{j1})}_{|V_{\pi,\mathbf{1}}|}(\mathbf{1}, a^{\circ}_{i_{j_1}}, \ldots, a^{\circ}_{i_{j_t}}) = 0,$$

because constants are free of non-constant elements in the original subalgebras $\mathcal{A}_l$.

Otherwise, $t > 1$ and there exist $1 \leq t' < t'' \leq t$ such that $i_{j_{t'}} \neq i_{j_{t''}}$. Then, by definition (13.15),

$$\kappa_{|V_{\pi,\mathbf{1}}|}(1, a^{\circ}_{i_{j_1}}, \ldots, a^{\circ}_{i_{j_t}}) = 0.$$

It follows that

$$\kappa_{\pi}[1, a^{\circ}_{i_1}, \ldots, a^{\circ}_{i_{s_i}}] = 0$$

for each π such that $\pi \vee \sigma = \mathbf{1}_{k+1}$.

Thus $\kappa_2(\mathbf{1}, a) = 0$, and, by a similar argument, so is $\kappa_2(a, \mathbf{1})$. ■

Lemma 13.3.3. Let $a = a^{\circ}_{i_1} \otimes \cdots \otimes a^{\circ}_{i_k} \in \mathcal{A}^{\circ}_{i_1} \otimes \cdots \otimes \mathcal{A}^{\circ}_{i_k}$, and $b = b^{\circ}_{j_1} \otimes \cdots \otimes b^{\circ}_{j_\ell} \in \mathcal{B}^{\circ}_{j_1} \otimes \cdots \otimes \mathcal{B}^{\circ}_{j_\ell}$. Then $\kappa_2(a^*, b)$ is non-zero only if $k = \ell$ and $i_u = j_u$ for all $1 \leq u \leq k$, in which case

$$\kappa_2(a^*, b) = \prod_{u=1}^{k} \kappa_2((a^{\circ}_{i_u})^*, b^{\circ}_{j_u}). \tag{13.21}$$

In particular, $\kappa_2(a^*, a) \geq 0$. ♦

Proof. We have, by (13.17), that

$$\kappa_2(a^*, b) = \sum_{\substack{\pi \in NC(k+\ell) \\ \pi \vee \sigma = \mathbf{1}_{k+\ell}}} \kappa_{\pi}[(a^{\circ}_{i_k})^*, \ldots, (a^{\circ}_{i_1})^*, b^{\circ}_{j_1}, \ldots, b^{\circ}_{j_\ell}] \tag{13.22}$$

where $\sigma = \{\{1, \ldots, k\}, \{k+1, \ldots, k+\ell\}\}$.

Suppose that $\kappa_2(a^*, b) \neq 0$. Consider a partition π appearing in (13.22) for which the corresponding $\kappa_\pi \neq 0$. Clearly π cannot be $\mathbf{1}_{k+\ell}$, for then κ_π would be zero since variables are centered.

So, assume that $\pi \neq \mathbf{1}_{k+\ell}$. Take two indices $i, j \in \{1, \ldots, k+\ell\}$ that are in the same block, with the smallest possible value for $j - i > 0$. Now, by freeness, we must have that the variables corresponding to i, j come from the same algebra in order for κ_π to be non-zero.

If $j - i > 1$, then, there cannot exist another index $i' \in \{i+1, \ldots, j-1\}$ which belongs to the same partition block as i and j, for then we would violate the minimality of $j-i$. For the same reason, no two $i', j' \in \{i+1, \ldots, j-1\}$ can belong to the same block of π. We conclude that all entries in $\{i+1, \ldots, j-1\}$ must constitute singleton blocks. But this would force κ_π to be zero by centeredness. It follows that we must have $j = i+1$.

Now, since consecutive elements of both a and b come from different algebras, in order for κ_π to be non-zero, i, j cannot both correspond to elements of a, or of b—the only possibility is that $i = k$ and $j = k+1$ and that $i_1 = j_1$.

Now, there cannot exist an $\bar{i} < i$ (resp. $\bar{j} > j$) such that $\bar{i}, i$ (resp. $j, \bar{j}$) are in the same block of π, for otherwise there would exist an $\bar{i}$ (resp. $\bar{j}$) with $i - \bar{i}$ (resp. $\bar{j} - j$) being the smallest, and then by the argument of the above paragraph, $i = \bar{i} + 1$ (resp. $\bar{j} = j + 1$) which would make k_π zero as $\bar{i}, i$ both correspond to elements of a (resp. $j, \bar{j}$ to elements of b). This means that $\{i, j\} = \{k, k+1\}$ is a block of π, and hence

$$\begin{aligned}\kappa_\pi[(a_{i_k}^\circ)^*, \ldots, (a_{i_1}^\circ)^*, b_{j_1}^\circ, \ldots, b_{j_\ell}^\circ] &= \kappa_2((a_{i_1}^\circ)^*, b_{j_1}^\circ) \times \\ &\qquad \kappa_{\pi \backslash \{k,k+1\}}[(a_{i_k}^\circ)^*, \ldots, (a_{i_2}^\circ)^*, b_{j_2}^\circ, \ldots, b_{j_\ell}^\circ].\end{aligned}$$

Noting that $\pi \backslash \{k, k+1\} \in NC(k+\ell-2)$, we can apply the same argument to conclude that $i_2 = j_2$ and $\{k-1, k+2\}$ is a partition block of π and so on. It is therefore clear that we must have $k = \ell$ and $i_u = j_u$ for all $1 \le u \le k$ and that only the partition $\pi_{2k} = \{\{1, 2k\}, \{2, 2k-1\}, \ldots, \{k, k+1\}\}$ has a non-zero κ_π. Since π_{2k} does appear in the sum in (13.22), (13.21) follows. The last assertion follows because

$$\kappa_2(a^*, a) = \prod_{u=1}^{k} \kappa_2((a_{i_u}^\circ)^*, a_{i_u}^\circ)) = \prod_{u=1}^{k} \kappa_2^{(i_u)}((a_{i_u}^\circ)^*, a_{i_u}^\circ)$$

and the variances $\kappa_2^{(i_u)}((a_{i_u}^\circ)^*, (a_{i_u}^\circ))$ are all non-negative. ■

Now we prove a lemma that handles elements a which are sums of elements of a particular $\mathcal{A}_{i_1}^\circ \otimes \cdots \otimes \mathcal{A}_{i_k}^\circ$.

Lemma 13.3.4. Let $a = \sum_{t=1}^{r} a_t$, where $a_t = a_{t,i_1}^\circ \otimes \cdots \otimes a_{t,i_k}^\circ \in \mathcal{A}_{i_1}^\circ \otimes \cdots \otimes \mathcal{A}_{i_k}^\circ$. Then $\kappa_2(a^*, a) \ge 0$. ♦

Proof. By multilinearity and Lemma 13.3.3,

$$\begin{aligned}\kappa_2(a^*, a) &= \sum_{1 \le s,t \le r} \kappa_2(a_s^*, a_t) \\ &= \sum_{1 \le s,t \le r} \prod_{u=1}^{k} \kappa_2((a_{s,i_u}^\circ)^*, a_{t,i_u}^\circ) =: u^\top M u,\end{aligned}$$

where u is the column vector of dimension r with all elements as 1, and M is the $r \times r$ matrix with entries $M_{st} = \prod_{u=1}^{k} \kappa_2((a_{s,i_u}^\circ)^*, a_{t,i_u}^\circ)$. It therefore suffices to show that M is positive semi-definite. But M can be written as

$$M = M_1 \odot \cdots \odot M_k,$$

where $(M_u)_{st} = \kappa_2((a_{s,i_u}^\circ)^*, a_{t,i_u}^\circ)$ and $\odot$ denotes the Schur-Hadamard product of matrices, where multiplication is element-wise.

The Schur product theorem (Schur (1911)[86]) says that the Schur-Hadamard product of positive semi-definite matrices is positive semi-definite.

So it is enough to show that each M_u is positive semi-definite. This follows since, for any $x' = (x_1, \ldots, x_r) \in \mathbb{C}^r$,

$$\begin{aligned}
x^* M_u x &= \sum_{s,t} \bar{x}_s x_t \kappa_2((a^\circ_{s,i_u})^*, a^\circ_{t,i_u}) \\
&= \kappa_2\Big(\big(\sum_s x_s a^\circ_{s,i_u}\big)^*, \sum_s x_s a^\circ_{s,i_u}\Big) \\
&= \kappa_2(w^*, w) \text{ (where } w = \sum_s x_s a^\circ_{s,i_u} \in \mathcal{A}^\circ_{i_u}) \\
&= \kappa_2^{(i_u)}(w^*, w) \text{ (since } \kappa_2 \text{ agrees with } \kappa_2^{(i_u)} \text{ on } \mathcal{A}^\circ_{i_u}) \\
&\geq 0 \text{ (since } \varphi^{(i_u)}(\cdot) \text{ is positive).}
\end{aligned}$$

This proves the lemma. ■

Now we consider a general element $a \in \mathcal{A}$, which can be expressed, for some $r \geq 0$, as

$$a = c\mathbf{1} \oplus \bigoplus_{i=1}^{r} a_i,$$

where $c \in \mathbb{C}$ and each a_i is a sum of some elements in $\mathcal{A}^\circ_{i_1} \otimes \cdots \otimes \mathcal{A}^\circ_{i_{s_i}}$, that is

$$a_i = \sum_{j=1}^{t_i} a_{ij},\ a_{ij} \in \mathcal{A}^\circ_{i_1} \otimes \cdots \otimes \mathcal{A}^\circ_{i_{s_i}}, 1 \leq j \leq t_i, 1 \leq i \leq r,$$

and the hosts $\mathcal{A}^\circ_{i_1} \otimes \cdots \otimes \mathcal{A}^\circ_{i_{s_i}}$ *are all different.*

Expanding κ_2 by multi-linearity,

$$\begin{aligned}
\kappa_2(a^*, a) \;\; = \;\; & |c|^2 \kappa_2(\mathbf{1}, \mathbf{1}) + \bar{c} \sum_i \kappa_2(\mathbf{1}, a_i) \\
& + c \sum_i \kappa_2(a_i^*, \mathbf{1}) + \sum_{i \neq i'} \kappa_2(a_i^*, a_{i'}) + \sum_i \kappa_2(a_i^*, a_i). \quad (13.23)
\end{aligned}$$

Note that $\kappa_2(\mathbf{1}, \mathbf{1}) = 0$ as κ_2 agrees with $\kappa_2^{(l)}$ for $l \in I$. By Lemmas 13.3.2 and 13.3.3 we see that all but the last term in the right side of (13.23) are zero. The last sum itself is non-negative by Lemma 13.3.4.

This completes the proof of the theorem. ■

Remark 13.3.1. Theorem 6.4.1 stated in Section 6.4 of Chapter 6 on the free product of C^*-algebras is more difficult to prove. See Nica and Speicher (2006)[74] for a detailed proof. ●

13.4 Exercises

1. Suppose $(P, \leq)$ is a POSET. Show that for any $\pi, \sigma \in P, \pi \leq \sigma$, $([\pi, \sigma], \leq)$ is a POSET and is a lattice if P is a lattice.

2. Suppose $P_1, \ldots, P_k$ are finite lattices with Möbius function $\mu_i, i = 1, \ldots, k$ respectively. Then verify that the Möbius function μ on the finite lattice $P_1 \times \cdots \times P_k$ is given by

$$\mu[(\pi_1, \ldots, \pi_k), (\sigma_1, \ldots, \sigma_k)] = \prod_{i=1}^{k} \mu_i[\pi_i, \sigma_i].$$

3. Consider the Möbius function $\mu[\cdot, \cdot]$ on $NC(n), n \geq 1$. Let $s_n = \mu_n[\mathbf{0}_n, \mathbf{1}_n], n \geq 1$. Show directly that $s_1 = 1, s_2 = -1, s_3 = 2$.

4. Verify that in the proof Theorem 13.3.1, the defined φ and κ satisfy the moment-free cumulant relation

$$\varphi(a_1, \ldots, a_n) = \sum_{\pi \in NC(n)} \kappa_\pi[a_1, \ldots, a_n].$$

5. Suppose in the statement of Theorem 13.3.1, we assume that all the states $\{\varphi^{(i)}, i \in I\}$ are faithful. Show that then φ is also faithful.

6. Suppose in the statement of Theorem 13.3.1, we assume that all the states $\{\varphi^{(i)}, i \in I\}$ are tracial. Show that then φ is also tracial.

Bibliography

[1] G. Anderson, A. Guionnet, and O. Zeitouni. *An Introduction to Random Matrices.* Cambridge University Press, Cambridge, UK, 2009. 15, 76, 107, 222

[2] O. Arizmendi and A Jaramillo. Convergence of the fourth moment and infinite divisibility: quantitative estimates. *Electron. Commun. Probab.*, 19(26):1–12, 2014. 81

[3] L. Arnold. On the asymptotic distribution of the eigenvalues of random matrices. *J. Math. Anal. Appl.*, 20:262–268, 1967. 146

[4] Z. D. Bai. Methodologies in spectral analysis of large-dimensional random matrices, a review. *Statist. Sinica*, 9(3):611–677, 1999. With comments by G. J. Rodgers and Jack W. Silverstein; and a rejoinder by the author. 146, 147

[5] Z. D. Bai and J. W. Silverstein. *Spectral Analysis of Large Dimensional Random Matrices.* Springer Series in Statistics. Springer, New York, second edition, 2010. 4, 67

[6] Z. D. Bai and Y. Q. Yin. Convergence to the semicircle law. *Ann. Probab.*, 16(2):863–875, 1988. 147

[7] Z. D. Bai and W. Zhou. Large sample covariance matrices without independence structures in columns. *Statist. Sinica*, 18(2):425–442, 2008. 147

[8] J. Baik, G. Akemann, and F. Di Francesco, Eds. *The Oxford Handbook of Random Matrix Theory.* Oxford University Press, 2011. 107

[9] T. Banica, S. Curran, and R. Speicher. de Finetti theorems for easy quantum groups. *Ann. Probab.*, 40(1):401–435, 2012. 168

[10] O. E. Barndorff-Nielsen and S. Thorbjørnsen. A connection between free and classical infinite divisibility. *Infin. Dimens. Anal. Qu.*, 7:573–590, 2004. 81

[11] S. Belinschi, P. Śniady, and R. Speicher. Eigenvalues of non-Hermitian random matrices and Brown measure of non-normal operators: Hermitian reduction and linearization method. *arXiv preprint arXiv:1506.02017*, 2015. 228

[12] F. Benaych-Georges. Rectangular random matrices, entropy, and Fisher's information. *J. Operator Theory*, 62(2):371–419, 2009. 223

[13] F. Benaych-Georges. Rectangular random matrices, related convolution. *Probab. Theory Related Fields*, 144(3-4):471–515, 2009. 223

[14] H. Bercovici and V. Pata. A free analogue of Khintchine's characterization of infinite divisibility. *Proc. Am. Math. Soc.*, 128:1011–1015, 2000. 81

[15] H. Bercovici and D. Voiculescu. Lévy-Hinčin type theorems for multiplicative and additive free convolution. *Pacific Journal of Mathematics*, 153(2), 1992. 83

[16] M. Bhattacharjee and A. Bose. Asymptotic freeness of sample covariance matrices via embedding. 2021. arXiv:2101.06481. 197

[17] M. Bhattacharjee, A. Bose, and A. Dey. Joint convergence of sample cross-covariance matrices. 2021. arXiv:2103.11946. 200, 206

[18] P. Biane. On the free convolution with a semi-circular distribution. *Indiana Univ. Math. J.*, 46(3):705–718, 1997. 222

[19] C. Bordenave and D. Chafai. Around the circular law. *Probability Surveys*, 9:1–89, 2012. 147, 228

[20] A. Bose. *Patterned Random Matrices.* CRC Press, Boca Raton, 2018. 4, 110, 121, 146, 147, 148, 176, 205, 211

[21] A. Bose and M. Bhattacharjee. *Large Covariance and Autocovariance Matrices.* CRC Press, Boca Raton, 2018. 186

[22] A. Bose, A. DasGupta, and H. Rubin. A contemporary review and bibliography of infinite divisible distributions. *Sankhyā, Special issue in memory of D.Basu*, 64(3):763–819, 2002. 80

[23] A. Bose and A. Dey. U-statistics clt using cumulants and a free version. *Statistics and Applications (New Series)*, 18(2):275–286, 2020. Special volume on the occasion of the 75th birthday of twin statisticians Bimal Sinha and Bikas Sinha. 60

[24] A. Bose, A. Dey, and W. Ejsmont. Characterization of non-commutative free gaussian variables. *ALEA: Latin American Journal of Probability and Mathematical Statistics*, 15(2):1241–1255, 2018. 48

[25] A. Bose and S. Mukherjee. Construction of product $*$-probability space via free cumulants, 2020. arXiv:2007.10989. 243

[26] A. Bose and K. Saha. *Random Circulant Matrices.* CRC Press, Boca Raton, 2018. 147

[27] A. Bose, K. Saha, A. Sen, and P. Sen. Random matrices with independent entries: beyond non-crossing partitions. 2021. arXiv:2103.09443. 146

[28] L. G. Brown. *Lidskii's Theorem in the Type II Case, Geometric Methods in Operator Algebras*, volume 123. Pitman Res. Notes Math. Ser Longman Sci. Tech., Harlow, 1986. xxii, 225

[29] W. Bryc, A. Dembo, and T. Jiang. Spectral measure of large random Hankel, Markov and Toeplitz matrices. *Ann. Probab.*, 34(1):1–38, 2006. 147

[30] M. Capitaine and M. Casalis. Asymptotic freeness by generalized moments for Gaussian and Wishart matrices. Application to beta random matrices. *Indiana Univ. Math. J.*, 53(2):397–431, 2004. 212, 222

[31] M. Capitaine and C. Donati-Martin. Strong asymptotic freeness for Wigner and Wishart matrices. *Indiana Univ. Math. J.*, 56(2):767–803, 2007. 222

[32] T. Carleman. *Les Fonctions Quasi Analytiques (in French).* Paris: Gauthier-Villars, 1926. 4

[33] B. Collins. Moments and cumulants of polynomial random variables on unitary groups, the Itzykson-Zuber integral, and free probability. *Int. Math. Res. Not.*, 2003(17):953–982, 2003. 212, 222, 223

[34] B. Collins, A. Guionnet, and E. Maurel-Segala. Asymptotics of unitary and orthogonal matrix integrals. *Adv. Math.*, 222(1):172–215, 2009. 222

[35] B. Collins and P. Śniady. Integration with respect to the Haar measure on unitary, orthogonal and symplectic group. *Comm. Math. Phys.*, 264(3):773–795, 2006. 222

[36] R. Couillet and M. Debbah. *Random Matrix Methods for Wireless Communications.* Cambridge University Press, Cambridge, UK, 2011. 186

[37] R. Couillet, M. Debbah, and J. W. Silverstein. A deterministic equivalent for the analysis of correlated MIMO multiple access channels. *IEEE Trans. Inform. Theory*, 57(6):3493–3514, 2011. 186

[38] P. Diaconis. Patterns in eigenvalues: the 70th Josiah Willard Gibbs Lecture. *Bull. Amer. Math. Soc.*, 40:155–178, 2003. 223

[39] K. Dykema. On certain free product factors via an extended matrix model. *J. Funct. Anal.*, 112(1):31–60, 1993. 222

[40] A. Edelman and N. R. Rao. Random matrix theory. *Acta Numerica*, pages 233–297, 2005. 223

[41] P. J. Forrester. *Log-gases and Random Matrices*, volume 34 of *London Mathematical Society Monographs Series.* Princeton University Press, Princeton, NJ, 2010. 107

[42] B. Fuglede and R. V. Kadison. Determinant theory in finite factors. *Ann. of Math.*, 55(2):520–530, 1952. 225

[43] W. Fulton. Eigenvalues of sums of Hermitian matrices (after A. Klyachko). *Astérisque*, 40:Exp. No. 845, 5, 255–269, 1998. Séminaire Bourbaki. 222

[44] V. L. Girko. *Theory of Stochastic Canonical Equations, Vol 1, 2.* Springer Science & Business Media, 2001. 107

[45] V.L. Girko. *An Introduction to Statistical Analysis of Random Arrays.* De Gruyter, 2018. 107

[46] B.V. Gnedenko. *Theory of Probability.* CRC Press, Boca Raton, sixth edition, 1997. 52

[47] F. Götze, A. Naumov, and A. Tikhomirov. On minimal singular values of random matrices with correlated entries. *Random Matrices Theory App*, 4(2):1550006, 30, 2015. 147

[48] R.M. Gray. *Toeplitz and Circulant Matrices: A Review.* Now Publishers, Inc., 2006. 52

[49] U. Grenander. *Probabilities on Algebraic Structures.* John Wiley & Sons, Inc., New York-London; Almqvist & Wiksell, Stockholm-Göteborg-Uppsala, 1963. 146

[50] U. Grenander and J. W. Silverstein. Spectral analysis of networks with random topologies. *SIAM J. Appl. Math.*, 32(2):499–519, 1977. 147

[51] U. Grenander and G. Szegő. *Toeplitz Forms and their Applications.* Chelsea Publishing Co., New York, second edition, 1984. 52, 113

[52] A. Guionnet. Large deviations and stochastic calculus for large random matrices. *Probability Surveys*, 1:72–172, 2004. 223

[53] A. Guionnet. *Large Random Matrices.* Springer Science & Business Media, 2009. 107

[54] A. Guionnet, M. Krishnapur, and O. Zeitouni. The single ring theorem. *Ann. of Math.*, 174(2):1189–1217, 2011. 228

[55] U. Haagerup and F. Larsen. Brown's spectral distribution measure for R-diagonal elements in finite von Neumann algebras. *J. Funct. Anal.*, 176(2):331–367, 2000. 227

[56] C. Hammond and S. J. Miller. Distribution of eigenvalues for the ensemble of real symmetric Toeplitz matrices. *J. Theoret. Probab.*, 18(3):537–566, 2005. 147

[57] F. Hiai and D. Petz. Asymptotic freeness almost everywhere for random matrices. *Acta Sci. Math. (Szeged)*, 66(3-4):809–834, 2000. 222, 223

[58] F. Hiai and D. Petz. *The Semicircle Law, Free Random Variables and Entropy*, volume 77 of *Mathematical Surveys and Monographs*. American Mathematical Society, Providence, RI, 2000. 15, 222

[59] L. Isserlis. On a formula for the product-moment coefficient of any order of a normal frequency distribution in any number of variables. *Biometrika*, 12(1-2):134–139, 1918. xviii, 11

[60] D. Jonsson. Some limit theorems for the eigenvalues of a sample covariance matrix. *J. Multivariate Anal.*, 12(1):1–38, 1982. 147

[61] H. Kesten. Symmetric random walks on groups. *Tran. Amer. Math. Soc.*, 92:336–354, 1959. 73

[62] B. Krawcyk and R. Speicher. Combinatorics of free cumulants. *J. Combin. Theory*, 90(2):267–292, 2000. 27

[63] F. Larsen. *Brown Measures and R-diagonal Elements in Finite von Neumann Algebras*. University of Southern Denmark, 1999. PhD Thesis, Department of Mathematics and Computer Science. 228

[64] G. Livan, M. Novaes, and P. Vibo. *Introduction to Random Matrices*. Springer, 2018. 107

[65] M. Loéve. *Probability Theory*. Dover Publications, 3rd edition, 1963. 26, 82

[66] E. Lukacs. *Characteristic Functions*. Griffin, London, 2nd edition, 1970. 3

[67] C. Male. The norm of polynomials in large random and deterministic matrices. *Probab. Theory Related Fields*, 154(3-4):477–532, 2012. With an appendix by Dimitri Shlyakhtenko. 186

[68] C. Male. Traffic distributions and independence: permutation invariant random matrices and the three notions of independence. *Memoirs of the American Mathematical Society*, 2020. To appear. 223

[69] V. A. Marčenko and L. A. Pastur. Distribution of eigenvalues in certain sets of random matrices. *Mat. Sb. (N.S.)*, 72 (114):507–536, 1967. 147

[70] J. Marcinkiewicz. Sur une propertie de la loi deGauss. *Math. Zeitschr.*, 44:612–618, 1938. Reprinted in *Collected Papers*, Pantswowe wydawnictwo, Naukowe, Warsawa, 1964. 3

[71] M. L. Mehta. *Random Matrices.* Pure and Applied Mathematics. Elsevier, 2004. 107

[72] J.A. Mingo and R. Speicher. *Free Probability and Random Matrices,* volume 35 of *Fields Institute Monographs.* Springer, 2017. xxii, 15, 225

[73] Hoi H. Nguyen and S. O'Rourke. The elliptic law. *Int. Math. Res. Not.*, 26(17):7620–7689, 2015. 147, 228

[74] A. Nica and R. Speicher. *Lectures on the Combinatorics of Free Probability,* volume 335 of *London Mathematical Society Lecture Note Series.* Cambridge University Press, Cambridge, 2006. xxii, 15, 58, 102, 212, 216, 226, 233, 241, 243, 248

[75] S. O'Rourke, D. Renfrew, A. Soshnikov, and V. Vu. Products of independent elliptic random matrices. *J. Stat. Phys.*, 160(1):89–119, 2015. 229, 230

[76] L. Pastur and M. Shcherbina. *Eigenvalue Distribution of Large Random Matrices.* Mathematical Surveys and Monographs. American Mathematical Society, 2011. 107

[77] L. A. Pastur and V. Vasilchuk. On the law of addition of random matrices. *Comm. Math. Phys.*, 214(2):249–286, 2000. 222

[78] N. Raj Rao and R. Speicher. Multiplication of free random variables and the S-transform: the case of vanishing mean. *Elect. Comm. in Probab.*, 12:248–258, 2007. 74

[79] R. Rashidi Far, T. Oraby, W. Bryc, and R. Speicher. On slow-fading MIMO systems with nonseparable correlation. *IEEE Trans. Inform. Theory*, 54(2):544–553, 2008. 186

[80] M. Riesz. Surle probleme des moments. Troisieme note (in French). *Ark.for matematik. Astronomi och Fysik*, 16:1–52, 1923. 4

[81] W. Rudin. *Real and Complex Analysis.* McGraw Hill Series in Higher Mathematics. McGraw Hill, third edition, 1987. 99

[82] Øyvind Ryan. On the limit distributions of random matrices with independent or free entries. *Comm. Math. Phys.*, 193(3):595–626, 1998. 222

[83] K. Sato. *Lévy Processes and Infinitely Divisible Distributions.* Cambridge University Press, London, 1999. 80

[84] J. H. Schenker and H. Schulz-Baldes. Semicircle law and freeness for random matrices with symmetries or correlations. *Mathematical Research Letters*, 12:531–542, 2005. 146

[85] H. Schultz. Non-commutative polynomials of independent Gaussian random matrices. The real and symplectic cases. *Probab. Theory Related Fields*, 131(2):261–309, 2005. 222

[86] J. Schur. Bemerkungen zur Theorie der beschränkten Bilinearformen mit unendlich vielen Veränderlichen. *J. Reine Angew. Math.*, 140:1–28, 1911. 248

[87] D. Shlyakhtenko. Random Gaussian band matrices and freeness with amalgamation. *Internat. Math. Res. Notices*, 1996(20):1013–1025, 1996. 223

[88] G. R. Shorack. *Probability for Statisticians.* Springer Texts in Statistics. Springer Verlag, New York, 2000. 56

[89] P. Śniady. Random regularization of Brown spectral measure. *J. Funct. Anal.*, 193(2):291–313, 2002. 228

[90] H. Sompolinsky, H. Sommers, A. Crisanti, and Y. Stein. Spectrum of large random asymmetric matrices. *Phys. Rev. Lett.*, 60(19):1895–1898, 1988. 147

[91] R. Speicher. Multiplicative functions on the lattice of non-crossing partitions and free convolution. *Math. Ann.*, 298:611–628, 1994. 27, 241

[92] R. Speicher. *Free Probability Theory and Non-Crossing Partitions.* Séminaire Lotharingien de Combinatoire B39c. Unpublished Lecture Notes, 1997. 15

[93] R. Speicher. On universal products. In *Free Probability Theory (Waterloo, ON, 1995)*, volume 12 of *Fields Inst. Commun.*, pages 257–266. Amer. Math. Soc., Providence, RI, 1997. 168

[94] R. Speicher. Free probability theory. In *The Oxford Handbook of Random Matrix Theory*, pages 452–470. Oxford Univ. Press, Oxford, 2011. 222

[95] R. P. Stanley. *Enumerative Combinatorics: Volume 2.* Cambridge University Press, 1999. Appendix by S. Fomin. 8

[96] R. P. Stanley. *Enumerative Combinatorics: Volume 1.* Cambridge University Press, 2nd edition, 2011. 8

[97] T. Tao. *Topics in Random Matrix Theory*, volume 132 of *Graduate Studies in Mathematics.* American Mathematical Society, Providence, RI, 2012. 107

[98] T. Tao and V. Vu. Random matrices: universality of ESDs and the circular law. *The Annals of Probability*, 38(5):2023–2065, 2010. With an appendix by M. Krishnapur. 228

[99] A. M. Tulino and S. Verdú. Random matrix theory and wireless communications. *Foundations and Trends in Communications and Information Theory*, 1(1):1–182, 2004. 186

[100] D. Voiculescu. Limit laws for random matrices and free products. *Invent. Math.*, 104(1):201–220, 1991. xviii, 31, 110, 155, 222, 223

[101] D. Voiculescu. A strengthened asymptotic freeness result for random matrices with applications to free entropy. *Internat. Math. Res. Notices*, 1998(1):41–63, 1998. 222

[102] D. Voiculescu, K. J. Dykema, and A. Nica. *Free Random Variables.* CRM Monograph Series, Volume 1. American Mathematical Society, 1992. 15

[103] K. W. Wachter. The strong limits of random matrix spectra for sample matrices of independent elements. *Ann. Probability*, 6(1):1–18, 1978. 147

[104] G. C. Wick. The evaluation of the collision matrix. *Physical Review*, 80(2):268–272, 1950. 12

[105] E. P. Wigner. Characteristic vectors of bordered matrices with infinite dimensions. *Ann. of Math.*, 62:548–564, 1955. xx, 110

[106] E. P. Wigner. On the distribution of the roots of certain symmetric matrices. *Ann. of Math. (2)*, 67:325–327, 1958. 146

[107] J. Wishart. The generalised product moment distribution in samples from a normal multivariate population. *Biometrika*, 20A(1-2):32–52, 1928. xx, 111

[108] Y. Q. Yin. Limiting spectral distribution for a class of random matrices. *J. Multivariate Anal.*, 20(1):50–68, 1986. 147

[109] Y. Q. Yin and P. R. Krishnaiah. Limit theorem for the eigenvalues of the sample covariance matrix when the underlying distribution is isotropic. *Teor. Veroyatnost. i Primenen.*, 30(4):810–816, 1985. 147

Index